DATA ANALYSIS FOR PHYSICAL SCIENTISTS: FEATURING EXCEL

The ability to summarise data, compare models and apply computer-based analysis tools are vital skills necessary for studying and working in the physical sciences. This textbook supports undergraduate students as they develop and enhance these skills.

Introducing data analysis techniques, this textbook pays particular attention to the internationally recognised guidelines for calculating and expressing measurement uncertainty. This new edition has been revised to incorporate Excel® 2010. It also provides a practical approach to fitting models to data using non-linear least squares, a powerful technique that can be applied to many types of model.

Worked examples using actual experimental data help students understand how the calculations apply to real situations. Over 200 in-text exercises and end of chapter problems give students the opportunity to use the techniques themselves and gain confidence in applying them. Answers to the exercises and problems are given at the end of the book.

LES KIRKUP is an Associate Professor in the School of Physics and Advanced Materials, University of Technology, Sydney. He is also an Australian Learning and Teaching Council National Teaching Fellow. A dedicated lecturer, many of his educational developments have focused on enhancing the laboratory experience of undergraduate students.

Data Analysis for Physical Scientists

Featuring Excel®

Les Kirkup

University of Technology, Sydney

CAMBRIDGE UNIVERSITY PRESS
Cambridge, New York, Melbourne, Madrid, Cape Town,
Singapore, São Paulo, Delhi, Tokyo, Mexico City

Cambridge University Press
The Edinburgh Building, Cambridge CB2 8RU, UK

Published in the United States of America by Cambridge University Press, New York

www.cambridge.org
Information on this title: www.cambridge.org/9780521883726

First published 2012

Printed in the United Kingdom at the University Press, Cambridge

A catalogue record for this publication is available from the British Library

Library of Congress Cataloguing in Publication data
Kirkup, Les.
Data analysis for physical scientists : featuring Excel / Les Kirkup. – 2nd ed.
  p.   cm.
Rev. ed. of: Data analysis with Excel. 2002.
Includes bibliographical references and index.
ISBN 978-0-521-88372-6
1. Research – Statistical methods – Data processing.   2. Electronic spreadsheets.   3. Microsoft
Excel (Computer file)   I. Kirkup, Les. Data analysis with Excel.   II. Title.
Q180.55.S7K57   2012
502.85'554–dc23

2011017662

ISBN 978-0-521-88372-6 Hardback

To Janet, Sarah and Amy
nee more late neets!

Contents

Preface to the second edition

I thank Cambridge University Press, and in particular Simon Capelin, for the opportunity to revisit *Data Analysis with Excel*. I have revised sections of the book to include topics of contemporary relevance to undergraduate students, particularly in the area of uncertainty in measurement. I hope the book will continue to assist in developing the quantitative skills of students destined to graduate in the physical sciences. There is little doubt that the demand for such skills will continue to grow in society in general and particularly within industry, research, education and commerce.

This edition builds on the first with a new chapter added and others undergoing major or minor modifications (for example, to remedy mistakes, update references or include more end of chapter exercises).

I have taken the opportunity to include topics requested by several readers of the first edition. In particular, feedback indicated that the inclusion of a chapter on non-linear least squares and Excel's Solver would be valued and broaden the appeal of the book.

The treatment of error and uncertainty in the first edition paid insufficient attention to the international guidelines for calculating and expressing uncertainty. I hope a major rewrite of chapter 5 has gone a long way to remedying this. The international guidelines on uncertainty deserve to be better known and I trust this book can contribute something to raising awareness of the guidelines within universities and colleges. Terms not usually found in a data analysis textbook for undergraduates, such as coverage factor, standard uncertainty and expanded uncertainty, have been introduced and their relationship to more familiar terms explained as the book progresses.

Microsoft's Excel features regularly throughout the book. References to Excel and the descriptions of its functions have been updated to be consistent with Excel 2010. While there have been several important changes to the look and feel of Excel over earlier versions, my main aim as in the first edition has been to describe features of most value to data analysis. There have been modifications to some of Excel's built-in functions and several new functions added. Also some of the statistical algorithms (which came in for criticism in earlier versions of Excel) have been improved.

I believe that the title of the first edition of this book *Data Analysis with Excel* was somewhat misleading as it was possible to interpret that the book was dominated by Excel, when this wasn't (isn't) the case. I hope the new title betters reflects the role of Excel within the book.

Many of the problems and exercises in the book are based on real, though unpublished, data. For this I thank colleagues from my institution and beyond who have been so generous with their data. These same colleagues have been equally generous with their encouragement throughout the writing of this edition and I thank them whole-heartedly. For the contribution of extra data to this edition, I would particularly like to thank Fraser Torpy, Anna Wilson, Mike Cortie, Andy Leigh (who also supplied the image used on the front cover) Jonathan Edgar, Alison Beavis, Greg Skilbeck and Francois Malan. I would also like to thank the following people for stimulating conversations on data analysis methods: Bob Frenkel, Kendal McGuffie, Michael Bulmer, Kelly Matthews, Andy Buffler, Paul Francis, Manju Sharma, Darren Pearce, Jo McKenzie and Kate Wilson-Goosens.

Preface to the first edition

Experiments and experimentation have central roles to play in the education of scientists. For many destined to participate in scientific enquiry through laboratory or field based studies, the ability to apply 'experimental methods' is a key skill that they rely upon throughout their professional careers. For others whose interests and circumstances take them into other fields upon completion of their studies, the experience of 'wrestling with nature' so often encountered in experimental work, offers enduring rewards: Skills developed in the process of planning, executing and deliberating upon experiments are of lasting value in a world in which some talents become rapidly redundant.

Laboratory and field based experimentation are core activities in the physical sciences. Good experimentation is a blend of insight, imagination, skill, perseverance and occasionally luck. Vital to experimentation is data analysis. This is rightly so, as careful analysis of data can tease out features and relationships not apparent at a first glance at the 'numbers' emerging from an experiment. This, in turn, may suggest a new direction for the experiment that might offer further insight into a phenomenon or effect being studied. Equally importantly, after details of an experiment are long forgotten, facility gained in applying data analysis methods remains as a highly valued and transferable skill.

My experience of teaching data analysis techniques at undergraduate level suggests that when the elements of content, relevance and access to contemporary analysis tools are sympathetically blended, students respond positively and enthusiastically. Believing that no existing text encourages or supports such a 'blend', I decided to write one. This text offers an introduction to data analysis techniques recognising the background and needs of students from the physical sciences. I have attempted to include those techniques most useful to students from the physical sciences and employ examples that have a physical sciences 'bias'.

It is natural to turn to the computer when the 'number crunching' phase of data analysis begins. Though many excellent computer based data analysis packages exist, I have chosen to exploit the facilities offered by spreadsheets throughout the text. In their own right, spreadsheets are powerful analysis tools which are likely to be familiar and readily accessible to students.

More specifically, my goals have been to,

- provide a readable text from which students can learn the basic principles of data analysis.
- ensure that problems and exercises are drawn from situations likely to be familiar and relevant to students from the physical sciences.
- remove much of the demand for manual data manipulation and presentation by incorporating the spreadsheet as a powerful and flexible utility.
- emphasise the analysis tools most often used in the physical sciences.
- focus on aspects often given little attention in other texts for scientists such as the treatment of systematic errors.
- encourage student confidence by incorporating 'worked' examples followed by exercises.
- provide access to extra topics not dealt with directly in the text through generally accessible Web pages.

Computers are so much a part of professional and academic life that I am keen to include their use, especially where this aids the learning and application of data analysis techniques. The Excel spreadsheet package by Microsoft has been chosen due to its flexibility, availability, longevity and the care that has been taken by its creators to provide a powerful yet 'user friendly' environment for the processing and presentation of data. This text does not, however, attempt a comprehensive coverage of the features of Excel. Anyone requiring a text focussing on Excel, and its many options, shortcuts and specialist applications must look elsewhere as only those features of most relevance to the analysis of experimental data are dealt with here.

While chapter 1 contains some material normally encountered at first year level, the text as a whole has been devised to be useful at intermediate and senior undergraduate levels. Derivations of formulae are mostly avoided in the body of the text. Instead, emphasis has been given to the assumptions underlying the formulae and range of applicability. Details of derivations may be found in the appendices. It is assumed that the reader is familiar with introductory calculus, graph plotting and the calculations of means and standard deviations. Experience of laboratory work at first year undergraduate level is also an advantage.

I am fortunate that many people have given generously of their time to help me during the preparation of this book. Their ideas, feedback and not least their encouragement are greatly appreciated. I also acknowledge many intense Friday night discussions with students and colleagues on matters relating to data analysis and their frequent pleadings with me to 'get a life'.

I would like to express my appreciation and gratitude to the following people:

From the University of Technology, Sydney (UTS):

Geoff Anstis, Mark Berkahn, Graziella Caprarelli, Bob Cheary, Michael Dawson, Chris Deller, Sherri Hilario, Suzanne Hogg, Ann–Marie Maher, Kendal McGuffie, Mary Mulholland, Matthew Phillips, Andrew Searle, Brian Stephenson, Mike Stevens, Paul Swift.

Formerly of UTS:

Andreas Reuben, Tony Fisher-Cripps, Patsy Gallagher, Gary Norton

Finally, I thank Janet Sutherland for her encouragement and support during the preparation of this text.

Chapter 1

Introduction to scientific data analysis

1.1 Introduction

'The principle of science, the definition almost, is the following: *The test of all knowledge is experiment. Experiment is the sole judge* of scientific "truth"'.

So wrote Richard Feynman, famous scientist and Nobel Prize winner, noted for his contributions to physics.[1]

It is possible that when Feynman wrote these words he had in mind elaborate experiments devised to reveal the 'secrets of the Universe', such as those involving the creation of new particles during high energy collisions in particle accelerators or others to determine the structure of DNA.[2] Experimentation encompasses an enormous range of more humble (but extremely important) activities such as testing the temperature of a baby's bath water by immersing an elbow into the water, or pressing on a bicycle tyre to establish whether it needs inflating. The absence of numerical measures of quantities distinguishes these experiments from those normally performed by scientists.

Many factors directly or indirectly influence the fidelity of data gathered during an experiment such as the quality of the experimental design, experimenter competence, instrument limitations and time available to perform the experiment. Identifying, appreciating and, where possible, accounting for, such factors are key tasks that must be carried out by an experimenter. After every care has been taken to acquire the best data possible, it is time to apply techniques of data analysis to extract the most from the data. The process of extraction requires qualitative as well as quantitative methods of analysis. The

[1] See Feynman, Leighton and Sands (1963).

[2] DNA stands for deoxyribonucleic acid.

first steps require consideration be given to how data may be summarised numerically and graphically.[3] This is the main focus of this chapter. Some of the ideas touched upon in this chapter, such as those relating to error and uncertainty, will be revisited in more detail in later chapters.

1.2 Scientific experimentation

To find out something about the world, we experiment. A child does this naturally, with no training or scientific apparatus. Through a potent combination of curiosity and trial and error, a child quickly creates a viable model of the 'way things work'. This allows the consequences of a particular action to be anticipated. Curiosity plays an equally important role in the professional life of a scientist who may wish to know the

- amount of contaminant in a pharmaceutical;
- concentration of CO_2 in the Earth's atmosphere;
- distribution of temperature across a leaf;
- stresses experienced by the wings of an aircraft;
- blood pressure of a person;
- frequency of electrical signals generated by the human brain.

Scientists look for relationships between quantities. For example, a scientist may wish to establish how the amount of energy radiated from a body each second depends on the temperature of that body. In formulating the problem, designing and executing the experiment and analysing the results, the intention may be to extend the domain of applicability of an established theory, or present convincing evidence of the limitations of that theory. Where results obtained conflict with accepted ideas or theories, a key goal is to provide a better explanation of the results. Before publishing a new and perhaps controversial explanation, the scientist needs to be confident in the data gathered and the methods used to analyse those data. This requires that experiments be well designed. In addition, good experimental design helps anticipate difficulties that may occur during the execution of the experiment and encourages the efficient use of resources.

Successful experimentation is often a combination of good ideas, good planning, perseverance and hard work. Though it is possible to discover something interesting and new 'by accident', it is usual for science to progress by small steps taken by many researchers. The insights gained by researchers (both experimentalists and theorists) combine to provide answers and explanations to some questions, and in the process create new questions that need to

---

[3] This is sometimes referred to as 'exploratory data analysis'.

be addressed. In fact, even if something new *is* found by chance, it is likely that the discovery will remain a curiosity until a serious scientific investigation is carried out to determine if the discovery or effect is real or illusory. While scientists are excited by new ideas, a healthy amount of scepticism remains until the ideas have been subjected to serious and sustained examination by others.

1.2.1 Aim of an experiment

An experiment needs a focus, more usually termed an 'aim', which is something the experimenter returns to during the design and analysis phases of the experiment. Essentially the aim embodies a question which can be expressed as 'what are we trying to find out by performing the experiment?'.

Expressing the aim clearly and concisely before the experiment begins is important, as it is reasonable to query as the experiment progresses whether the steps taken are succeeding in addressing the aim, or whether the experiment has deviated 'off track'. Deviating from the main aim is not necessarily a bad thing. After all, if you observe an interesting and unexpected effect during the course of an experiment, it would be quite natural to want to know more, as rigidly pursuing the original aim might cause an important discovery to be overlooked. Nevertheless, it is likely that if a new effect *has* been observed, this effect deserves its own separate and carefully planned experiment.

Implicit in the aim of the experiment is an idea or hypothesis that the experimenter wishes to promote or test, or an important question that requires clarification. Examples of questions that might form the basis of an experiment include the following.

- Is a new spectroscopic technique better able to detect impurities in silicon than existing techniques?
- Does heating a glass substrate during vacuum deposition of a metal improve the quality of films deposited onto the substrate?
- To what extent does a reflective coating on windows reduce the heat transfer into a motor vehicle?
- In what way does the efficiency of a thermoelectric cooler depend on the size of the electrical current supplied to the cooler?
- How does the flow rate of fluid through a hollow tube depend on the internal diameter of that tube?

Such questions can be restated explicitly as aims of a scientific investigation. It is possible to express those aims in a number of different, but essentially equivalent, ways. For example:

(a) the aim of the experiment is to determine the change in heat transfer to a motor vehicle when a reflective coating is applied to the windows of that vehicle;

(b) the aim of the experiment is to test the hypothesis that a reflective coating applied to the windows of a motor vehicle reduces the amount of heat transferred into that vehicle.

Most physical scientists and engineers would recognise (a) as a familiar way in which an aim is expressed in their disciplines. By contrast, the explicit inclusion of a hypothesis to be tested, as stated in (b) is often found in studies in the biological, medical and behavioural sciences. The difference in the way the aim is expressed is largely due to the conventions adopted by each discipline, as all have a common goal of advancing understanding and knowledge through experimentation, observation and analysis.

1.2.2 Experimental design

Deciding the aim or purpose of an experiment at an early stage is important, as precious resources (including the time of the experimenter) are to be devoted to the experiment. Experimenting is such an absorbing activity that it is possible for the aims of an experiment to become too ambitious. For example, the aim of an experiment might be to determine the effect on the thermal properties of a ceramic when several types of atoms are substituted for (say) atoms of calcium in the ceramic. If a month is available for the study, careful consideration must be given to the number of samples of ceramic that can be prepared and tested and whether a more restricted aim, perhaps concentrating on the substitution of just one type of atom, would be more judicious.

Once the aim of an experiment is decided, a plan of how that aim might be achieved is devised. Matters that must be considered include the following.

- What quantities are to be measured during the experiment?
- Over what ranges should the controllable quantities be measured?
- What are likely to be the dominant sources of error, and how can the errors be minimised?
- What equipment is needed and what is its availability?
- In what ways are the data to be analysed?
- Does the experimenter need to become skilled in new techniques (say, how to operate an electron microscope) in order to complete the experiment?
- Does new equipment need to be designed/constructed/acquired or does existing equipment require modification?
- Is there merit in developing a computer controlled acquisition system to gather the data?

- How much time is available to carry out the experiment?
- Are the instruments to be used performing within their specifications?

A particularly important aspect of experimentation is the identification of influences that can affect any result obtained through experiment or observation. Such influences are regarded as sources of 'experimental error' and we will have cause to consider these in this text. In the physical sciences, many of the experimental variables that would affect a result are readily identifiable and some are under the control of the experimenter. Identifying sources that would adversely influence the outcomes of an experiment may lead to ways in which the influence might be minimised. For example, the quality of a metal film deposited onto a glass substrate may be dependent upon the temperature of the substrate during the deposition process. By improving the temperature control of the system, so that the variability of the temperature of the substrate is reduced to (say) less than 5 °C, the quality of the films may be enhanced.

Despite the existence of techniques that allow us to draw out much from experimental data, a good experimenter does not rely on data analysis to compensate for data of dubious quality. If large scatter is observed in data, a sensible option is to investigate whether improved experimental technique can reduce the scatter. For example, incorporating electromagnetic shielding as part of an experiment requiring the measurement of extremely small voltages can improve the quality of the data dramatically and is preferred to the application of sophisticated data analysis techniques which attempt to compensate for shortcomings in the data.

An essential feature of experiments in the physical sciences is that the measurement process yields numerical values for quantities such as temperature, pH, strain, pressure and voltage. These numerical values (often referred to as *experimental data*) may be algebraically manipulated, graphed, compared with theoretical predictions or related to values obtained by other experimenters who have performed similar experiments.

1.3 The vocabulary of measurement

Scientists draw on statistical methods as well as those deriving from the science of measurement (termed *metrology*) when analysing their data. A consequence is that sometimes there is inconsistency between the way terms, such as error and uncertainty, are used in texts on the treatment of data written by statisticians and by those written by metrologists. The diversity of terms can be a distraction. In this text we will tend to rely on the internationally recognised explanation of terms found in the 'International vocabulary of metrology' (usually abbreviated to the VIM).[4]

1.4   Units and standards

Whenever a value is recorded in a table or plotted on a graph, the unit of measurement must be stated, as numbers by themselves have little meaning. To encompass all quantities that we might measure during an experiment, we need units that are:

- comprehensive;
- clearly defined;
- internationally endorsed;
- easy to use.

Reliable and accurate standards based on the definition of a unit must be available so that instruments designed to measure specific quantities may be compared against those standards. Without agreement between experimenters in, say, Australia, the United Kingdom and the United States, as to what constitutes a metre or a second, a comparison of values obtained by each experimenter would be impossible.

A variety of instruments may be employed to measure quantities in the physical sciences. These range from a simple hand-held stopwatch for timing a body in free-fall, to a state of the art HPLC[5] to determine the concentration of contaminant in a pharmaceutical. Whatever the particular details of a scientific investigation, we generally attach much importance to the 'numbers' that emerge from an experiment as they may provide support for a new theory of the origin of the Universe, assist in monitoring the concentration of CO_2 in the Earth's atmosphere, or help save a life. Referring to the outcome of a measurement as a 'number' is rather vague and misleading. Through experiment we obtain *values*. A value is the product of a number and the unit in which the measurement is made. The distinction in scientific contexts between number and value is important. Table 1.1 includes definitions of number, value, and other important terms as they are used in this text.

---

[4] ISO/IEC Guide 99:2007, International vocabulary of metrology – Basic and general concepts and associated terms (VIM). Available as a free download from http://www.bipm.org/en/publications/guides/vim.html [accessed 30/6/2011].

[5] HPLC stands for High Performance Liquid Chromatography.

Table 1.1. *Definitions of commonly used terms in data analysis.*

Term	Definition
Quantity	An attribute or property of a body, phenomenon or material. Examples of quantities are: the temperature, mass or electrical capacitance of a body, the time elapsed between two events such as starting and stopping a stopwatch, or the resistivity of a metal.
Unit	An amount of a quantity, suitably defined and agreed internationally, against which some other amount of the same quantity may be compared. As examples, the kelvin is a unit of temperature, the second is a unit of time and the ohm-metre is a unit of resistivity.
Value	The product of a number and a unit. As examples, 273 K is a value of temperature, 0.015 s is a value of time interval and 1.7×10^{-8} $\Omega \cdot$m is a value of resistivity.
Measurement	A process by which a value of a quantity is determined. For example, the measurement of water temperature using an alcohol-in-glass thermometer entails immersing a thermometer into the water followed by estimating the position of the top of a narrow column of alcohol against an adjacent scale.
Data	Values obtained through measurement or observation.

1.4.1 Units

The most widely used system of units in science is the SI system[6] and has been adopted officially by most countries around the world. Despite strongly favouring SI units in this text, we will also use some non-SI units such as the minute and the degree, as these are likely to remain in widespread use in science for the foreseeable future.

The origins of the SI system can be traced to pioneering work done on units in France in the late eighteenth century. In 1960 the name 'SI system' was adopted and at that time the system consisted of six fundamental or 'base' units. Since 1960 the system has been added to and refined and remains constantly under review. From time to time suggestions are made regarding how the definition of a unit may be improved. If this allows for easier or more accurate realisation of the

[6] SI stands for Système International. An authoritative document on the SI system prepared by the Bureau International des Poids et Mesures (custodians of the SI system) is freely available as a download from www.bipm.org/utils/common/pdf/si_brochure_8_en.pdf [accessed 9/11/2010].

Table 1.2. *SI base units, symbols and definitions.*

Quantity	Unit	Symbol	Definition
Mass	kilogram	kg	The kilogram is equal to the mass of the international prototype of the kilogram. (The prototype kilogram is made from an alloy of platinum and iridium and is kept under very carefully controlled environmental conditions by the Bureau International des Poids et Mesures (BIPM) in Sèvres near Paris, France.)
Length	metre	m	The metre is the length of the path travelled by light in a vacuum during a time interval of $\frac{1}{299792458}$ of a second.
Time	second	s	The second is the duration of 9192631770 periods of the radiation corresponding to the transition between the two hyperfine levels of the ground state of the caesium 133 atom.
Thermodynamic temperature	kelvin	K	The kelvin is the fraction $\frac{1}{273.16}$ of the thermodynamic temperature of the triple point of water.
Electric current	ampere	A	The ampere is that current which, if maintained between two straight parallel conductors of infinite length, of negligible cross-section and placed one metre apart in a vacuum, would produce between these conductors a force of 2×10^{-7} newton per metre of length.
Luminous intensity	candela	cd	The candela is the luminous intensity, in a given direction, of a source that emits monochromatic radiation of frequency 540×10^{14} hertz and that has a radiant intensity in that direction of $\frac{1}{683}$ watt per steradian.
Amount of substance	mole	mol	The mole is the amount of substance of a system which contains as many elementary entities as there are atoms in 0.012 kilogram of carbon 12.

unit as a standard (permitting, for example, improvements in instrument calibration), then appropriate modifications are made to the definition of the unit.

Currently the SI system consists of seven base units as defined in table 1.2.

Other quantities may be expressed in terms of the base units. For example, energy can be expressed in units $kg \cdot m^2 \cdot s^{-2}$ and electric potential difference in units $kg \cdot m^2 \cdot s^{-3} \cdot A^{-1}$. The cumbersome nature of units expressed this way is such that other, so called *derived* units, are introduced which are formed from products of the base units. Some familiar quantities with their units expressed in derived and base units are shown in table 1.3.

Table 1.3. *Symbols and units of some common quantities.*

Quantity	Derived unit	Symbol	Unit of quantity expressed in base units
Energy, work	joule	J	$\text{kg·m}^2\text{·s}^{-2}$
Force	newton	N	kg·m·s^{-2}
Power	watt	W	$\text{kg·m}^2\text{·s}^{-3}$
Potential difference, electromotive force (emf)	volt	V	$\text{kg·m}^2\text{·s}^{-3}\text{·A}^{-1}$
Electrical charge	coulomb	C	s·A
Electrical resistance	ohm	Ω	$\text{kg·m}^2\text{·s}^{-3}\text{·A}^{-2}$

Example 1

The farad is the SI derived unit of electrical capacitance. With the aid of table 1.3, express the unit of capacitance in terms of the base units, given that the capacitance, C, may be written

$$C = \frac{Q}{V}, \tag{1.1}$$

where Q represents electrical charge and V represents potential difference.

ANSWER

From table 1.3, the unit of charge expressed in base units is s·A and the unit of potential difference is $\text{kg·m}^2\text{·s}^{-3}\text{·A}^{-1}$. It follows that the unit of capacitance can be expressed with the aid of equation 1.1 as,

$$\text{unit of capacitance } = \frac{\text{s} \cdot \text{A}}{\text{kg} \cdot \text{m}^2 \cdot \text{s}^{-3} \cdot \text{A}^{-1}} = \text{kg}^{-1} \cdot \text{m}^{-2} \cdot \text{s}^4 \cdot \text{A}^2.$$

Exercise A

The henry is the derived unit of electrical inductance in the SI system of units. With the aid of table 1.3, express the unit of inductance in terms of the base units, given the relationship

$$E = -L\frac{dI}{dt}, \tag{1.2}$$

where E represents emf, L represents inductance, I represents electric current, and t represents time.

1.4.2   **Standards**

How do the definitions of the SI units in table 1.2 relate to measurements made in a laboratory? For an instrument to measure a quantity in SI units, the definitions need to be made 'tangible' so that an example or *standard* of the unit is made available. Only when the definition is realised as a practical and maintainable standard, can values obtained by an instrument designed to measure the quantity be compared against that standard. Where a difference is established between standard and instrument, that difference is stated as a correction to the instrument. The process by which the comparison is made and the issuing of a statement of discrepancy is referred to as *calibration*.

Accurate standards based on the definitions of some of the units appearing in table 1.2 are realised in specialist laboratories. For example, a clock based on the properties of caesium atoms can reproduce the second to high accuracy.[7] By comparison, creating an accurate standard of the ampere based directly on the definition of the ampere appearing in table 1.2 is more challenging. In this case it is common for laboratories to maintain standards of related derived SI units such as the volt and the ohm, which can be implemented to high accuracy.

Most countries have a 'national standards laboratory', or equivalent, which maintains the most accurate standards achievable, referred to as *primary* standards. From time to time each national laboratory compares its standards with other primary standards held in other national laboratories around the world. In addition, a national laboratory creates and calibrates secondary standards by reference to the primary standard. Such secondary standards are found in government, industrial and university laboratories. Secondary standards in turn are used to calibrate and maintain working standards and eventually a working standard may be used to calibrate (for example) a hand-held voltmeter used in an experiment.

The result of a measurement is said to be *traceable* if, by a documented chain of comparisons involving secondary and working standards, the results can be compared with a primary standard. Traceability is very important in some situations, particularly when the 'correctness' of a value indicated by an instrument is in dispute.

---

[7] See appendix 2 of The International System of Units (English translation) 8th edition, 2006, published by BIPM. Appendix 2 is freely available as a download from http://www.bipm.org/utils/en/pdf/SIApp2_s_en.pdf [accessed 2/11/2010].

Table 1.4. *Prefixes used with the SI system of units.*

Factor	Prefix	Symbol	Factor	Prefix	Symbol
10^{-24}	yocto	y	10^1	deka	da
10^{-21}	zepto	z	10^2	hecto	h
10^{-18}	atto	a	$\mathbf{10^3}$	**kilo**	**k**
10^{-15}	femto	f	$\mathbf{10^6}$	**mega**	**M**
$\mathbf{10^{-12}}$	**pico**	**p**	$\mathbf{10^9}$	**giga**	**G**
$\mathbf{10^{-9}}$	**nano**	**n**	$\mathbf{10^{12}}$	**tera**	**T**
$\mathbf{10^{-6}}$	**micro**	**μ**	10^{15}	peta	P
$\mathbf{10^{-3}}$	**milli**	**m**	10^{18}	exa	E
$\mathbf{10^{-2}}$	**centi**	**c**	10^{21}	zetta	Z
10^{-1}	deci	d	10^{24}	yotta	Y

1.4.3 Prefixes and scientific notation

Values obtained through experiment are often much larger or much smaller than the base (or derived) SI unit in which the value is expressed. In such situations there are two widely used methods by which the value of the quantity may be specified. The first is to choose a multiple of the unit and indicate that multiple by attaching a *prefix* to the unit. So, for example, we might express the value of the capacitance of a capacitor as 47 μF. The letter μ is the symbol for the prefix 'micro' which represents a factor of 10^{-6}. A benefit of expressing a value in this way is the conciseness of the representation. A disadvantage is that many prefixes are required in order to span the orders of magnitude of values that may be encountered in experiments. As a result several unfamiliar prefixes exist. For example, the size of the electrical charge carried by an electron is about 160 zC. Only dedicated students of the SI system would instantly recognise z as the symbol for the prefix 'zepto' which represents the factor 10^{-21}.

Table 1.4 includes the prefixes currently used in the SI system. The prefixes shown in bold are the most commonly used.

Another way of expressing the value of a quantity is to give the number that precedes the unit in scientific notation. To express any number in scientific notation, we separate the first non-zero digit from the second digit by a decimal point, so for example, the number 1200 becomes 1.200. So that the number remains unchanged we must multiply 1.200 by 10^3 so that 1200 is written as 1.200×10^3. Scientific notation is preferred for very large or very small numbers.

For example, the size of the charge carried by the electron is written[8] as 1.60×10^{-19} C.

Though any value may be expressed using scientific notation, we should avoid taking this approach to extremes. For example, suppose the mass of a body is 1.2 kg. This *could* be written as 1.2×10^0 kg, but this is arguably going too far.

Example 2

Rewrite the following values using (a) commonly used prefixes and (b) scientific notation:

(i) 0.012 s; (ii) 601 A; (iii) 0.00064 J.

ANSWER

(i) 12 ms or 1.2×10^{-2} s; (ii) 0.601 kA or 6.01×10^2 A; (iii) 0.64 mJ or 6.4×10^{-4} J.

Exercise B

(1) Rewrite the following values using prefixes:

(i) 1.38×10^{-20} J in zeptojoules; (ii) 3.6×10^{-7} s in microseconds; (iii) 43258 W in kilowatts; (iv) 7.8×10^8 m/s in megametres per second.

(2) Rewrite the following values using scientific notation:

(i) 0.650 nm in metres; (ii) 37 pC in coulombs; (iii) 1915 kW in watts; (iv) 125 μs in seconds.

1.4.4 Significant figures

In a few situations, a value obtained in an experiment can be exact. For example, in an experiment to determine the wavelength of light by using Newton's rings,[9] the number of rings can be counted exactly. By contrast, the temperature of an object cannot be known exactly and so we must be careful when we interpret values of temperature. Presented with the statement that '*the temperature of the water bath was 21 °C*' it is unreasonable to infer that the temperature was 21.0000000 °C. It is more likely that the temperature of the water was closer to 21 °C than it was to either 20 °C or 22 °C. By expressing

---

[8] To three significant figures (see section 1.4.4).

[9] For a description of Newton's rings, see Bennett (2008).

the temperature as 21 °C, the implication is that the value of temperature obtained by a single measurement is known to two figures, often referred to as *two significant figures.*

Inferring how many figures are significant simply by the way a number is written can sometimes be difficult. If we are told that the mass of a body is 1200 kg, how many figures are significant? If the instrument measures mass to the nearest 100 kg, then the mass of the body lies between 1150 kg and 1250 kg, such that only the first two figures are significant. On the other hand, if the measuring instrument is capable of measuring to the nearest kilogram, then all four figures are significant. The ambiguity can be eliminated if we express the value using *scientific notation.* If the mass of the body, m, is correct to two significant figures we would write

$$m = 1.2 \times 10^3 \text{ kg.}$$

When a value is written using scientific notation, every figure preceding the multiplication sign is regarded as significant. If the mass is correct to four significant figures then we write

$$m = 1.200 \times 10^3 \text{ kg.}$$

Though it is possible to infer something about a value by the way it is written, it is better to state explicitly the uncertainty in a value. For example, we might write

$$m = (1200 \pm 12) \text{ kg,}$$

where 12 kg is the uncertainty in the value of the mass. Estimating uncertainty is considered in chapter 5.

In some circumstances, it is required to round a value to a specified number of significant figures. For example, we might want to round 1.752×10^{-7} m to three significant figures. To do this, we consider the fourth significant figure (which in this example is a '2'). If this figure is equal to or greater than 5, we increase the third significant figure by one, otherwise we leave the figure unchanged. So, for example, 1.752×10^{-7} m becomes 1.75×10^{-7} m to three significant figures. Using the same convention, a mass of 3.257×10^3 kg becomes 3.3×10^3 kg to two significant figures.

1.5 Picturing experimental data

Our ability to recognise patterns and trends is so good that it makes sense to exploit this talent when analysing experimental data. Though a table of experimental values may contain the same information as appears on a graph, it is often difficult to extract useful information from a table 'by eye'. Comparison

---

Exercise C

(1) How many significant figures are implied by the way each of the following values is written:

(i) 1.72 m; (ii) 0.00130 mol/cm^3; (iii) 6500 kg; (iv) 1.701 × 10^{-3} V; (v) 100 °C; (vi) 100.0 °C; (vii) 0.04020 g; (viii) 1.30 × 10^{-8} lx?

(2) Express the following values using scientific notation to two, three and four significant figures.

(i) 775710 m/s^2; (ii) 0.001266 s; (iii) −105.4 °C; (iv) 14000 nH in henrys; (v) 12.400 kJ in joules; (vi) 101.56 nm in metres.

---

between two or more data sets is generally much easier when the data are presented in graphical form. To appreciate the 'big picture' it is helpful to devise ways of graphically representing the values.

When values are obtained through repeat measurements of a single quantity, then the histogram is used extensively to display data. Values obtained through repeat measurements of a single quantity are often referred to as 'univariate' data. By contrast, if an experiment involves investigating the relationship between two quantities, then the x-y graph is a preferred way of displaying the data (such data are often referred to as 'bivariate' data).

1.5.1 Histograms

The histogram is a pictorial representation of data that is regularly used to reveal the scatter or distribution of values obtained from repeat measurements of a single quantity. For example, we might measure the diameter of a wire many times in order to establish the variation of the diameter along the length of the wire. A table containing the values is a convenient and compact way to present the numerical information. However, we are usually happy (at least in the early stages of analysis) to forego knowledge of individual values in the table for a broader overview of the whole data. This should help indicate whether some values are much more common than others and whether there are any that appear to differ greatly from the others. These 'extreme' values are usually termed *outliers*.

To illustrate the histogram, let us consider data gathered in a radioactive decay experiment. In an experiment to study the emission of beta particles from a strontium 90 source, measurements were made of the number of particles

Table 1.5. *Counts from a radioactivity experiment.*

1265	1196	1277	1320	1248	1245	1271	1233	1231	1207
1240	1184	1247	1343	1311	1237	1255	1236	1197	1247
1301	1199	1244	1176	1223	1199	1211	1249	1257	1254
1264	1204	1199	1268	1290	1179	1168	1263	1270	1257
1265	1186	1326	1223	1231	1275	1265	1236	1241	1224
1255	1266	1223	1233	1265	1244	1237	1230	1258	1257
1252	1253	1246	1238	1207	1234	1261	1223	1234	1289
1216	1211	1362	1245	1265	1296	1260	1222	1199	1255
1227	1283	1258	1199	1296	1224	1243	1229	1187	1325
1235	1301	1272	1233	1327	1220	1255	1275	1289	1248

emitted from the source over 100 consecutive periods of 1 minute. The data gathered are shown in table 1.5.

Inspection of table 1.5 indicates that the values lie between about 1100 and 1400, but little else can be discerned. Do some values occur more often than others and if so which values? A good starting point for establishing the distribution of the data is to count the number (referred to as the *frequency*) of values which occur in predetermined intervals of equal width. The next step is to plot a graph consisting of frequency on the vertical axis, versus interval on the horizontal axis. In doing this we create a histogram.

Table 1.6, created using the data in table 1.5, shows the number of values which occur in consecutive intervals of 20 counts beginning with the interval 1160 to 1180 counts and extending to the interval 1360 to 1380 counts. This table is referred to as a *grouped frequency distribution*.

The distribution of counts is shown in figure 1.1. We note that most values are clustered between 1220 and 1280 and that the distribution is approximately symmetric, with the hint of a longer 'tail' at larger counts.

Other methods by which univariate data can be displayed include stem and leaf plots and pie charts,[10] though these tend to be used less often than the histogram in the physical sciences.

There are no strict rules about choosing the width of intervals for a histogram, but a good histogram:

- is easy to construct, so intervals are chosen to reduce the risk of mistakes when preparing a grouped frequency distribution. For example, an interval between 1160 and 1180 is preferable to one from (say) 1158 to 1178;
- reveals the distribution of the data clearly. If too many intervals are chosen then the number of values in each interval is small and the histogram

[10] See Blaisdell (1998) for details of alternate methods of displaying univariate data.

Table 1.6. *Grouped frequency distribution of the data shown in table 1.5.*

Interval (counts)	Frequency
$1160 < x \leq 1180$	3
$1180 < x \leq 1200$	10
$1200 < x \leq 1220$	7
$1220 < x \leq 1240$	24
$1240 < x \leq 1260$	25
$1260 < x \leq 1280$	16
$1280 < x \leq 1300$	6
$1300 < x \leq 1320$	4
$1320 < x \leq 1340$	3
$1340 < x \leq 1360$	1
$1360 < x \leq 1380$	1

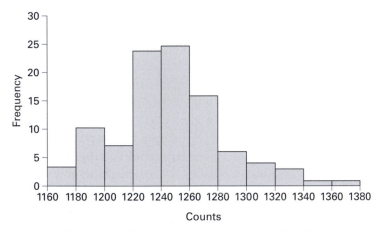

Figure 1.1. Histogram showing the frequency of counts in a radioactivity experiment.

appears 'flat' and featureless. At the other extreme, if the histogram consists of only two or three intervals, then all the values will lie within those intervals and the shape of the histogram reveals little.

In choosing the total number of intervals[11] to be used, a useful rule of thumb is to make the number of intervals, N, equal to[12]

---

[11] Histogram intervals are sometimes referred to as *bins*.

[12] There are other ways of determining the number of intervals, N. Sturge's formula, as described by DeCoursey (2003), gives N as, $N = 1.443 \times \ln(n) + 1$. For the number of data in table 1.5 ($n = 100$), gives N as ≈ 7.6, which rounds up to 8.

$$N = \sqrt{n}. \tag{1.3}$$

where n is the number of values. Once N has been rounded to a whole number, the interval width, w, can be calculated by using

$$w = \frac{\text{range}}{N}, \tag{1.4}$$

where range is defined as,

$$\text{range} = \text{maximum value} - \text{minimum value}. \tag{1.5}$$

We should err on the side of selecting 'easy to work with' intervals, rather than holding rigidly to the value of w given by equation 1.4. If, for example, w were found using equation 1.4 to be 13.357, then a value of w of 10 or 15 should be considered, as this would make tallying up the number of values in each interval less prone to mistakes.

Preparing a grouped frequency distribution and plotting a histogram 'by hand' is tedious if there are many data. Happily, there are many computer based analysis packages, such as spreadsheets (discussed in chapter 2) which reduce the effort that would otherwise be required.

Exercise D

Table 1.7 shows the values of 52 'weights' of nominal mass 50 g used in an undergraduate laboratory.

Using the values in table 1.7, construct:

(i) a grouped frequency distribution;
(ii) a histogram.

Table 1.7. *Values of 52 weights.*

Mass (g)								
50.42	50.09	49.98	50.16	50.10	50.18	50.12	49.95	50.05
50.14	50.07	50.15	50.06	50.22	49.90	50.09	50.18	50.04
50.02	49.81	50.10	50.16	50.06	50.14	50.20	50.06	49.84
50.07	50.08	50.19	50.05	50.13	50.13	50.08	50.05	50.01
49.84	50.11	50.11	50.05	50.15	50.17	50.05	50.12	50.30
49.97	50.05	50.09	50.17	50.08	50.21	50.21		

1.5.2   Relationships and the *x–y* graph

A preoccupation of many scientists is to discover, and account for, the relationship between quantities. Experiment and theory combine in often complex and unpredictable ways before any relationship can be said to be accounted for in a quantitative as well as qualitative manner. Examples of relationships that may be studied through experiment include how the:

- intensity of light emitted from a light emitting diode (LED) varies as the temperature of the LED is reduced;
- power output of a solar cell changes as the angle of orientation of the cell with respect to the Sun is altered;
- electrical resistance of a humidity sensor depends on humidity;
- flow rate of a fluid through a pipe increases as the pressure difference between the ends of the pipe increases;
- acceleration caused by gravity varies with depth below the Earth's surface.

Let us consider the last example, in which the acceleration caused by gravity varies with depth below the Earth's surface. Based upon considerations of the gravitational attraction between bodies, it is possible to predict a relationship between acceleration and depth when a body has uniform density. By gathering 'real data' this prediction can be examined. Conflict between theory and experiment might suggest modifications are required to the theory or perhaps indicate that some anomaly, such as the existence of large deposits of a dense mineral close to the site of the measurements, has influenced the values of acceleration.

As the acceleration in the example above depends on depth, we often refer to the acceleration as the *dependent* variable, and the depth as the *independent* variable. The independent and dependent variables are sometimes referred to as the predictor and response variables respectively.

A convenient way to record values of the dependent and independent variables is to construct a table. Though concise, a table of data is fairly dull and cannot assist efficiently with the identification of trends or patterns in data or allow for easy comparison between data sets. A revealing and very popular way to display bivariate data is to plot an *x–y* graph (sometimes referred to as a scatter graph). The '*x*' and the '*y*' are the symbols used to identify the horizontal and vertical axes respectively of a Cartesian co-ordinate system.[13]

---

[13] The horizontal and vertical axes are also referred to as the abscissa and ordinate respectively.

Properly prepared, a graph is a potent summary of many aspects of an experiment.[14] It can reveal:

- the quantities being investigated;
- the number and range of values obtained;
- gaps in the measurements;
- a trend between the x and y quantities;
- values that conflict with the trend followed by the majority of the data;
- the extent of uncertainty in the values (sometimes indicated by 'error bars').

If a graph is well constructed, this qualitative information can be digested in a few seconds. To construct a good graph we should ensure that:

- a caption describing the graph is included;
- axes are clearly labelled (and the label includes the unit of measurement);
- the scales for each axis are chosen so that plotting, if done by hand, is made easy. Choosing sensible scales also allows for values to be easily read from the graph;
- the graph is large enough to allow for the efficient extraction of information 'by eye';
- plotted values are clearly marked with a conspicuous symbol such as a circle or a square.

An x–y graph is shown in figure 1.2 constructed from data gathered in an experiment to establish the cooling capabilities of a thermoelectric cooler (TEC).[15]

Attached to each point in figure 1.2 are lines which extend above and below the point. These lines are generally referred to as *error bars* and in this example they are used to indicate the uncertainty in the values of temperature.[16] The 'y' error bars attached to the points in figure 1.2 indicate that the uncertainty in the temperature values is about 2 °C. As 'x' error bars are absent we infer that the uncertainty in values of time is too small to plot on this scale.

If an x–y graph is used to present many values, a convenient way to plot the values and the error bars is to use a computer based spreadsheet (see section 2.7.1).

---

[14] Cleveland (1994) discusses what makes 'good practice' in graph plotting.

[15] A thermoelectric cooler is a device made from junctions of semiconductor material. When a current passes through the device, some of the junctions expel thermal energy (causing a temperature rise) while others absorb thermal energy (causing a temperature drop).

[16] Chapter 5 considers uncertainties in detail.

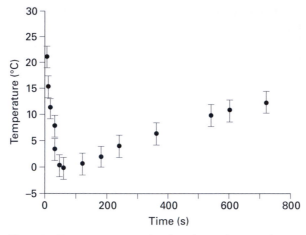

Figure 1.2. Temperature versus time for a thermoelectric cooler.

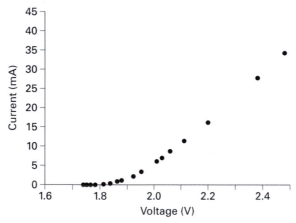

Figure 1.3. Variation of current with voltage for an LED.

1.5.3   Logarithmic scales

The scales on the graph in figure 1.2 are linear. That is, each division on the x axis corresponds to a time interval of 200 s and each division on the y axis corresponds to a temperature interval of 5 °C. In some situations important information can be obscured if linear scales are employed. As an example, consider the current–voltage relationship for a light emitting diode (LED) as shown in figure 1.3.

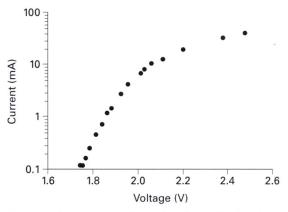

Figure 1.4. Current versus voltage values for an LED plotted on semi-logarithmic scales.

It is difficult to determine the relationship between current and voltage for the LED in figure 1.3 for values of voltage below about 1.9 V. As the current data span several orders of magnitude, the distribution of values can be more clearly discerned by replacing the linear y scale in figure 1.3 by a logarithmic scale. Though graph paper is available that has logarithmic scales, many computer based graph plotting routines, including those supplied with spreadsheet packages, allow easy conversion of the y or x or both axes from linear to logarithmic scales. Figure 1.4 shows the data from figure 1.3 re-plotted using a logarithmic y scale. As one of the axes remains linear, this type of graph is sometimes referred to as *semi-logarithmic*.

Exercise E

The variation of current through a Schottky diode[17] is measured as the temperature of the diode increases. Table 1.8 shows the data gathered in the experiment. Choosing appropriate scales, plot a graph of current versus temperature for the Schottky diode.

1.6 Key numbers summarise experimental data

Scientists strive to express data concisely so that important features are not obscured. The histogram can give us the 'big picture' regarding the distribution of values and can alert us to important features such as lack of symmetry in the

---

[17] A Schottky diode is a device consisting of a junction between a metal and a semiconductor. Such diodes are used extensively in power supplies.

Table 1.8. *Variation of current with temperature for a Schottky diode.*

Temperature (K)	Current (A)
297	2.86×10^{-9}
317	1.72×10^{-8}
336	6.55×10^{-8}
353	2.15×10^{-7}
377	1.19×10^{-6}
397	3.22×10^{-6}
422	1.29×10^{-5}
436	2.45×10^{-5}
467	9.97×10^{-5}
475	1.41×10^{-4}

distribution, or the existence of outliers. This information, though vital, is essentially qualitative. What quantitative measures can we use to summarise all the data?

1.6.1 The mean and the median

It might seem surprising that many values may be usefully summarised by a single number, better referred to as a 'statistic'. But this is exactly what is done on a routine basis. Suppose that, as part of an experiment, we are required to measure the diameter of a wire. Upon making the measurements of diameter with a micrometer we find small variations in the diameter along the wire (these could be due to 'kinks', bends or scratches in the wire, lack of experience in using the measuring instrument, or variations in diameter that occurred during the manufacturing process). Whatever the cause of the variations, there is unlikely to be any reason for favouring one particular value over another. What is required is to determine an 'average' of the values which is regarded as representative of all the values. Several types of average may be defined, but the most frequently used in the physical sciences is the *mean*, $\bar{x}$, which is defined as

$$\bar{x} = \frac{x_1 + x_2 + x_3 + \cdots + x_n}{n} = \frac{\sum_{i=1}^{i=n} x_i}{n}, \tag{1.6}$$

where x_i denotes the ith value and n is the number of values.[18]

---

[18] The limits of the summation are often not shown explicitly, and we write $\bar{x} = \frac{\sum x_i}{n}$.

Table 1.9. *Resonance frequency in an a.c circuit.*

Frequency (Hz)	2150	2120	2134	2270	2144	2156	2139	2122

Table 1.10. *Resonance frequency in ascending order from left to right.*

Frequency (Hz)	2120	2122	2134	2139	2144	2150	2156	2270

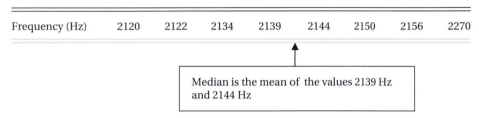

Median is the mean of the values 2139 Hz and 2144 Hz

If values have been grouped so that the value, x_i, occurs f_i times, then the mean is given by

$$\bar{x} = \frac{\sum f_i x_i}{\sum f_i}. \tag{1.7}$$

Another 'average' is the *median* of a group of values. Suppose data are ordered from the smallest to the largest value. The median separates the ordered values into two halves. As an example, consider the data in table 1.9 which shows values of the resonance frequency of an a.c. circuit.

Table 1.10 shows the resonant frequencies arranged in ascending order.

As the median divides the ordered values into two halves, it must lie between 2139 Hz and 2144 Hz. The median is taken to be the mean of these two values, i.e. $\frac{2139+2144}{2} = 2141.5$ Hz.

If n is an even number, the median is the mean of the $(n/2)$th value and the $(n/2+1)$th value. If n is an odd number, the median is the $(n/2 + \frac{1}{2})$th value.

Although the mean and median differ little when there is a symmetric spread of data, they do differ considerably when the spread is asymmetric, or if the group contains an outlier. As an example, for the data in table 1.9, the mean is 2154.4 Hz and the median is 2141.5 Hz. The difference between the mean and median is 12.9 Hz. The 'discrepancy' between mean and median is largely due to the outlier with value 2270 Hz. If the outlier is discarded, then the mean is 2137.9 Hz and the median is 2139 Hz, representing a difference of just over 1 Hz.

We are not suggesting that outliers should be discarded, as this is a matter requiring very careful consideration. However, this example does illustrate that the mean is more sensitive to outliers than the median. Despite this sensitivity,

Table 1.11. *Capacitance values.*

Capacitance (pF)	103.7	100.3	98.4	99.3	101.0	106.1	103.9	101.5	100.9	105.3

the mean is more widely used than the median in the physical sciences for characterising the average of a group of values.

While the mean is arguably the most important number, or statistic, that can be derived from a group of repeat measurements, by itself it tells us nothing of the spread of the values. We now seek a number that is representative of the spread of the data.

Exercise F

Determine the mean and the median of the values of capacitance in table 1.11.

1.6.2 Variance and standard deviation

The starting point for finding a number which usefully describes the spread of values is to calculate the deviation from the mean of each value. If the ith value is written as x_i, then the deviation, d_i, is defined as[19]

$$d_i = x_i - \bar{x}, \tag{1.8}$$

where $\bar{x}$ is the mean of the values.

At first inspection it appears plausible to use the mean of the sum of the deviations as representative of the spread of the values. In this case,

$$\text{mean deviation} = \frac{\sum d_i}{n} = \frac{1}{n}\sum (x_i - \bar{x}), \text{ and expanding the brackets gives}$$

$$\text{mean deviation} = \frac{\sum x_i}{n} - \frac{\sum \bar{x}}{n}.$$

Now $\frac{\sum x_i}{n}$ is the mean, $\bar{x}$, and $\sum \bar{x} = \bar{x} + \bar{x} + \bar{x} + \cdots = n\bar{x}$, so

$$\text{mean deviation} = \bar{x} - \frac{n\bar{x}}{n} = 0.$$

As the mean deviation is always zero, it is not a promising candidate as a number useful for describing the amount of spread in a group of values.

---

[19] Note that d_i is sometimes referred to as the *residual*.

As a useful measure of spread we introduce the *variance*, σ^2, which is defined as the mean of the sum of the square of the deviations, so that

$$\sigma^2 = \frac{\sum (x_i - \bar{x})^2}{n}. \tag{1.9}$$

One of the difficulties with using variance as a measure of spread of values is that its units are the square of the units in which the measurements were made. A new quantity based on the variance is therefore defined which is the *standard deviation* and is equal to the square root of the variance. Representing the standard deviation by σ, we have

$$\sigma = \left[\frac{\sum (x_i - \bar{x})^2}{n}\right]^{1/2}. \tag{1.10}$$

Except in situations where we retain extra figures to avoid rounding errors in later calculations, we will express standard deviations to two significant figures.[20]

Example 3

A rare earth oxide gains oxygen when it is heated to high temperature in an oxygen-rich atmosphere. Table 1.12 shows the mass gain from twelve samples of the oxide which were held at 600 °C for 10 h.

Calculate the (i) mean, (ii) standard deviation and (iii) variance of the values in table 1.12.

ANSWER

(i) The mean of the values in table 1.12 = 5.9083 mg.

(ii) The standard deviation, as defined by equation 1.10, can be found on many scientific pocket calculators, such as those made by CASIO and Hewlett Packard. An alternative is to use a computer based spreadsheet, as most have built in functions for calculating σ. If neither of these options is available then it is possible to use equation 1.10 directly. To assist computation, equation 1.10 is rearranged into the form

$$\sigma = \left(\frac{\sum x_i^2}{n} - (\bar{x})^2\right)^{1/2}. \tag{1.11}$$

For the data in table 1.12, $\sum x_i^2 = 422.53 (\text{mg})^2$ and $\bar{x} = 5.9083$ mg, so that

$$\sigma = \left(\frac{422.53}{12} - (5.9083)^2\right)^{1/2} = 0.55 \text{ mg}.$$

(iii) The variance = σ^2 = 0.30 $(\text{mg})^2$ = 0.30 mg^2

---

[20] See Barford (1985) for a discussion of rounding standard deviations.

Rosewarne Learning Centre

Table 1.12. *Mass gain of twelve samples of ceramic.*

Mass gain (mg)	6.4	6.3	5.6	6.8	5.5	5.0	6.2	6.1	5.5	5.0	6.2	6.3

Table 1.13. *Heights to which water rises in a capillary tube.*

Height (cm)	4.15	4.10	4.12	4.12	4.32	4.20	4.18	4.13	4.15

Exercise G

(1) Show that equation 1.10 may be rewritten in the form given by equation 1.11.

(2) When a hollow glass tube of narrow bore is placed in water, the water rises up the tube due to capillary action. The values of height reached by the water in a small bore glass tube are shown in table 1.13.

For the values in table 1.13 determine the:
  (i) range;
  (ii) mean;
  (iii) median;
  (iv) variance;
  (v) standard deviation.

1.7 Population and sample

In an experiment we must make decisions regarding the amount of time to devote to gathering data. This likely means that fewer measurements are made than would be 'ideal'. But what *is* ideal? This depends on the experiment being performed. As an example, suppose we want to know the mean and standard deviation of the resistance of a batch of 100000 newly manufactured resistors. Ideally, we would measure the resistance, R_i, of every resistor then calculate the mean resistance, $\bar{R}$, using

$$\bar{R} = \frac{\sum_{i=1}^{i=100\,000} R_i}{100\,000}. \tag{1.12}$$

If the resistance of every resistor *is* measured, we regard the totality of values produced as the *population*. The standard deviation of the resistance values may be determined using equation 1.10.

Measuring the resistance of every resistor is costly and time consuming. Realistically, measurements are made of n resistors drawn at random from the population, where $1 < n << 100000$. The values of resistance obtained are regarded as a *sample* taken from a larger population. We hope (and anticipate) that the sample is representative of the whole population, so that the mean and the standard deviation of the sample are close to that of the population mean and standard deviation.

The population of resistors in the previous example, though quite large, is finite. There are other situations in which the size of the population is regarded as infinite. Suppose, for example, we choose a single resistor and measure its resistance many times. The values of resistance obtained are not constant but vary due to many factors including ambient temperature fluctuations, 50 Hz electrical interference, stability of the measuring instrument and (if we carried on for a very long time) ageing of the resistor. The number of repeat measurements of resistance that *could* be made on a single resistor is infinite and so we should regard the population as infinite. No matter the size of the population, we can estimate the mean and standard deviation by considering a sample drawn from the population. We will discover in chapter 3 that the larger the sample, the better are these estimates.

1.7.1 Population parameters

A population of values has a mean and a standard deviation. Any number that is characteristic of a population is referred to as a *population parameter*. One such parameter, the population mean, is usually denoted by the Greek symbol, μ. In situations in which the population is infinite, μ is defined as[21]

$$\mu = \lim_{n \to \infty} \frac{\sum x_i}{n},$$ (1.13)

where x_i is the ith value and n is the number of values.

Similarly, for an infinite population, the standard deviation of the population, σ (often referred to as the 'population standard deviation') is given by

$$\sigma = \lim_{n \to \infty} \left[\frac{\sum (x_i - \mu)^2}{n} \right]^{1/2}.$$ (1.14)

---

[21] Essentially, equation 1.13 says that the population mean is equal to $\frac{\sum x_i}{n}$, in the situation where n tends to infinity.

1.7.2 True value and population mean

The term 'true value' is often used to express the value of a quantity that would be obtained if no influences, such as shortcomings in an instrument used to measure the quantity, existed to affect the measurement. In order to determine the true value of a quantity, we require that the following conditions hold.

- The quantity being measured does not change over the time interval in which the measurement is made.
- External influences that might affect the measurement, such as power supply fluctuations, or changes in room temperature and humidity, are absent.
- The instrument used to make the measurement is 'ideal'.[22]

As an example, suppose an experiment is devised to measure the charge carried by an electron. If only experimental techniques were good enough, and measuring instruments ideal, we would be able to know the value of the charge exactly. As the ideal instrument has yet to be devised, we must make do with values obtained using the best instruments and experimental techniques available. By making many repeat measurements and taking the mean of the values obtained, we might expect that the spread in values due to imperfections in the measurement process would cancel out, in which case the mean would be close to the true value. We might even go one step further and suggest that if we make an infinite number of measurements, the mean of the values (i.e. the population mean, μ) will coincide with the true value. Unfortunately, due to systematic errors,[23] not all imperfections in the measurement process 'average out' by taking the mean, and so we must be very careful when equating the population mean with the 'true value'.

There are situations in which using the term 'true value' may be misleading. Returning to our example in which a population consists of the resistances of 100000 resistors, there is no doubt that this population has a mean, but in what sense, if any, is this population mean the 'true value'? Unlike the example of determination of the charge on an electron, in which variability in values is due to inadequacies in the measurement process, the variability in resistance values is mainly due to variations *between* resistors introduced during manufacture. Therefore no true value, in the sense used to describe an attribute of a single entity, such as the charge on an electron, exists for the group of resistors.

---

[22] An ideal instrument would have many (non-attainable) attributes such as freedom of influence on the quantity being measured (see section 5.6.5), zero drift over time, and the capability of infinitely fine resolution.

[23] Sections 1.8.2 and 5.6.2 consider systematic errors.

1.7.3 Sample statistics

As values of a quantity usually show variability, we take the mean, $\bar{x}$, of values obtained through repeat measurement as the best estimate of the true value (or the population mean). $\bar{x}$ is referred to as the *sample mean*, where

$$\bar{x} = \frac{\sum x_i}{n}, \tag{1.15}$$

and n is the number of values in the sample.

As $\bar{x}$ is determined using a sample of values drawn from a population, it is an example of a *sample statistic*. As n tends to a large value, then $\bar{x} \rightarrow \mu$, as given by equation 1.13.

Another extremely important statistic determined using sample data is the estimate of the population standard deviation, s, given by[24]

$$s = \left[\frac{\sum (x_i - \bar{x})^2}{n - 1} \right]^{1/2}. \tag{1.16}$$

As n tends to a very large number then $\bar{x} \rightarrow \mu$ and the difference between n and $n - 1$ becomes negligible, so that $s \rightarrow \sigma$.

Subtracting 1 from n in the denominator of equation 1.16 is sometimes referred to as the 'Bessel correction'.[25] It arises from the fact that before we can use data to calculate s using equation 1.16, we must first use the same data to determine $\bar{x}$ using equation 1.15. In doing this we have reduced the number of *degrees of freedom* by 1. The degrees of freedom[26] is usually denoted by the symbol, ν.

Example 4

In a fluid flow experiment, the volume of water flowing through a pipe is determined by collecting water at one minute intervals in a measuring cylinder. Table 1.14 shows the volume of water collected in ten successive 1 minute intervals.

(i) Determine the mean of the values in table 1.14.
(ii) Estimate the standard deviation and the variance of the population from which these values were drawn.

---

[24] In some texts, s as defined in equation 1.16 is referred to as the *sample standard deviation*. We will not use this term in this text.

[25] See Ghilani (2010).

[26] In general, ν is equal to the number of observations n, minus the number of constraints (in this situation there is one constraint, given by equation 1.15).

ANSWER

With only ten values to deal with, it is quite easy to find $\bar{x}$ using equation 1.15. Determining s using equation 1.16 requires more effort. Most scientific calculators allow you to enter data and will calculate $\bar{x}$ and s. A better alternative is to use a spreadsheet (see chapter 2) as values entered can be inspected before proceeding to the calculations.

  (i) Using equation 1.15, $\bar{x} = 251.3$ cm^3.
  (ii) Using equation 1.16, the estimate of population standard deviation, $s = 6.1$ cm^3. The variance, $s^2 = 37$ (cm^3)$^2 = 37$ cm^6.

Exercise H

In an experiment to study electrical signals generated by the human brain, the time elapsed for a particular 'brain' signal to double in size when a person closes both eyes was measured. The values obtained for 20 successive eye closures are shown in table 1.15.

   Using the values in table 1.15, determine the sample mean and estimate the standard deviation of the population.

1.7.4   Which standard deviation do we use?

A reasonable question to ask is 'which equation should be used to determine standard deviation?' (After all, we have already introduced three, equations 1.10, 1.14 and 1.16.)

Table 1.14. *Volume of water collected in 1 minute intervals.*

Volume (cm^3)	256	250	259	243	245	260	253	254	244	249

Table 1.15. *Times for electrical signal to increase by a factor of two.*

				Time (s)					
4.51	2.33	1.51	1.91	2.54	1.91	1.51	1.52	2.71	3.03
2.12	2.61	0.82	2.51	2.07	1.73	2.34	1.82	2.32	1.92

The first equation we can 'discard' is equation 1.14. The reason for this is that the parameter μ appears in the equation, and there is no way we can determine this parameter when the population is infinite. The best we can do is to estimate μ. As our goal is to identify the characteristics of the population from which the values are drawn, we choose equation 1.16 in preference to equation 1.10. This is because equation 1.16 *is* an estimate of the population standard deviation. By contrast, equation 1.10 gives the standard deviation of a population when *all* the values that make up that population are known. Since this rarely, if ever, applies to data gathered in experiments in the physical sciences, equation 1.10 is not favoured.

Though equation 1.16 is preferred to 1.10, it may be demonstrated that, as the number of values, n, becomes large, the difference between the standard deviation calculated using each becomes negligible. To show this, we write the percentage difference, d, between the standard deviations given by equations 1.10 and 1.16 as

$$d = \left(\frac{s - \sigma}{s}\right) \times 100\%. \tag{1.17}$$

By substituting equations 1.16 and 1.10 for s and σ respectively, we obtain

$$d = \left[1 - \left(\frac{n - 1}{n}\right)^{1/2}\right] \times 100\%. \tag{1.18}$$

Figure 1.5 shows the variation of d with n as given by equation 1.18, for n between 3 and 100. Figure 1.5 indicates that as n exceeds 10, the percentage difference between standard deviations falls below 5%.

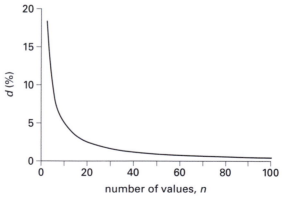

Figure 1.5. Percentage difference between s and σ as a function of n.

1.7.5 Approximating *s*

Calculating the standard deviation using equation 1.16 is time consuming, though using a spreadsheet or pocket calculator reduces the effort considerably. As the standard deviation, *s*, is usually not required to more than two significant figures (and often one figure is sufficient), we can use an approximation which works well so long as the number of values, *n*, does not exceed 12. The approximation is[27]

$$s \approx \frac{\text{range}}{\sqrt{n}}, \tag{1.19}$$

where the range is the difference between the maximum value and the minimum value in any given set of data (see equation 1.5). We can regard equation 1.19 as a 'first order approximation' to equation 1.16 for small *n*. Equation 1.19 is useful for determining *s* rapidly and with the minimum of effort.[28]

Example 5

Table 1.16 contains values of repeat measurements of the distance between a lens and an image produced using the lens.

Use equations 1.16 and 1.19 to determine the standard deviation of the values in table 1.16.

ANSWER

Using equation 1.16, $s = 2.5$ cm.

Using equation 1.19,

$$s \approx \frac{(42 - 35)\text{cm}}{\sqrt{7}} = 2.6 \text{ cm}.$$

The difference between the two values of *s* is about 5% and would be regarded as unimportant for most applications.

Exercise I

The period, *T*, of oscillation of a body on the end of a spring is measured and values obtained are given in table 1.17.

  (i) Use equation 1.16 to find the standard deviation, *s*.

 (ii) Use equation 1.19 to find an approximate value for *s*.

(iii) What is the percentage difference between the standard deviations calculated in parts (i) and (ii)?

---

[27] See Lyon (1980) for a discussion of equation 1.19.

[28] In situations in which *n* exceeds 12, the value of *s* given by equation 1.19 is consistently less than *s* as calculated using equation 1.16.

Table 1.16. *Image distances.*

Image distance (cm)	35	42	40	38	39	36	36

Table 1.17. *Period of oscillation of a body on the end of a spring.*

T (s)	2.53	2.65	2.67	2.56	2.56	2.60	2.67	2.64	2.63

1.8 Experimental error

The elusive true value of a quantity sought through experiment could be established with a single measurement so long as 'interfering' factors (such as human reaction time when measuring an interval of time using a stopwatch) could be eliminated. Such factors are the cause of *errors* in values obtained through measurement. If a scale is misread on a meter, or a number is transcribed incorrectly, this is not regarded as an error in the context of data analysis, but as a mistake or blunder which (in principle) can be avoided. Errors exist in all experiments and though their effects may be minimised in a number of ways, such as controlling the physical environment in which an experiment is performed, they cannot be eliminated completely. Chapter 5 deals with the identification and treatment of errors, but it is appropriate to introduce here two important categories of errors: random and systematic.

1.8.1 Random error

The International Vocabulary of Metrology (VIM) defines random measurement error as the:

> component of measurement error that in replicate measurements varies in an unpredictable manner.

As a consequence of random errors, measured values are either above or below the true value. In fact, the most obvious 'signature' of random errors is a scatter or variability of values obtained through experiment.

If we know the size of the error, we can correct the measured value by a amount equal to the error and hence obtain the true value. Regrettably, the size of the error is as elusive to determine as the true value. Nevertheless, by gathering many values we can quantify the scatter caused by random errors through determining the standard deviation of the values. The existence of random error introduces some *uncertainty* in the estimate of the true value of a quantity. Though the error cannot be determined, the uncertainty can. This is very important since, by calculating the uncertainty, we can express limits between which the true value of a quantity is expected to lie. When repeat measurements of a quantity are made, we can write:

interval containing the true value = sample mean ± uncertainty.

We consider uncertainty in measurement in some detail in chapter 5.

1.8.2 Systematic error

The International Vocabulary of Metrology (VIM) defines systematic measurement error as the:

> *component of measurement error that in replicate measurements remains constant or varies in a predictable manner.*

The existence of random errors is revealed through the variability in measured values. By contrast, systematic errors give no such clues. In an experiment, a poorly calibrated or faulty thermometer may consistently indicate water temperature as 25 °C, when the true value is close to 20 °C. The consistent *offset* or *bias* which causes the measured value to be consistently above or below the true value is most often referred to as a *systematic* error. From this example it might seem that all we need do is check the calibration of the instrument, correct the values appropriately and thereby eliminate the systematic error. Though we might reduce the systematic error in this way it cannot be eliminated as it requires the calibration procedure to be 'perfect' and that relies on some superior instrument determining the true value. No matter how 'superior' the instrument is, it still cannot determine the true value of the quantity, so some systematic error must remain. Further systematic error may be introduced when dynamic measurements are made. For example, a thermometer may be used to measure the temperature rise of a water bath. Irrespective of how quickly a thermometer responds to a temperature rise, there will be some small difference between the water temperature and the temperature indicated by the thermometer.

1.8.3 Repeatability and reproducibility

A particular quantity, for example the time for a ball to fall through a viscous liquid, may be measured a number of times by the same observer using the same equipment where the conditions of the measurement (including the environmental conditions) remain unchanged. If the scatter of values produced is small, we say the measurement is *repeatable*. By contrast, if measurements are made of a particular quantity by various workers using a variety of instruments and techniques in different locations, we say that the measurements are *reproducible* if the values obtained by the various workers are in close agreement. The terms repeatability and reproducibility are qualitative only and in order to assess the degree of repeatability or reproducibility we must consider quantitative estimates of scatter such as the standard deviation.

1.9 Modern tools of data analysis – the computer based spreadsheet

The availability of powerful computer based data analysis packages has reduced the effort required to plot histograms and graphs, calculate standard deviations and perform statistical analysis of data. A data set consisting of hundreds or thousands of values is no more difficult to summarise graphically and numerically than one containing 10 or 20 values. Though statistical analysis tools are available in dedicated statistics packages, many spreadsheets incorporate analysis features useful to the scientist. Excel by Microsoft is a powerful and widely available spreadsheet package that is employed in this text where it complements the discussion of methods of analysis and assists in computation.

It is worth sounding a note of caution here. A spreadsheet (or any other computer based package) can return the results of thousands of calculations almost instantly. It is important to ask 'are the results reasonable?'. The meaning and validity of the results deserve some consideration. Using a spreadsheet places a responsibility on the user to understand what is going on 'behind the scenes' when a calculation is performed. A preliminary evaluation of data (such as roughly estimating the mean of many values) can often help anticipate the number that the spreadsheet will return. If the number differs greatly from that expected, further investigation is warranted. Despite this concern, we recommend the use of spreadsheets for data analysis in science and recognise that they have become so well established that their use is taken as much for granted as the pocket calculator.

1.10 Review

Prior to the analysis of data, it is proper to consider issues relevant to good experimentation. Good experimentation begins with careful consideration of such matters as:

- what is the purpose of the experiment;
- is the experimental design feasible;
- what factors might constrain the experiment (such as the availability of equipment);
- which 'exploratory' data analysis techniques should be used (such as the determination of means and standard deviations) and what display methods will allow the most insight to be drawn from the data?

In this chapter we have considered issues associated with experimentation and data analysis, such as how to express values determined through experiment. Of special importance is the capacity to present data in ways that reveal characteristic features or trends in the data, as well as exposing anomalies or outliers. Graphs and histograms provide excellent visual summaries of data and, if carefully constructed, these can reveal features of interest that would be extremely difficult to identify through inspection of a table of 'raw' data.

Complementary to summarising data visually is to summarise data numerically. To this end we introduced the mean and standard deviation as the most frequently used measures of 'average' and 'scatter' respectively. In addition, we briefly considered units, standards, errors and uncertainty.

The vocabulary used when discussing measurement and data analysis can be confusing owing to the everyday usage of terms such as accuracy and error, as well as the sometimes inconsistent way terms are used. In this text we give preference to the definition of terms such as error and accuracy as expressed in the International Vocabulary of Metrology (VIM) so that this text is consistent with international usage of the terms.

In chapter 2 we will consider the spreadsheet as a powerful computer based tool that can assist in the efficient analysis of experimental data.

End of chapter problems

(1) What are the derived SI units of the quantities:

  (i)   specific heat capacity;
  (ii) pressure;
 (iii) thermal conductivity;
  (iv) impulse;
   (v) electrical resistivity;
  (vi) magnetic flux;
 (vii) heat flux density?

(2) Express the units for the quantities in question (1) in terms of the base units of the SI system.

(3) (i) When the temperature of a metal rod of length, l_0, rises by an amount of temperature, ΔT, the rod's length increases by an amount, Δl, where

$$\Delta l = l_0 \; \alpha \Delta T,$$

and α is known as the temperature coefficient of expansion.
Use this equation to express the unit of α in terms of the base units of the SI system.

(ii) The force, F, between two point charges Q_1 and Q_2, separated by a distance, r, in a vacuum is given by

$$F = \frac{Q_1 Q_2}{4\pi\varepsilon_0 r^2}, \quad \text{where } \varepsilon_0 \text{ is the permittivity of free space.}$$

Use this equation and the information in table 1.3 to express the unit of ε_0 in terms of the base units in the SI system.

(4) Write the following values in scientific notation to two significant figures:

(i) 0.0000571 s;
(ii) 13700 K;
(iii) 1387.5 m/s;
(iv) 101300 Pa;
(v) 0.001525 Ω.

(5) The lead content of river water was measured five times each day for 20 days. Table 1.18 shows the values obtained in parts per billion (ppb).

(i) Construct a histogram beginning at 30 ppb and extending to 70 ppb with interval (bin) widths of 5 ppb.
(ii) Determine the median value of the lead content.

Table 1.18. *Lead content of river water (in ppb).*

43	58	53	49	60	48	48	49	49	45
57	35	51	67	49	51	59	55	62	59
40	52	53	70	48	51	44	51	46	47
42	41	53	56	40	42	47	54	54	56
46	40	57	57	47	54	48	61	51	56
56	57	45	42	54	56	66	48	48	52
64	56	49	54	66	45	63	49	65	40
54	54	50	56	51	49	51	46	52	41
43	46	57	51	46	68	58	69	52	49
52	53	46	53	44	36	54	60	61	67

(6) Table 1.19 contains values of the retention time (in seconds) for pseudo-ephedrine using high performance liquid chromatography (HPLC).

    (i) Construct a histogram using the values in table 1.19.
    (ii) Determine the mean retention time.
    (iii) Estimate the population standard deviation of the values in table 1.19.

Table 1.19. *Retention times for pseudoephedrine (in seconds).*

6.035	6.049	6.032	6.065	6.057	6.069	6.084	6.110
6.122	6.066	6.072	6.046	6.100	6.262	6.262	6.252
6.276	6.042	6.067	6.054	6.098	6.093	6.072	6.124
6.085	6.076	6.045	6.067	6.220	6.223	6.271	6.219

(7) The variation of electrical resistance of a humidity sensor with relative humidity is shown in table 1.20.

    (i) Plot a graph of resistance versus relative humidity (consider: should linear or logarithmic scales be employed?).
    (ii) Use the graph to estimate the relative humidity when the resistance of the sensor is 5000 Ω.

Table 1.20. *Variation of the resistance of a sensor with humidity.*

Resistance (Ω)	Relative humidity (%)
90.5	98
250	90
945	80
3250	70
8950	60
22500	50
63500	40
82500	30
124000	20

(8) Consider table 1.21 which contains 45 values of nitrate ion concentration.

    (i) Calculate the mean, $\bar{x}$, and standard deviation, s, of the nitrate concentration.
    (ii) Draw up a histogram for the data in table 1.21.

(9) Using accurate clocks in satellites orbiting the Earth, the global positioning system (GPS) may establish the position of a GPS receiver on the Earth.

Table 1.21. *Nitrate ion concentration (in µmol/mL).*

0.481	0.462	0.495	0.493	0.501	0.481	0.479	0.503	0.497
0.506	0.512	0.457	0.521	0.474	0.487	0.504	0.445	0.490
0.455	0.480	0.509	0.484	0.475	0.511	0.481	0.509	0.500
0.490	0.550	0.493	0.513	0.483	0.505	0.503	0.501	0.506
0.480	0.491	0.493	0.509	0.475	0.482	0.486	0.479	0.488

Table 1.22 shows 15 values of the vertical displacement, d (relative to mean sea level), of a fixed receiver, as determined using the GPS.

Using the values in table 1.22:

   (i)   calculate the mean displacement and median displacement;
   (ii) calculate the range of values;
   (iii) estimate the population standard deviation and the population variance.

Table 1.22. *Vertical displacement of a GPS receiver.*

d (m)	120	108	132	125	118	106	115	103
	117	120	135	129	123	127	128	

(10) The masses of 120 'Ownbrand' tablets containing paracetemol are shown in table 1.23.

Plot a histogram of the data in table 1.23. Determine the mean mass, and estimate the population variance of the tablets.

Table 1.23. *Masses of 120 Ownbrand paracetemol tablets (g).*

0.5429	0.5447	0.5460	0.5470	0.5480	0.5489	0.5492	0.5498	0.5503	0.5506
0.5430	0.5448	0.5461	0.5472	0.5480	0.5489	0.5492	0.5498	0.5503	0.5506
0.5430	0.5449	0.5461	0.5472	0.5480	0.5489	0.5493	0.5499	0.5503	0.5507
0.5433	0.5453	0.5463	0.5474	0.5481	0.5490	0.5493	0.5499	0.5504	0.5508
0.5433	0.5453	0.5463	0.5474	0.5485	0.5490	0.5493	0.5500	0.5504	0.5508
0.5436	0.5455	0.5465	0.5475	0.5485	0.5490	0.5493	0.5500	0.5505	0.5508
0.5436	0.5455	0.5465	0.5477	0.5486	0.5490	0.5495	0.5500	0.5505	0.5509
0.5438	0.5455	0.5465	0.5477	0.5486	0.5490	0.5495	0.5500	0.5505	0.5510
0.5438	0.5456	0.5466	0.5477	0.5486	0.5491	0.5495	0.5501	0.5505	0.5511
0.5439	0.5456	0.5466	0.5478	0.5487	0.5491	0.5497	0.5502	0.5505	0.5511
0.5440	0.5458	0.5468	0.5479	0.5488	0.5492	0.5497	0.5502	0.5505	0.5511
0.5446	0.5460	0.5468	0.5479	0.5488	0.5492	0.5497	0.5503	0.5506	0.5512

Chapter 2

Excel and data analysis

2.1 Introduction

Thorough analysis of experimental data frequently requires extensive numerical manipulation. Many tools exist to assist in the analysis of data, ranging from the pocket calculator to specialist computer based statistics packages. Despite limited editing and display options, the pocket calculator remains a well-used tool for basic analysis due to its low cost, convenience and reliability. Intensive data analysis may require a statistics package such as Systat or Origin.[1] As well as standard functions, such as those used to determine means and standard deviations, these packages possess advanced features routinely required by researchers and professionals. Between the extremes of the pocket calculator and specialised statistics package is the spreadsheet. While originally designed for business users, spreadsheet packages are popular with other users due to their accessibility, versatility and ease of use. The inclusion of advanced features into spreadsheets means that, in many situations, a spreadsheet is a viable alternative to a statistics package. The most widely used spreadsheet available for personal computers (PCs) is Excel by Microsoft. Excel appears within this book in the role of convenient data analysis tool with short sections within most chapters devoted to describing specific features. Its clear layout, extensive help facilities, range of in-built statistical functions and availability for both PCs and Mac computers make Excel a popular choice for data analysis. This chapter introduces Excel and describes some of its basic features using examples drawn from the physical sciences. Some familiarity with using a PC is assumed, to the extent that terms such as 'mouse', 'pointer', 'Enter key' and 'save' are assumed understood in the context of using a program such as Excel.

---

[1] Systat is a product of Systat Inc, Illinois. Origin is a product of OriginLab Corporation, Massachusetts.

2.2 What is a spreadsheet?

A computer based spreadsheet is a sophisticated and versatile analysis and display tool for numeric and text based data. As well as the usual arithmetic and mathematical functions found on pocket calculators, spreadsheets offer other features such as data sorting and display of data in the form of an x–y graph. Some spreadsheet packages include more advanced analysis options such as linear regression and hypothesis testing. An attractive feature of many spreadsheets is the ability to accept data directly from other computer based applications, simplifying and speeding up data entry as well as avoiding mistakes caused by faulty transcription.

A spreadsheet consists of a two-dimensional array of rectangular boxes, usually referred to as *cells*, into which text, symbols, numbers or formulae can be typed. An example of such an array is shown in sheet 2.1.[2] The data appearing in the array were gathered during a study of the performance of a vacuum system. The letters A and B in sheet 2.1 serve to identify the columns in the spreadsheet, and the numbers 1 to 5 identify the rows. So, for example, the number 95 appears in cell B3.

Sheet 2.1. *Spreadsheet containing data on the variation of pressure in a vacuum system with time after switching on a vacuum pump.*

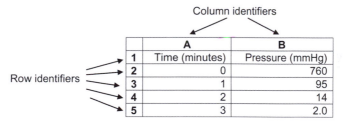

	A	B
1	Time (minutes)	Pressure (mmHg)
2	0	760
3	1	95
4	2	14
5	3	2.0

As it stands, sheet 2.1 is little more than a table constructed on paper into which values have been entered.[3] Where a computer based spreadsheet differs from a piece of paper is the dynamic way in which data in the spreadsheet can be manipulated, linked together and presented pictorially, for instance as an x–y graph. For example, the analysis of the data in sheet 2.1 may require that the pressure be expressed in pascals instead of mmHg, the time be expressed in

---

[2] We will use the term 'sheet', rather than 'table', in order to indicate that we are dealing with a spreadsheet.

[3] In chapter 1 we defined a 'value' as consisting of the product of number and a unit. When we refer in this and other chapters to 'entering a value' into a spreadsheet, it is taken for granted that we are considering only the numerical part of the value.

seconds, and that a graph of pressure versus time be plotted. It is not a difficult task to perform the conversions or plot the graph manually, but if the experiment consists of gathering pressure data at one minute intervals for an hour then the task of manipulating so many data and plotting a graph certainly becomes laborious and time consuming. This is where the spreadsheet is able to assist. With a few key strokes, formulae can be entered into the spreadsheet to perform the conversions. A highly valued feature is the ability to link cells together dynamically. For example, where values have to be corrected, all calculations and graphs using those values are updated as soon as they are typed in and the Enter key pressed. This facility is powerful and allows us to investigate the consequences, for example, of omitting, adding or changing one or more values in a set of data. This is sometimes referred to as a 'What if' calculation.

2.3 Introduction to Excel

Designed originally with business users in mind, Excel has evolved since the late 1980s into a powerful spreadsheet package able to support users from science, mathematics, statistics, engineering and business disciplines. With regard to the analysis of experimental data, Excel possesses 100 or so built in statistical functions which will evaluate such things as the standard deviation, maximum, minimum and mean of values. More advanced analysis facilities are available such as linear and multiple regression, hypothesis testing, histogram plotting and random number generation. Graphing options include pie and bar graphs and, perhaps the most widely used graph in the physical sciences, the 'x–y' graph.

At the time of writing, the latest version of Excel for PCs is Excel 2010. While Excel 2010 and its predecessors offer many shortcuts to setting up and using a spreadsheet,[4] they are not vital for solving data analysis problems and we will use them sparingly.

We will describe briefly some of the features of Excel before moving on to use it. Excel 2010 commonly runs under Windows XP, Windows Vista or Windows 7 operating systems. The screen looks slightly different depending on the operating system, but the differences are minor and do not affect the functionality of Excel.

2.3.1 Starting Excel

If a PC is running under Windows 7 operating system, then one way to start Excel is as follows.

---

[4] For a comprehensive text on the features of Excel 2010, see Walkenbach (2010).

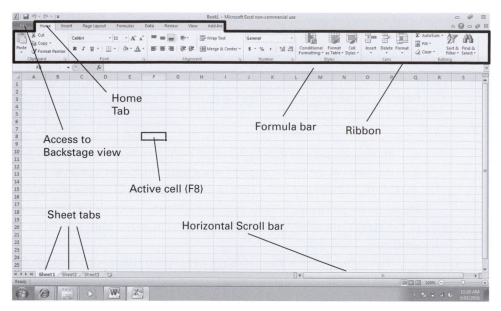

Figure 2.1. Appearance of screen after starting Excel.

(1) Using the left hand mouse button, click on at the bottom left hand corner of the screen.

(2) Click on ▶ All Programs then click on Microsoft Office , then click on X Microsoft Excel 2010 (a shortcut to Excel might exist on your computer, for example on the desktop, which will allow you to start Excel more quickly).

(3) After a few seconds the screen appears which should be similar to that shown in figure 2.1.

Excel commands are available on what Microsoft refers to as the 'Ribbon'. The Ribbon is indicated on figure 2.1 within a thick black outline. The contents of the Ribbon depend on which tab has been clicked. If the Home tab has been clicked, then the Ribbon appears as in figure 2.1. There are seven or more tabs, but we will mainly consider the commands appearing on the Ribbon when the Home, Insert or Data tabs are pressed.

Similar commands are grouped together in the Ribbon to make them easier to find. Figure 2.2 shows a group of commands that appear on the Home Ribbon which allow you to modify font size, type, colour and so on. The commands themselves are accessed by clicking on the buttons.

Figure 2.2. Example of grouped commands on the Home Ribbon.

Moving the pointer over each button highlights that button. A moment later the name of the button appears along with a brief description of what it does. At the bottom and to the right of the screen there are scroll bars which are useful for navigating around large spreadsheets. Towards the bottom left of the screen are sheet tabs which allow you to move from one sheet to another.

New to Excel 2010 is the 'Backstage view' which is accessible through the File tab at the top left of the screen. This tab allows access to features such as 'Save' and 'Print' and makes available other options that you might want to add in to Excel.

2.3.2 Worksheets and workbooks

When starting Excel, a screen of empty cells appears as shown in figure 2.1. Sheet tabs, Sheet1 Sheet2 Sheet3, are visible near to the bottom of the screen. Switching between sheets is accomplished by clicking on each tab. Excel refers to each of the sheets as a worksheet.[5] These sheets (and others, if they are added) constitute a workbook. A workbook can contain many worksheets, and incorporating several sheets into a workbook is useful if interrelated data or calculations need to be kept together. For example, a workbook might consist of three worksheets, where:

- worksheet 1 is used to store the original data as obtained from an experiment;
- worksheet 2 contains values after conversion of units and other data manipulation steps have occurred;
- worksheet 3 contains the values displayed in graphical form.

Worksheets can be renamed by moving the pointer to the worksheet tab and double clicking on it. At this stage it is possible to overwrite (for example) 'Sheet1' with something more meaningful or memorable.

---

[5] Other sheets can be added by clicking on the tab.

2.3.3 Entering and saving data

After starting Excel, the majority of the computer screen is filled with cells. Sheet 2.2 shows cells which contain data from an experiment in which repeat measurements were made in order to establish the potassium concentration in a sample of human blood.

Sheet 2.2. *Potassium concentration in a blood sample.*

	A	B	C	D	E
1	Concentration (mmol/L)				
2	5.2				
3	4.7				
4	4.9				
5	4.6				
6	4.3				
7	3.9				
8	4.2				
9	5.2				
10	5.1				
11	4.6				
12					

To begin entering text and data into the cells, do the following.

(1) Move the pointer, ⊕ , to cell A1.
(2) Click the left hand mouse button[6] to make the cell active. A conspicuous border appears around the active cell.
(3) Type[7] **Concentration (mmol/L)**. Once the Enter key has been pressed, the active cell becomes A2.
(4) Type **5.2** into cell A2 and press the Enter key. Continue this process until all the data are entered. If a transcription mistake occurs, use the mouse to activate the cell containing the mistake then retype the text or number.

Before moving on to manipulate the data, it is prudent to save the data. One way to do this is as follows.

(1) In Excel 2010, click on the ▢ File ▢ tab. (In Excel 2007, click on the Office button, ▢)
(2) Click on the Save As option. At this point you can choose the format in which the file is saved. If compatibility with previous (but fairly recent) versions of

---

[6] There are several alternatives to using a mouse for 'pointing and clicking', including trackballs and touchpads. In the interests of brevity, we will assume that the pointing device being used is a mouse, and that the mouse incorporates left and right buttons.
[7] We adopt the convention that anything to be typed into a cell appears in **bold**.

Excel is important, then choose Excel 97 – 2003 Workbook,[8] otherwise save it as an Excel workbook.

(3) Save the sheet by giving it an easy to remember name such as 'Conc1', then clicking on the <u>S</u>ave button. The extension .xlsx will automatically be attached to the file and identifies the file as having been created in Excel 2007 or Excel 2010.

2.3.4   Rounding, range and the display of numbers

The values appearing in sheet 2.2 are likely to have been influenced by several factors: the underlying variation in the quantity being measured, the resolution of the instrument used to make the measurement, as well as the experience, care and determination of the experimenter. As we begin to use a spreadsheet to 'manipulate' the values, we need to be aware that the spreadsheet introduces its own influences, such as those due to rounding errors or the accuracy of algorithms used in the calculation of the statistics.[9] As with a pocket calculator, a spreadsheet is only able to store a value to a certain number of digits. However, in contrast to a calculator, which might display up to 10 digits and hold two 'in reserve' to ensure that rounding errors do not affect the least significant digit displayed, Excel holds 15 digits internally. It is unlikely you will come across a situation in which 15 digits is insufficient.[10]

Most pocket calculators cannot handle values with magnitudes that fall outside the range 1×10^{-99} to $9.99999 \times 10^{+99}$. A value with magnitude less than 1×10^{-99} is rounded to zero, and a value with magnitude $10 \times 10^{+99}$ (or greater) causes an overflow error (displayed as -E-, or something similar, depending on the type of calculator).

Excel is able to cope with much larger and smaller values than the average scientific calculator, but cannot handle values with magnitudes that fall outside the approximate range 2.3×10^{-308} to $1.8 \times 10^{+308}$. A value with magnitude smaller than 2.3×10^{-308} is rounded to zero. A value typed into a cell which is equal to or in excess of $1 \times 10^{+308}$ is regarded by Excel as a string of characters and not a number. The limitation regarding the size of a value is unlikely to be of concern unless calculations are performed in which a divisor is close to zero. A division by zero causes Excel to display the error message, #DIV/0!. If the result

---

[8] This could be important if you wish to distribute your Excel file to someone who does not have the latest version of Excel.

[9] For a forthright account of the statistical limitations of versions of Excel prior to Excel 2010, see McCullough and Heiser (2008).

[10] It is possible that if a process requires a very large number of iterations, rounding errors will become significant, especially if the calculations require the subtraction of two nearly equal numbers.

of a calculation is a value in excess of $\approx 1.8 \times 10^{+308}$, then the error message #NUM! appears in the cell. Another way to cause the #NUM! message to appear is to attempt to use Excel to calculate the square root of a negative number.

Values appear in cells as well as in the formula bar[11] as you type. Values such as 0.0023 or 1268 remain unchanged once they have been entered. However, if a value is very small, say, 0.0000000000123, or very large, say, 165000000000, then Excel automatically switches to scientific notation. So 0.0000000000123 is displayed as 1.23E-11 and 165000000000 as 1.65E+11. The 'E' notation is interpreted as follows:

$$1.23\text{E-}11 = 1.23 \times 10^{-11}; \text{ and } 1.65\text{E+}11 = 1.65 \times 10^{11}.$$

How a value appears in a cell can be modified. To illustrate some of the display options, consider sheet 2.3 containing data from an experiment in which a capacitor is discharged through a resistor. The voltage across the resistor is measured as a function of time. Cells A1 and B1 contain text indicating the quantities in the A and B columns. Neglecting to include such information risks the consequence of returning to the spreadsheet some time later with no idea of what the numbers relate to. Clear headings and formatting enhances the readability and enduring usefulness of any spreadsheet.

Scientific notation is often preferred for very small or very large values. We can present values in this manner using Excel. As an example, consider the values in column B of sheet 2.3(a). These values are shown in scientific notation in column B of sheet 2.3(b).

Sheet 2.3. *Data from discharge of a 0.47 μF capacitor through a 12 MΩ resistor.*

(a) *Voltage values in column B in 'general' format*

(b) *Voltage values in column B in scientific notation*

	A	B
1	t(s)	V(volts)
2	0	3.98
3	5	1.58
4	10	0.61
5	15	0.24
6	20	0.094
7	25	0.035
8	30	0.016
9	35	0.0063
10	40	0.0031
11	45	0.0017
12	50	0.0011
13	55	0.0007
14	60	0.0006

	A	B
1	t(s)	V(volts)
2	0	3.98E + 00
3	5	1.58E + 00
4	10	6.10E − 01
5	15	2.40E − 01
6	20	9.40E − 02
7	25	3.50E − 02
8	30	1.60E − 02
9	35	6.30E − 03
10	40	3.10E − 03
11	45	1.70E − 03
12	50	1.10E − 03
13	55	7.00E − 04
14	60	6.00E − 04

---

[11] The formula bar is indicated in figure 2.1.

To convert values in column B to scientific notation, do the following.

(1) Highlight cells B2 to B14 in sheet 2.3(a). Do this by moving the pointer to cell B2, pressing down on the left mouse button, then (with the button held down) 'dragging' the mouse from B2 to B14. On reaching cell B14, release the button. Every cell from B2 to B14 now lies within a clearly defined border.

(2) Bring up the Home Ribbon by clicking on the Home tab. In the Cells group (near to the right of the screen) click on Format then click on Format Cells.

(3) A box (referred to as a *dialog* box) appears headed Format Cells. Click on the Number tab. From the list that appears, choose the Scientific option.

(4) The dialog box shows how the first value in the column of highlighted numbers will appear in scientific notation. The number of decimal places to which the value is expressed can be modified at this stage. The default number is two decimal places. Click on the OK button. The values in column B now appear as in sheet 2.3(b).

Another option in the Format Cells dialog box, useful for science applications, is the Number category. With this option the number of decimal places to which a value is displayed may be modified (but the value is not forced to appear in scientific notation). Irrespective of how values are displayed by Excel on the screen, they are stored internally to 15 digits.

2.3.5 Entering formulae

After entering data into a spreadsheet, the next step is usually to perform a mathematical, statistical or other operation on the data. This may be accomplished by entering a formula into one or more cells. Though Excel provides many advanced functions, often only simple arithmetic operations such as multiplication or division are required. As an example, consider the capacitor discharge data in sheet 2.3. Suppose at each point in time we require both the current through the discharge resistor and the charge remaining on the capacitor. The equations for the current, I, and charge, Q, are

$$I = \frac{V}{R} \tag{2.1}$$

$$Q = CV, \tag{2.2}$$

where V is the voltage across the parallel combination of a resistor and a capacitor. R is the resistance and C is the capacitance.

To enter a formula into a cell, we begin the formula with an equals sign. Sheet 2.4(a) shows a formula entered into cell C2. To calculate the current corresponding to the voltage value in cell B2, we divide the value in B2 (which is 3.98 V) by the resistance, 12 MΩ. To do this, do the following.

(1) Make cell C2 active by moving the pointer to C2 and click on the left hand mouse button.

(2) Type =**B2/12E6**.

(3) Press the Enter key.

(4) The value 3.31667E-07 is returned[12] in cell C2, as shown in sheet 2.4(b).

Sheet 2.4. *Calculation of current through a 12 MΩ resistor.*

(a) *Formula typed into cell C2*

(b) *Value returned in C2 when Enter key is pressed*

	A	B	C
1	t(s)	V(volts)	I(amps)
2	0	3.98	=B2/12E6
3	5	1.58	
4	10	0.61	
5	15	0.24	
6	20	0.094	
7	25	0.035	
8	30	0.016	
9	35	0.0063	
10	40	0.0031	
11	45	0.0017	
12	50	0.0011	
13	55	0.0007	
14	60	0.0006	

	A	B	C
1	t(s)	V(volts)	I(amps)
2	0	3.98	3.31667E−07
3	5	1.58	
4	10	0.61	
5	15	0.24	
6	20	0.094	
7	25	0.035	
8	30	0.016	
9	35	0.0063	
10	40	0.0031	
11	45	0.0017	
12	50	0.0011	
13	55	0.0007	
14	60	0.0006	

If the current is required at other times between $t = 5$ s to $t = 60$ s, then it is possible to type =B3/12E6 into cell C3, =B4/12E6 into cell C4, and so on down to =B14/12E6 in cell C14. This is a lot of (unnecessary) work. Spreadsheets are designed to reduce the effort required to carry out such a task. Enter the required formulae as follows.

(1) Move the pointer to cell C2. With the left hand mouse button pressed down, move to cell C14 and release the button. The cells from C2 to C14 should be highlighted as shown in sheet 2.5(a).

(2) Go to the Editing group on the Home Ribbon and click on Fill, Fill ▾.

(3) Click on Down.[13]

(4) Values now appear in cells C3 to C14, as shown in sheet 2.5(b).

---

[12] When Excel performs a calculation, it 'returns' the result of the calculation into the cell in which the formula was typed.

[13] For the sake of brevity from this point on we will write the steps given by steps 2 and 3 as Home>Editing>Fill>Down.

Sheet 2.5. *Entering formulae into a cell.*

(a) *Highlighting cells*

(b) *Contents of cells after choosing* 🔽 *Fill ▾, then* 🔽 Down

	C
1	I(amps)
2	3.31667E – 07
3	
4	
5	
6	
7	
8	
9	
10	
11	
12	
13	
14	

	C
1	I(amps)
2	3.31667E – 07
3	1.31667E – 07
4	5.08333E – 08
5	0.00000002
6	7.83333E – 09
7	2.91667E – 09
8	1.33333E – 09
9	5.25E – 10
10	2.58333E – 10
11	1.41667E – 10
12	9.16667E – 11
13	5.83333E – 11
14	5E – 11

Clicking Home>Editing>Fill>Down, copies the formula into the high-lighted cells and automatically increments the cell reference in the formula so that the calculation is carried out using the value in the cell in the adjacent B column. Cell referencing is discussed in the section 2.3.6.

Another common arithmetic operation is to raise a number to a power. If it is required that we find the square of the contents of, say, cell C2 in sheet 2.5, then we would type[14] in another cell =C2^2. To illustrate this, suppose we wish to calculate the power, P, dissipated in the 12 MΩ resistor. The equation required is

$$P = I^2 R, \tag{2.3}$$

where I is the current flowing through the resistance, R. The formula used to calculate the power dissipated in the resistor is shown in cell D2 of sheet 2.6.

Sheet 2.6. *Calculation of power dissipated in a resistor.*

(a) *Formula to calculate power typed into cell D2*

(b) *Value returned in D2 when the Enter key is pressed*

	C	D
1	I(amps)	P(watts)
2	3.31667E – 07	=C2^2*12E6
3	1.31667E – 07	
4	5.08333E – 08	

	C	D
1	I(amps)	P(watts)
2	3.31667E – 07	1.32003E – 06
3	1.31667E – 07	
4	5.08333E – 08	

---

[14] The ^ symbol is found by holding down the shift key and pressing the '6' key.

Exercise A

(1) Column B of sheet 2.4 shows the voltage across a 0.47 μF capacitor at times $t = 0$ to $t = 60$ s. Calculate the charge remaining on the capacitor (as given by equation 2.2) at times $t = 0$ to $t = 60$ s. Tabulate values of charge in column D of the spreadsheet.

(2) Enter a formula into cell E2 of sheet 2.5 to calculate the square root of the value of current in cell C2. Click Home>Editing>Fill>Down to calculate the square root of the other values of current in sheet 2.5.

(3) The energy, E, stored in a capacitor, C, is given by $E = \frac{1}{2}CV^2$. Given a 'supercapacitor' with capacitance 55 F, use Excel to tabulate the energy stored in the capacitor for voltages from 0.2 V to 2.8 V in steps of 0.2 V.

2.3.6 Cell references and naming cells

Relative referencing

How cells are referenced within other cells affects how calculations are performed. For example, consider the formula, =A1*B1, appearing in cell C1 in sheet 2.7.

Sheet 2.7. *Formula incorporating relative referencing of cells.*

	A	B	C
1	20	6.5	=A1*B1
2	30	7.2	
3	40	8.5	

Excel interprets the formula in cell C1 as, 'starting from the current cell (C1) multiply the contents of the cell two to the left (the value in A1) by the contents of the cell one to the left (the value in B1)'. This is referred to as relative referencing of cells. When the Enter key is pressed, the value 130 appears in cell C1. By clicking Home>Editing>Fill>Down, cells C2 and C3 are filled with formulae. Relative referencing assures that the correct cells in the A and B columns are used in the calculations. Specifically, =A2*B2 appears in cell C2, and =A3*B3 in cell C3. If cells are moved around, for example by 'cutting and pasting',[15] Excel automatically updates the references to ensure that calculations return the correct values irrespective of which cells contain the raw data.

---

[15] To 'cut and paste', first highlight the cells containing the values to be moved. Click on the Cut symbol, ✂, which is in the Clipboard group within the Home Ribbon. Move the cursor to the cell where you want the first value to appear. Click on the arrow below Paste, ▼, (which is also found within the Clipboard group) then choose .

Exercise B

Highlight cells C1 to C3 in sheet 2.7. Click Home>Editing>Fill>Down to calculate the product of values in adjacent cells in the A and B columns.

Absolute referencing

Another way in which cells may be referenced is shown in sheet 2.8.

Sheet 2.8. *Formula using absolute referencing of cells.*

	A	B	C
1	20	6.5	= A1*B1
2	30	7.2	
3	40	8.5	

The formula in cell C3 is interpreted as 'multiply the value in cell A1 by the value in cell B1'. This is referred to as absolute referencing. On the face of it, that is not very different from relative referencing. Certainly, when the Enter key is pressed, the value 130 appears in cell C1, just as in the previous example. The difference becomes more obvious by highlighting cells C1 to C3 and choosing from the Editing group, Fill then Down. The consequences of these actions are shown in sheet 2.9.

Sheet 2.9. *Using Fill then Down with absolute referenced cells.*

	A	B	C
1	20	6.5	130
2	30	7.2	130
3	40	8.5	130

Cells C1 to C3 each contain 130. This is because the formulae in cells C1 to C3 use the contents of the cells which have been absolutely referenced, in this case cells A1 and B1, and no incrementing of references occurs when Fill> Down is used. This can be very useful, for example, if we wish to multiply values in a row or column by a constant.

As an example, consider the values of distance, h, shown in sheet 2.10. The time, t, for an object to fall freely through a distance, h, is given by

$$t = \left(\frac{2h}{g}\right)^{1/2},$$
(2.4)

where g is the acceleration due to gravity (= $9.81 \, \text{m/s}^2$ on the Earth's surface).

Sheet 2.10. *Calculation of time for object to fall a distance, h.*

	A	B	C	D
1	h (m)	t (s)		
2		2	=(2*A2/D3)^0.5		
3		4		g	9.81
4		6			
5		8			
6		10			

Cell B2 contains a relative reference to cell A2. When cells B2 to B6 are highlighted and Fill > Down chosen, the cell references are automatically incremented. Cell B2 contains an absolute reference to cell D3, so that the value 9.81 is used in the formulae contained in cells in B2 to B6.

Exercise C

(1) Complete sheet 2.10 using Home>Editing>Fill>Down to find the time of fall for all the heights given in the A column.
(2) Use the spreadsheet to calculate the times of fall when the acceleration due to gravity is $1.6\,\mathrm{m/s^2}$.

Naming cells

The use of absolute referenced cells for values that we might want to use repeatedly is fine, but it is possible to incorporate constants into a formula in a way that makes the formula easier to read. That way is to give the cell a name. Consider sheet 2.11.

Sheet 2.11. *Time of fall calculated using a named cell.*

	A	B	C	D
1	h (m)	t (s)		
2		2	=(2*A2/g)^0.5		
3		4		g	9.81
4		6			
5		8			
6		10			

Sheet 2.11 is similar to sheet 2.10, the difference being that the absolute reference, D3, in cell B2 has been replaced by the symbol, g. Before proceeding, we must allocate the name g to the contents of cell D3. One way to allocate the name is as follows.

(i) Highlight cell D3.
(ii) Click Formulas>Define Name. A dialog box opens indicating that the intended new name for the cell D3 is 'g' (Excel 'guesses' that the symbol entered into cell C3 is the preferred name). By clicking OK, g becomes the new name for cell D3.

Whenever we need to use the value of g in a formula, we use the symbol g instead of giving an absolute reference to the cell containing the value.[16]

A word of caution: some words or letters are reserved by Excel and using them as a name will cause difficulties. For example, trying to use the letter c (or C) as a name prompts Excel to report that the name entered is not valid.

Exercise D

The heat emitted each second, H, from a blackbody of surface area, A, at temperature, T, is given by

$$H = \sigma A T^4, \qquad\qquad (2.5)$$

where σ is the Stefan–Boltzmann constant ($= 5.67 \times 10^{-8}$ W/(m^2·K^4)). In this problem take $A = 0.062$ m^2.

Create a spreadsheet with cells containing 5.67×10^{-8} and 0.062 and name the cells SB and A, respectively. Use Excel to calculate H, for $T = 1000$ K to 6000 K in steps of 1000 K.

2.3.7 Operator precedence and spreadsheet readability

Care must be taken when entering formulae, as the order in which calculations are carried out affects the final values returned by Excel. For example, suppose Excel is used to calculate the equivalent resistance of two resistors of values 4.7 kΩ and 6.8 kΩ connected in parallel. The formula for the equivalent resistance, R_{eq}, of two resistors R_1 and R_2 in parallel is

$$R_{eq} = \frac{R_1 R_2}{R_1 + R_2}. \qquad\qquad (2.6)$$

Sheet 2.12 shows the resistor values entered into cells A1 and A2 (with cells formatted to represent numbers in scientific notation to two decimal places). The equation to calculate R_{eq} is entered in cell A3.

Sheet 2.12. *Calculation of parallel equivalent resistance.*

(a) *Formula entered into cell A3*

	A
1	4.70E+03
2	6.80E+03
3	=A1*A2/A1+A2
4	

(b) *Value returned after pressing the Enter key*

	A
1	4.70E+03
2	6.80E+03
3	1.36E+04
4	

---

[16] We have not included the units of g. If we type g(m/s^2) in cell C3, then g(m/s^2) becomes the name which we would need to type out in full when incorporating it into other formulae.

When the Enter key is pressed, the value 1.36E+04 (making R_{eq} equal to 13.6 kΩ) appears in cell A3 as indicated in sheet 2.12(b). This value is incorrect, as R_{eq} should be 2.779 kΩ. Excel interprets the formula in cell A3 of sheet 2.12 as

$$= (A1 * A2/A1) + A2, \text{instead of} = (A1 * A2)/(A1 + A2).$$

To avoid such mistakes, it is advisable to include parentheses in formulae, as Excel carries out the operations within the parentheses first. Difficulties with calculations can be lessened if a formula is divided into a number of smaller formulae, with each entered into a different cell. As an example, consider an equation relating the velocity of water through a tube to the cross-sectional area of the tube:

$$v = \sqrt{\frac{2gh}{(A_1/A_2)^2 - 1}}. \tag{2.7}$$

Here, v is the velocity of water through a tube, g is the acceleration due to gravity, h is the height through which the water falls as it moves through the tube whose cross-sectional area changes from A_1 to A_2. Values for these quantities are shown in sheet 2.13.

The formulae in cells B5, B6 and B7 return values which are subsequently used in the calculation of v in cell B8. This step by step approach is also useful if a mistake is suspected in the final value and troubleshooting is required. For completeness, symbols used for the quantities along with the appropriate units are shown in column A of the spreadsheet. The corresponding values and formulae are shown in column B of sheet 2.13(a). The values returned by the formulae are shown in sheet 2.13(b).

Sheet 2.13. *Calculation of the velocity of water.*

(a) *Formulae entered into cells B5 to B8*

	A	B
1	g (m/s^2)	9.81
2	h (m)	0.15
3	A_1 (m^2)	0.062
4	A_2 (m^2)	0.018
5	2gh (m^2/s^2)	= 2*B1*B2
6	$(A_1/A_2)^2$	= (B3/B4)^2
7	$(A_1/A_2)^2-1$	= B6−1
8	v (m/s)	= (B5/B7)^0.5

(b) *Values returned in cells*

	A	B
1	g (m/s^2)	9.81
2	h (m)	0.15
3	A_1 (m^2)	0.062
4	A_2 (m^2)	0.013
5	2gh (m^2/s^2)	2.943
6	$(A_1/A_2)^2$	11.8642
7	$(A_1/A_2)^2-1$	10.8642
8	v (m/s)	0.520471

Subscripts and superscripts have been added to quantities and units in column A. This further improves the readability of the spreadsheet. To add a super or subscript, do the following.

(1) Click on the cell containing the symbol to be made into a super or subscript.
(2) Highlight the symbol to be super or subscripted. This may be done in the formula bar. Hold down the left hand mouse button and drag across the symbol or number you wish to make a super or subscript.
(3) Go to the Home ribbon. Go to the Font group and click on ⃞ which is in the bottom right hand corner of the group. A dialog box opens.
(4) Click the Superscript or Subscript box. A tick should appear in the box selected.
(5) Click on OK. The chosen symbol or number should appear as a superscript or a subscript.

Exercise E

(1) The radius of curvature of a spherical glass surface may be found using the Newton's rings method.[17] If the radius of the mth ring is r_m and the radius of the nth ring is r_n, then the radius, R, of the spherical glass surface is given by

$$R = \frac{r_n^2 - r_m^2}{(n-m)\lambda},$$ (2.8)

where λ is the wavelength of the light incident on the surface. Table 2.1 contains data from an experiment carried out to determine R.

Using the data in table 2.1, create a spreadsheet to calculate R, as given by equation 2.8.

(2) The velocity of sound, v_d, in a tube depends on the diameter of the tube, the frequency of the sound, and the velocity of sound in free air. The equation relating the quantities, when the walls of the tube are made from smooth glass, is

$$v_d = v\left(1 - \frac{3 \times 10^{-3}}{d\sqrt{f}}\right),$$ (2.9)

where v is the velocity of sound in free air in m/s, d is the diameter of the tube in metres and f is the frequency of the sound in Hz.

Taking $v = 344$ m/s and $f = 5$ Hz, use Excel to tabulate values of v_d when d varies from 0.1 m to 1 m in steps of 0.1 m.

2.3.8 Verification and troubleshooting

The previous section revealed the ease with which a formula may be entered into a cell that 'looks right', but without the appropriate parenthesis the formula

---

[17] For details on Newton's rings, see Kraftmakher (2007).

Table 2.1. *Values used to determine the radius of a spherical surface.*

Quantity	Value
r_m (m)	6.35×10^{-3}
r_n (m)	6.72×10^{-3}
m	52
n	86
λ (m)	6.02×10^{-7}

returns values inconsistent with the equation upon which it is based. Establishing or verifying that a spreadsheet is returning the correct values can sometimes be difficult, especially if there are many steps in the calculation. While no single approach can assure that the spreadsheet is behaving as intended, there are a number of actions that can be taken which minimise the chance of a mistake going unnoticed. If a mistake is detected, a natural response is to suspect some logic error in the way the spreadsheet has been assembled. However, it is easy to overlook the possibility that a transcription error has occurred when entering data into the spreadsheet. In this situation little is revealed by stepping through calculations performed by the spreadsheet in a 'step by step' manner. Table 2.2 offers some general advice intended to help reduce the occurrence of mistakes.

When a possible error in a formula is detected by Excel, a small green triangle appears in the top left hand corner of the cell containing the error. By clicking on the cell, a 'smart tag' icon, ⬦ , appears beside the cell. Clicking on the icon brings up a menu which offers information and options. Excel identifies the nature of the possible error at the top of the menu and offers assistance to fix the error. Sometimes the error is correctly identified, such as when trying to divide the contents of a cell by zero. When this occurs, a message 'Divide by Zero Error' appears when clicking on the smart tag. On other occasions closer inspection is required.

For example, suppose the mean of the contents of several adjacent cells in the A column of a worksheet is required. To do this we would use Excel's Average() function. If the Average() function has been entered into a cell as =Average(A1:A9) then the small green triangle will appear if all the adjacent cells in the A column containing numbers have not been included in the cell range. When the smart tag icon is clicked, the message 'Formula Omits Adjacent Cells' appears. This may or may not be what was intended. If it *was* intended then after the smart tag icon has been clicked, the 'Ignore Error' option can be chosen. If this was *not* intended, then the range of cells appearing in the function will need to be corrected.

Table 2.2. *Troubleshooting advice.*

Suggestion	Explanation/Example
Make the spreadsheet do the work	Enter 'raw' values into a spreadsheet in the form in which they emerge from an experiment. For example, if the diameter of a ball bearing is measured using a micrometer, then it is unwise to convert the diameter to a radius 'in your head'. It is better to add an extra column (with a clear heading) and to calculate the radius in that column. This makes backtracking to find mistakes much easier.
Perform an order of magnitude calculation	Having a feel for the size and sign of numbers emerging from the spreadsheet means we are alerted when those numbers do not appear. For example, if the volume of a small ball bearing is calculated to be roughly $150 \times 10^{-9} \, \mathrm{m}^3$ but the value determined by the spreadsheet is $143.8 \, \mathrm{m}^3$, this might point to an inconsistency in the units used for the volume calculation or that the formula entered is incorrect.
Use data that have already been analysed	On some occasions 'old' data are available that have already been analysed 'by hand' or using another computer package. The purpose of the spreadsheet might be to analyse similar data in a similar manner. By repeating the analysis of the 'old' data using the spreadsheet it may be established whether the analysis is consistent with that performed previously.
Choose the appropriate built in function	Many built in functions in Excel appear to be very similar, for example when calculating a standard deviation it is possible to use[18] STDEV.A(), STDEV.P(), STDEVP() or STDEVPA(). Knowing the definition of each function (by consulting the help available for each function if necessary) the appropriate function may be selected.
Graph the data and scan for outliers	If many values are entered by hand into a spreadsheet it is easy for a transcription mistake to occur. By using the x–y graphing feature of Excel we are able to plot the values in the order in which they appear in a column or row of the spreadsheet. Any gross mistakes can usually be identified 'by eye' within a few moments. As an example, most y values in figure 2.3 are

[18] In Excel 2010 the functions STDEV.A() and STDEV.P() replace STDEV() and STDEVP() respectively found in earlier version of Excel. The names have been changed to make the naming of functions more consistent. Many other functions in Excel have been changed for the same reason. For compatibility with earlier versions of Excel the functions STDEV() and STDEVP() are still available in Excel 2010.

Table 2.2. (*cont.*)

Suggestion	Explanation/Example
	about 3.6. As one value is close to 6.3 we should consider the possibility that a transcription error has occurred.

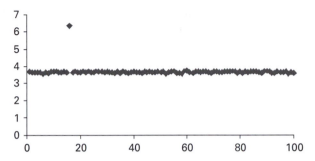

Figure 2.3. An *x–y* graph of data indicating a possible transcription error.

Suggestion	Explanation/Example
Be alert to error messages	#DIV/0!, #NAME?, #REF!, #NUM! and #N/A! are some of the error messages that Excel may return into a cell due to a variety of causes. As examples:

- A cell containing the #DIV/0! error indicates that the calculation being attempted in that cell requires Excel to divide a value by zero. A common cause of this error is that a cell such as B1 contains no value, yet this cell is referenced in a formula such as =32.45/B1.
- The #NAME? error occurs when a cell contains reference to an invalid name. For example, typing =averag(A1:A10) into a cell causes the #NAME? error to appear. Excel does not recognise averag() as a function (most likely average(A1:A10) should have been entered, which is a valid Excel function, but it was spelled incorrectly). Excel assumes that averag(A1:A10) is a name that has not been defined and so returns the error message.
- A cell containing the #NUM! error indicates that some invalid mathematical operation is being attempted. For example, if a cell contains =LN(-6) then the #NUM! error is returned into that cell as it is not possible to take the logarithm of a negative number.

A more thorough consideration of how the contents of various cells are 'brought together' to calculate, for example, a mean and standard deviation is aided by using some of Excel's in-built tools. These are the 'Auditing' tools and can assist in identifying problems in a spreadsheet.

2.3.9 Formula auditing

Sheet 2.14 contains nine values of temperature as well as the mean and the standard deviation of the values.

Sheet 2.14. *Calculation of mean and standard deviation.*

	A	B
1		Temperature (C)
2		23.5
3		24.2
4		26.4
5		23.1
6		22.8
7		22.5
8		25.4
9		28.3
10		26.5
11	mean	24.74444444
12	standard deviation	1.893328117

Calculating the mean and standard deviation using a pocket calculator we find that the mean = 24.74 °C (to four significant figures), consistent with the mean calculated in cell B11 of sheet 2.14. The standard deviation found using a pocket calculator is 2.008 °C which is quite different from 1.893 °C which appears in sheet 2.14. To trace the calculation of the standard deviation appearing in sheet 2.14, we can use Excel's auditing commands which give a graphical representation how values in cells are calculated. Specifically, by using the auditing commands, we can establish which cells contribute to the calculation of a value in a cell. The appropriate auditing command can be accessed by going to the Formulas Ribbon, then to the Formula Auditing group. The Formula Auditing group is shown in figure 2.4.

To audit the formula in B12 we proceed as follows.

(1) Click on the cell which contains the calculation we wish to trace (in the example shown in sheet 2.14, this would be cell B12).

Figure 2.4. The Formula Auditing group.

◢	A	B	C
1		Temperature (°C)	
2		23.5	
3		24.2	
4		26.4	
5		23.1	
6		22.8	
7		22.5	
8		25.4	
9		28.3	
10		26.5	
11	mean	24.74444444	
12	standard deviation	1.893323117	
13			

Figure 2.5. Tracing a calculation using a formula auditing tool.

(2) Click on 🔲 **Trace Precedents** in the Formula Auditing group. A blue line with an arrow appears which indicate the cells used in the calculation of the value in cell B12. If a range of cells contribute to the calculation, then this range is outlined in blue.

Figure 2.5 shows Trace Precedents used to trace the calculation of the standard deviation in sheet 2.14.

Trace Precedents indicates that cells in the range B2 to B11 have been used in the calculation of the standard deviation in cell B12. The mistake that has been made has been the inclusion of the cell B11 in the range, as B11 does not contain 'raw data', but the mean of the contents of cells B2 to B10. This mistake could have been detected by examining the range appearing in the formula in cell B12. However, as the relationships between cells becomes more complex, the pictorial representation of those relationships as provided by the formula auditing tools can help identify mistakes that would be otherwise difficult to find. The Formula Auditing group contains other facilities to assist in tracking down mistakes, details of which can be found either by using Excel's Help or by referring to a standard text.[19]

---

[19] See, for example, Jelen (2010).

2.4　Built in mathematical functions

Excel possesses the mathematical and trigonometrical functions present on most scientific pocket calculators. In addition, Excel possesses other functions rarely found on a calculator such as a function to calculate the sum of a geometric series. A full list of mathematical and trigonometrical functions provided by Excel is found by clicking in the Formulas tab, then clicking on Math & Trig. When entering any function into a cell, Excel anticipates which function you want to use and shows a list of functions to choose from consistent with the letters you have typed. The list and a description of the highlighted function, appear adjacent to the cell selected to contain the function.

In many situations in the physical sciences, data may need to be transformed before further analysis or graphing can take place. For example, in the capacitor discharge experiment discussed in section 2.3.4, it is likely that the decay of the current with time follows a logarithmic relationship. The first step that would normally be taken to verify this would be to calculate the natural logarithm of all of the values of current. In Excel, the natural logarithm of a number is calculated using the LN() function. To calculate the natural logarithm of the value appearing in cell C2.

(1) type the function =**LN(C2)** into cell D2, as shown in sheet 2.15(a) (LN can either be in upper or lower case letters) and press the Enter key. The number -14.9191 is returned in cell D2 as indicated in sheet 2.15(b);

Sheet 2.15. *Calculation of natural logarithm of current.*

(a) *Formula entered into cell D2*　　(b) *Value returned in cell D2*

	C	D		C	D
	I(amps)	ln(I)		I(amps)	ln(I)
1	I(amps)	ln(I)	1	I(amps)	ln(I)
2	3.31667E − 07	=LN(C2)	2	3.31667E − 07	−14.9191
3	1.31667E − 07		3	1.31667E − 07	
4	5.08333E − 08		4	5.08333E − 08	
5	0.00000002		5	0.00000002	
6	7.83333E − 09		6	7.83333E − 09	
7	2.91667E − 09		7	2.91667E − 09	
8	1.33333E − 09		8	1.33333E − 09	
9	5.25E − 10		9	5.25E − 10	
10	2.58333E − 10		10	2.58333E − 10	
11	1.41667E − 10		11	1.41667E − 10	
12	9.16667E − 11		12	9.16667E − 11	
13	5.83333E − 11		13	5.83333E − 11	
14	5E − 11		14	5E − 11	

(2) highlight cells D2 to D14 then click Home>Editing>Fill>Down to calculate the natural logarithms of all the values in the C column.

Exercise F

(1) Calculate the logarithms to the base 10 of the values shown in column C of sheet 2.15.

(2) An equation used to calculate atmospheric pressure at height h above sea level, when the air temperature is T, is

$$P = P_0 \exp\left(\frac{-3.39 \times 10^{-2}h}{T}\right), \tag{2.10}$$

where P is the atmospheric pressure in pascals, P_0 is equal to 1.01×10^5 Pa, h is the height in metres above sea level and T is the temperature in kelvins.

Using equation 2.10 and assuming $T = 273$ K, use the EXP() function to calculate P for $h = 1 \times 10^3$ m, 2×10^3 m, and so on, up to 9×10^3 m. Express the values of P in scientific notation to three significant figures.

(3) When a cable is held at each end, the cable sags and forms a curve called a catenary. The shape of the catenary can be described by the following relation:

$$y = c \cosh\left(\frac{x}{c}\right), \tag{2.11}$$

where c is a constant, cosh is a hyperbolic function, and x and y are the coordinates of a point on the catenary.

Taking $c = 250$ m, use equation 2.11 to calculate y for $x = -100$ m to 100 m in steps of 20 m. Express the values of y to four significant figures.

2.4.1 Trigonometrical functions

The usual sine, cosine and tangent functions are available in Excel, as are their inverses. Note Excel expects angles to be entered in radians and not degrees. An angle in degrees can be converted to radians by using the RADIANS() function. Sheet 2.16 shows a range of angles and the functions for calculating the sine, cosine and tangent of each angle.

Sheet 2.16. *Calculation of sine, cosine and tangent.*

	A	B	C	D	E
1	x (degrees)	x (radians)	sin(x)	cos(x)	tan(x)
2	10	= RADIANS(A2)	= SIN(B2)	= COS(B2)	= TAN(B2)
3	30				
4	56				
5	125				

Table 2.3. *Data for exercise G.*

x	0.0	0.1	0.2	0.3	0.4	0.5	0.6	0.7	0.8	0.9	1.0

In section 2.3.5 we used Home>Editing>Fill>Down to fill a single column of Excel with formulae. We can complete sheet 2.16 by filling all the cells from B2 to E5 with the required formulae. To do this,

(1) enter the numbers, headings and formulae as shown in sheet 2.16;

(2) move the mouse pointer to cell B2 and hold down the left mouse button;

(3) with the button held down, drag across and down to cell E5, then release the button;

(4) click on the Home tab then click Editing>Fill>Down.

(5) The cells from B2 to E5 should now contain the values shown in sheet 2.17.

Sheet 2.17. *Values returned by Excel's trigonometrical functions.*

	A	B	C	D	E
1	x(degrees)	x (radians)	sin(x)	cos(x)	tan(x)
2	10	0.174533	0.173648	0.984808	0.176327
3	30	0.523599	0.5	0.866025	0.57735
4	56	0.977384	0.829038	0.559193	1.482561
5	125	2.181662	0.819152	−0.57358	−1.42815

Exercise G

Use the Excel functions ASIN(), ACOS() and ATAN(), to calculate the inverse sine, inverse cosine and inverse tangent of the values of x in table 2.3. Express the inverse values in degrees using Excel's DEGREES() function.

2.5 Built in statistical functions

In addition to mathematical and trigonometrical functions, Excel possesses about 100 functions directly related to statistical analysis of data. Some of these functions are found on scientific pocket calculators. For example, the AVERAGE() function calculates the arithmetic mean of a list of values and so performs the same function as the $\bar{x}$ button appearing on most pocket calculators. Other statistical functions are more specialised, such as NORM.DIST(), which is related to the normal distribution. We consider such statistical functions in later chapters.

2.5.1 SUM(), MAX() and MIN()

The summation of values is required frequently in data analysis, for example it is a necessary step when finding the mean of a column of numbers. Summation can be accomplished by Excel in several ways. Consider the values in sheet 2.18.

Sheet 2.18. *Example of summing numbers.*

	A	B
1		y
2		23
3		34
4		42
5		42
6		65
7		87
8		

To sum the values in the B column,

(1) click on the cell B8;

(2) click on the Home tab, then move the pointer to the Editing group and click on Σ AutoSum ▾;

(3) a flashing dotted line appears around cells B2 to B7 and =SUM(B2:B7) appears in cell B8. Excel has made an 'intelligent guess' that we want to sum the numbers appearing in cells B2 to B7 inclusive. =SUM(B2:B7) can be read as 'sum the values in cells B2 through to B7'.

(4) Pressing the Enter key causes the number 293 to be returned in cell B8.

Exercise H

Consider the x and y values in sheet 2.19.

Sheet 2.19. *The x and y values for exercise H.*

	A	B	C	D
1	x	y	xy	x2
2	1.75	23		
3	3.56	34		
4	5.56	42		
5	5.85	42		
6	8.76	65		
7	9.77	87		
8				

(1) Use Excel to calculate the product of x and y in column C.

(2) Calculate x^2 in column D.

(3) Use the SUM() function in cells A8 to D8 to calculate Σx, Σy, Σxy and Σx^2

Another way to calculate the sum of the values in cells B2 to B7 is to make cell B8 active then type =**SUM(B2:B7)**. On pressing the Enter key, the number 293 appears in cell B8.

The MAX() and MIN() functions return the maximum and minimum values respectively in a list of values. Such functions are useful, for example, if we wish to scale a list of values by dividing each value by the maximum or minimum value in that list, or if we require the range (i.e. the maximum value – minimum value) perhaps as the first step to plotting a histogram.[20] To illustrate these functions, consider the values in sheet 2.20. The formula for finding the maximum value in cells A1 to E6 is shown in cell A8.

Sheet 2.20. *Identifying the maximum value in a group of values.*

	A	B	C	D	E
1	23	23	13	57	29
2	65	22	45	87	76
3	34	86	76	79	35
4	45	55	89	34	43
5	45	61	56	43	12
6	98	21	87	56	34
7					
8	= MAX(A1:E6)				

When the Enter key is pressed, 98 is returned in cell A8.

Exercise I

Incorporate the MIN() function into sheet 2.20 so that the smallest value appears in cell B8. Show the range of the values in cell C8.

2.5.2 AVERAGE(), MEDIAN() and MODE.SNGL()

The AVERAGE(), MEDIAN() and MODE.SNGL() functions calculate three numerical descriptions of the centre of a set a values. Briefly, we can describe them as follows.

AVERAGE() returns the arithmetic mean of a group or list of values.

MEDIAN() orders values in a list of n values from lowest to highest and returns the middle value in the list. The middle value is the $\left(\frac{n}{2}+\frac{1}{2}\right)$th value if n is an odd number of values. If n is an even number then the median value is the mean of the $\left(\frac{n}{2}\right)$th value and the $\left(\frac{n}{2}+1\right)$th value.

---

[20] See section 2.8.1 for a description of how to use Excel to plot a histogram.

MODE.SNGL() determines the frequency of occurrence of values in a list and returns the value which occurs most often.[21] If two or more values tie for first place with regard to the frequency with which they appear in a list then Excel returns only the first value that it completed tallying.[22] Of the three measures of centre of data discussed in this section, the mode is used least frequently in the physical sciences.

Consider sheet 2.21 containing 48 integer values. The mean of the values is found by entering the AVERAGE() function into cell A9.

Sheet 2.21. *Use of the AVERAGE() function.*

	A	B	C	D	E	F
1	27	1	49	2	39	11
2	27	29	40	8	28	5
3	18	5	25	0	4	33
4	26	30	20	14	22	10
5	23	28	33	30	28	16
6	23	5	27	9	48	4
7	39	41	46	22	25	25
8	35	49	13	8	40	43
9	=AVERAGE(A1:F8)					

When the Enter key is pressed, the value 23.60417 is returned in cell A9.

Exercise J

Calculate the median and mode of the values in sheet 2.21 by entering the MEDIAN() and MODE.SNGL() functions into cells B9 and C9 respectively.

2.5.3 Other useful statistical functions

A full list of statistical functions in Excel may be obtained by making any cell active then clicking on the Formulas>Function Library>Insert Function. A function may be chosen from the list that appears in the dialog box. The scroll bar is used to scroll down to the desired function. Clicking on the function causes a brief description to appear. Further information can be obtained by clicking on 'Help on this function' appearing in the bottom left corner of the dialog box.

---

[21] In Excel 2010 MODE.SNGL() replaces the MODE() function found in earlier versions of Excel (and operates identically to MODE()). The MODE() function is available in Excel 2010 for backward compatibility with earlier versions of Excel.

[22] Another function new to Excel 2010, MODE.MULT(), returns all the values that tie for first place with respect to the frequency of occurrence in a list. MODE.MULT() is a function that returns numbers into an array of cells. We consider array functions in Excel in chapter 6.

Table 2.4. *Useful statistical functions in Excel.*

Function	What it does	Defining equation	Example of use	Value returned
AVEDEV() (abbreviated to A.D.)	Calculates the mean of the absolute deviations of values from the mean.	$\text{A.D.} = \frac{1}{n}\sum |x - \bar{x}|$	=AVEDEV (14, 23, 12, 34, 36)	8.96
HARMEAN() (abbreviated to H.M.)	Calculates the reciprocal of the mean of reciprocals of values.	$\text{H.M.} = \frac{n}{\sum\left(\frac{1}{x}\right)}$	=HARMEAN (2.4, 4.5, 6.7, 6.4, 4.3)	4.248266
STDEV.S() (symbol, s)	Calculates s which is the estimate of population standard deviation.	$s = \left[\dfrac{\sum (x - \bar{x})^2}{n-1}\right]^{1/2}$	=STDEV.S (2.1, 3.5,4.5, 5.6)	1.488568
STDEV.P() (symbol, σ)	Calculates the population standard deviation, σ (all values in population are required).	$\sigma = \left[\dfrac{\sum (x - \mu)^2}{n}\right]^{1/2}$	=STDEV.P (2.1, 3.5,4.5, 5.6)	1.289137

Other useful statistical functions are given in table 2.4 along with brief descriptions of each.

Exercise K

Repeat measurements made of the pH of river water are shown in sheet 2.22. Use the built in functions in Excel to find the mean, harmonic mean, average deviation and estimate of the population standard deviation for these data.

Sheet 2.22. *Data for exercise K.*

	A	B	C	D	E	F	G	H	I	J
1	6.6	6.4	6.8	7.1	6.9	7.4	6.9	6.4	6.3	7.0
2	6.8	7.2	6.4	6.7	6.8	6.1	6.9	6.7	6.4	7.1
3	6.8	6.7	6.3	6.6	7.0	6.7	6.4	6.7	6.7	6.4
4	6.6	6.7	7.4	7.1	7.0	6.8	7.0	6.8	6.7	6.2
5	7.1	6.4	6.7	6.9	6.9	6.6	7.2	6.8	6.4	6.5

2.6 Presentation options

A spreadsheet consisting largely of numbers and formulae can appear incomprehensible to the user, even if that user is the person who created it. For clarity, it is important to consider the layout of the spreadsheet carefully.

It may be necessary to move the contents of cells to allow space for titles or labels. This does not pose a problem, as whole sections of a spreadsheet can be moved without 'messing up' the calculations. For example, when columns containing values are moved, the formulae in cells that use those values are automatically updated to include the references to the new cells containing the values.

Sheet 2.23 contains data from the measurement of the coefficient of static friction between two flat wooden surfaces.

Sheet 2.23. *Coefficient of static friction data.*

	A	B	C	D	E	F
1	0.34	0.34	0.40	0.49	0.42	0.33
2	0.41	0.39	0.38	0.39	0.44	0.49
3	0.42	0.50	0.31	0.36	0.48	0.26
4	0.46	0.42	0.32	0.37	0.40	0.36
5	0.44	0.40	0.36	0.38	0.45	0.37
6	0.38	0.30	0.29	0.39	0.49	0.55

To improve the readability of the spreadsheet we might:

- give the sheet a clear heading;
- indicate what quantities (with units, if any) the values in the cells represent.

Allow some space at the top of sheet 2.23 for labels and titles to be inserted as follows.

(1) Highlight the cells containing values by moving the pointer to cell A1 and, with the left hand mouse button held down, drag down and across to cell F6. Release the mouse button.

(2) Move the pointer to the outline of the highlighted cells, The pointer should change from an open cross, ⬚, to an arrow with a four-headed arrow, ✛. Holding the left hand mouse button pressed down, move the mouse around the screen. The outline of the cells should move with the mouse.

(3) Release the mouse button to transfer the contents of the cells to a new location.

By moving the contents of sheet 2.23 down by three rows, and adding labels, the spreadsheet can be made to look like sheet 2.24.

Sheet 2.24. *Improved spreadsheet layout for static friction data.*

	A	B	C	D	E	F
1	Coefficient of static friction for two wooden surfaces in contact					
2						
3		Coefficient of static friction (no units)				
4	0.34	0.34	0.40	0.49	0.42	0.33
5	0.41	0.39	0.38	0.39	0.44	0.49
6	0.42	0.50	0.31	0.36	0.48	0.26
7	0.46	0.42	0.32	0.37	0.40	0.36
8	0.44	0.40	0.36	0.38	0.45	0.37
9	0.38	0.30	0.29	0.39	0.49	0.55

A title (in bold font) has been added to the sheet and the quantity associated with the values in the cells is included.

The impact of a spreadsheet can be improved by formatting cells. This is done by clicking on the Home>Cells>Format>Format Cells. With this option, the colour of digits or text contained within cells may be changed, borders can be drawn around one or more cells, fonts can be altered and background shading added. A word of caution: too much formatting and bright colours can be distracting and, instead of making the spreadsheet easier to read, actually have the opposite effect.

2.7 Charts in Excel

A spreadsheet populated solely with numbers is sometimes difficult to make sense of, even when it is well laid out. Often the most efficient way to absorb data and to identify trends, anomalies and features of interest is to create a picture or graph. The graphical features of Excel are extensive and include options such as bar, pie and x–y graph. Excel refers to graphs as *charts*. Graphs may be constructed by going to the Insert Ribbon then to the Charts group. Among the many graphing alternatives is the x–y scatter graph. This is extensively used in the physical sciences and we will concentrate on this type of graph.

Another Excel graphing option called 'Line chart' seems similar to the x–y scatter graph, save for the fact that a line is automatically drawn through the data points. However the difference goes deeper than that, and if we wish to plot x–y data, the Line chart option should be avoided.[23]

2.7.1 The x–y graph

The x–y graphing facility in Excel (referred to in Excel as a Scatter chart) offers many options, including:

- conversion of scales from linear to logarithmic;
- addition of error bars to data points;
- addition of 'line of best fit' using a variety of models;
- automatic updating of points and trendline(s) on the graph when values within cells are changed.

---

[23] For information on Excel's Line chart, see Walkenbach (2010).

The Scatter chart is accessed by clicking on Insert>Charts>Scatter. We illustrate x-y graphing features using data gathered from an experiment in which the electrical resistance of a tungsten wire is measured as a function of temperature. We consider:

- plotting an x-y graph;
- adding a line of best fit (referred to in Excel as a 'trendline') to the graph;
- adding error bars to each point.

We begin by entering the resistance–temperature data as shown in sheet 2.25. To simplify the construction of an x-y graph, values to be plotted are placed in adjacent columns. The values in the left of the two columns are interpreted by Excel as the x values and those in the other column as the y values.

Sheet 2.25. *Resistance versus temperature data for a tungsten wire.*

	A	B
1	t (C)	R (ohms)
2	20	15.9
3	30	15.1
4	40	16.3
5	50	17.5
6	60	18.2
7	70	18.5
8	80	18.9
9	90	19.7
10	100	20.1
11	110	20.2

To plot the data shown in sheet 2.25, take the following steps.

(1) Highlight cells A2 to B11 by moving the pointer to cell A2. With the left hand mouse button held down, drag across and down from cell A2 to cell B11. Release the mouse button.

(2) Click on Insert>Charts>Scatter.

(3) A list of charts appears from which you can choose one. Select .

(4) An x-y graph appears. The axes of the graph are not labelled and it has no title. Excel anticipates that the graph will need more added to it and so it automatically activates the Design Ribbon.

(5) From the Chart Layouts group, select the top left layout, . Default labels appear for the title and the axes which can be modified by clicking on each label then overwriting it. Note that the ° symbol appearing in the x axis label and the Ω symbol appearing in the y axis label can be found by going to Insert>Symbol then isolating each symbol in the Symbol dialog box.

(6) The graph can be enlarged by moving to a corner of the frame in which it is enclosed, holding down the left mouse button, then dragging the corner.

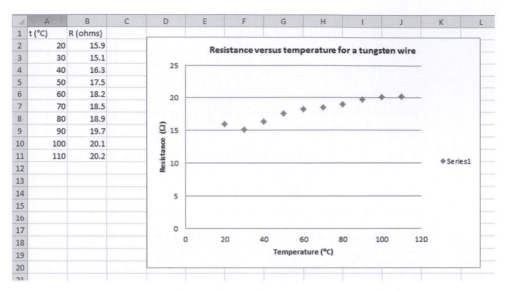

	A	B	C	D	E	F	G	H	I	J	K	L
1	t (°C)	R (ohms)										
2	20	15.9										
3	30	15.1										
4	40	16.3										
5	50	17.5										
6	60	18.2										
7	70	18.5										
8	80	18.9										
9	90	19.7										
10	100	20.1										
11	110	20.2										
12												
13												
14												
15												
16												
17												
18												
19												
20												
21												

Figure 2.6. Section of screen of a spreadsheet containing an x–y graph of resistance versus temperature of a tungsten wire.

Figure 2.6 shows the graph with an appropriate title and axes labels.

The graph in figure 2.6 is missing some important features. For example it would be routine to include tick marks on the x and y axes of a graph. Tick marks can be inserted by clicking on any number, for example on the x axis, then choose Format>Current Selection>Format Selection. The Format Axis dialog will appear which allows an appropriate Axis Option (for example adding a major tick mark) to be chosen.

Adding a trendline

If there is evidence to suggest that there is a linear relationship between quantities plotted on the x and y axes, as there is in this case, then a 'line of best fit', or trendline may be added to the graph.[24] To add a trendline to the resistance versus temperature data shown in sheet 2.25.

(1) Single click on the chart to make it active.
(2) Click on the Layout tab. In the Analysis group click on Trendline
(3) In the dialog box, click on More Trendline Options, then click on Linear, Display Equation on chart.
(4) A line of best fit should now appear, along with the equation of the line $(y = 0.0577x + 14.29)$.
(5) Click on Close in the Format Trendline dialog box.

---

[24] Fitting a line to data is dealt with in detail in chapter 6.

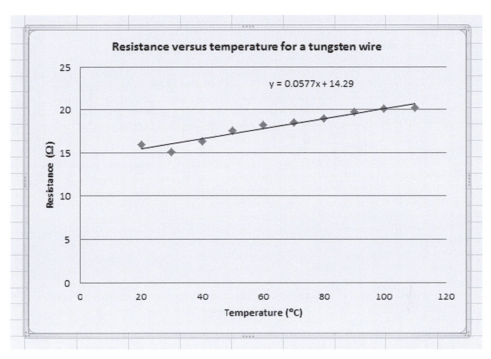

Figure 2.7. Screen of resistance–temperature graph with trendline and trendline equation added.

Figure 2.7 shows the line of best fit attached to the resistance versus temperature data.[25]

If data appearing in the cells A2 to B11 are changed, then the points on the line, the trendline and the trendline equation are instantly updated.

Adding error bars[26]

When drawing x–y graphs by hand, error bars are often omitted due to the effort required to plot them, especially if the graph contains many points. This effort can be considerably reduced by using Excel to add error bars to all points.

We add error bars to data shown in figure 2.7.

(1) Make the chart active.
(2) Move the pointer to one of the data points and click on the point.
(3) Click on Layout>Analysis>Error Bars>More Error Bars Options. A dialog box appears.

---

[25] The legend to the right of the screen can be removed by making the chart active, clicking on the legend, then pressing the delete key.

[26] Error bars are discussed briefly in section 1.5.2.

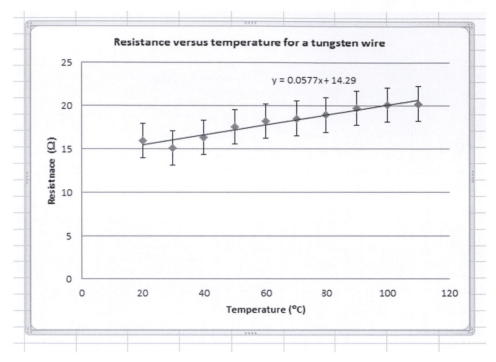

Figure 2.8. Resistance–temperature graph with error bars attached.

(4) If the uncertainty in the resistance is estimated to be, say ± 2 Ω, then click on Custom followed by Specify Value. A Custom Error Bars dialog box appears. Type 2 within the {} brackets for both the Positive Error Value and Negative Error Value boxes

(5) Click on OK. Error bars appear attached to each data point as shown in figure 2.8.

Exercise L

(1) The graph in figure 2.7 would be improved by better filling the graph with the resistance–temperature data. This can be accomplished if the y axis were to begin at 14 Ω, rather than at zero. Click on any number on the y axis, then use the Format>Current Selection>Format Selection to change the y axis minimum to 14.

(2) The intensity of light emitted from a red light emitting diode (LED) is measured as a function of current through the LED. Table 2.5 shows data from the experiment.

　Use Excel's x–y graphing facility to plot a graph of Intensity versus Current. Attach a straight line to the data using the Trendline option, and show the equation of the line on the graph.

Table 2.5. *Variation of light intensity with current for an LED.*

Current (mA)	1	2	3	4	5	6	7	8	9	10
Light intensity (arbitrary units)	71	97	127	134	159	175	197	203	239	251

2.7.2 Plotting multiple sets of data on an *x–y* graph

There are situations in which it is helpful to show more than one set of data on an *x–y* graph, as this allows comparisons to be made between data. Two basic approaches may be adopted.

Several sets of data can be plotted 'simultaneously' by arranging the leftmost column of the spreadsheet to contain the *x* values and adjacent columns to contain the corresponding *y* values for each set of data.

If an *x–y* graph has already been created with one set of data, then another set of data may be added to the same graph.

For example, consider the data in sheet 2.26.

Sheet 2.26. *Multiple data sets to be plotted on an x–y graph.*

	A	B	C	D	E	F
1	x	y1	y2	y3	y4	y5
2	0.2	71	77	83	89	96
3	0.4	53	61	70	81	93
4	0.6	42	50	61	74	91
5	0.8	36	44	54	69	89
6	1.0	34	40	50	65	87
7	1.2	33	39	48	62	86
8	1.4	34	39	47	61	85
9	1.6	36	40	47	60	85
10	1.8	39	42	48	60	85
11	2.0	42	44	49	60	85

Sheet 2.26 contains five sets of data. Columns B to F in contain *y* values for each set corresponding to the *x* values in column A. To plot the data, highlight the contents of cells A1 through to F11, then follow the instructions in section 2.7.1. The graph produced is shown embedded in the worksheet in figure 2.9.

Note that by including the headings of the columns when selecting the range of data to be plotted, Excel has included those headings in the legend at the right hand side of the graph.

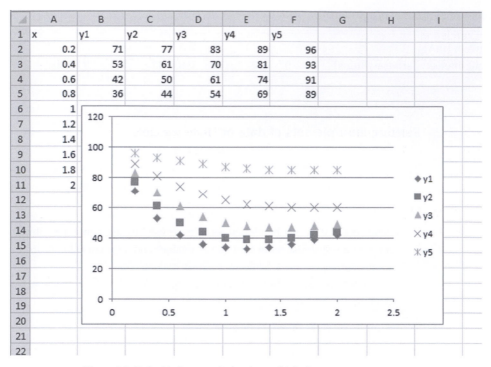

Figure 2.9. Embedded *x-y* graph showing multiple data sets.

If another set of data is to be added to an existing Excel *x-y* graph, then:

(1) Place the new *x-y* data in adjacent columns in a Worksheet;
(2) Highlight the *x-y* data, including the heading of the columns, and click on Home>Clipboard>🖺 (which copies the *x-y* data);
(3) Click on the existing *x-y* graph to make it active;
(4) Click on Home<Clipboard<Paste<Paste Special;
(5) In the dialog box choose 'Add cells as New series';
(6) Make sure that the box described as 'Categories (X Values) in First Column' has been ticked as well as the 'Series Names in First Row' and that the option 'Values (Y) in Columns' has been selected;
(7) Click on OK.

The new data should be added to the graph and the legend updated.

Exercise M

The intensity of radiation (in units W/m^3) emitted from a body can be written

$$I = \frac{A}{\lambda^5 \left(e^{\frac{B}{\lambda T}} - 1 \right)}, \tag{2.12}$$

where T is the temperature of the body in kelvins, λ is the wavelength of the radiation in metres, $A = 3.75 \times 10^{-16}$ W·m^2 and $B = 1.443 \times 10^{-2}$ m·K.

(1) Use Excel to determine I at $T = 1250$ K for wavelengths between 0.2×10^{-6} m to 6×10^{-6} m in steps of 0.2×10^{-6} m.

(2) Repeat part (1) for temperatures 1500 K, 1750 K and 2000 K.

(3) Plot I versus λ at temperatures 1250 K, 1500 K, 1750 K and 2000 K on the same x-y graph.

2.8 Data analysis tools

Besides many built in statistical functions, there exists an 'add in' within Excel that offers powerful data analysis facilities. This is referred to as the Analysis ToolPak. To establish whether the ToolPak is available, click on the Data tab. If the option 'Data Analysis' does not appear on the Data Ribbon at the far right, then the Analysis ToolPak must be added.

To add the Analysis ToolPak, do the following.

(1) Click on the **File** tab in Excel 2010 (in Excel 2007 click on the Office button,).

(2) Click on Options.

(3) Click on Add-Ins. Near to the bottom of the dialog box, you should see Manage: Excel Add-ins ▾ Go... .

(4) Click on Go.

(5) A dialog box appears. Tick the Analysis ToolPak. Click OK.

Returning to the Data Ribbon and clicking on Data Analysis should cause the dialog box in figure 2.10 to appear.[27]

Other analysis tools are discussed in chapter 10, but for the moment we consider just two; Histogram and Descriptive Statistics.

---

[27] Most dialog boxes in the Analysis ToolPak allow for the easy entry of cell ranges into the tool and often allow you to modify parameters.

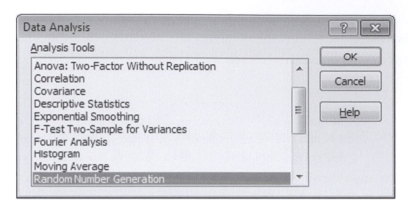

Figure 2.10. Dialog box for Data Analysis tools.

2.8.1 Histograms

A histogram is a revealing way to display repeated values of a single quantity or variable. The Histogram tool within the Analysis ToolPak takes much of the tedium out of creating a histogram. To illustrate this tool, consider the data shown in sheet 2.27 consisting of 64 repeat measurements of the lead content (in ppb) in river water.[28]

Sheet 2.27. *Lead content of river water.*

	A	B	C	D	E	F	G	H
1	Lead content of river water (in ppb)							
2	45	71	47	55	44	59	46	49
3	46	45	57	44	53	46	45	66
4	52	54	48	53	49	40	41	46
5	43	63	35	42	33	33	44	57
6	51	41	40	43	37	40	32	37
7	39	60	34	47	58	37	40	57
8	48	50	32	40	47	37	41	61
9	49	53	37	58	40	45	21	43
10								

Before using the Histogram tool, consideration should be given to what intervals or 'bins' we want to appear in the histogram. If this is not specified,

---

[28] The abbreviation ppb stands for 'parts per billion'.

Excel divides up the data range into bins of equal width, but may not choose a sensible number of bins. For example, applying the histogram tool to the data in sheet 2.27 without specifying the bins produces a histogram with only three bins. Such a histogram reveals little about the distribution of the data.

Before selecting the bins, it is helpful to find the range of the data by using the MAX() and MIN() functions on the data in sheet 2.27. The maximum value is 71 ppb and the minimum value is 21 ppb. It seems reasonable to choose the limits of the bins to be 20, 30, 40, 50, 60, 70, 80. These numbers are entered into the spreadsheet as shown in sheet 2.28.

Sheet 2.28. *Bin limits for the lead content histogram.*

	A	B
11	Bin limits	
12	20	
13	30	
14	40	
15	50	
16	60	
17	70	
18	80	

We are now ready to create the histogram.

(1) Click on the Data tab, then choose Data Analysis>Histogram.
(2) A dialog box appears into which we must enter references to the cells containing the values, reference to the cells containing the bin information and indicate where we want the histogram frequencies to appear. Figure 2.11 shows the values entered into the dialog box. Sheet 2.29 shows the numerical output from the Histogram tool. See also table 2.6.
(3) Enter the cell references shown in figure 2.11 into the dialog box.
(4) To obtain a graphical output of the histogram, tick the Chart output box in the dialog box. Click on the OK button.

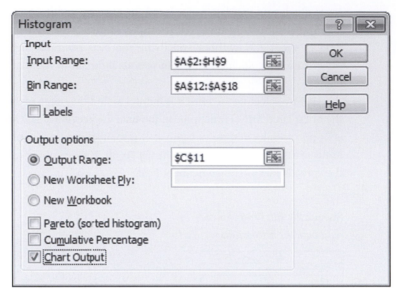

Figure 2.11. Dialog box for the Histogram tool.

The values returned in columns C and D are shown in sheet 2.29.

Sheet 2.29. *Numerical output from the Histogram tool.*

	A	B	C	D
11	Bin limits		Bin	Frequency
12	20		20	0
13	30		30	1
14	40		40	18
15	50		50	27
16	60		60	14
17	70		70	3
18	80		80	1
19			More	0

It is not immediately clear how the bin limits in the C column of sheet 2.29 relate to the frequencies appearing in the D column. The way to interpret the values in cell D12 to D18 is shown in table 2.6, where x represents the lead concentration in ppb.

Figure 2.12 shows the chart created using the bins and frequencies appearing in sheet 2.29.

Inspection of figure 2.12 reveals that the horizontal axis is not labelled correctly. The labels, 20 and 30 and so on, have been placed between the tick

Table 2.6. *Relating frequencies to intervals in the Histogram output shown in sheet 2.29.*

Excel bin label	Actual interval (ppb)	Frequency
20	$x \leq 20$	0
30	$20 < x \leq 30$	1
40	$30 < x \leq 40$	18
50	$40 < x \leq 50$	27
60	$50 < x \leq 60$	14
70	$60 < x \leq 70$	3
80	$70 < x \leq 80$	1
More	$x > 80$	0

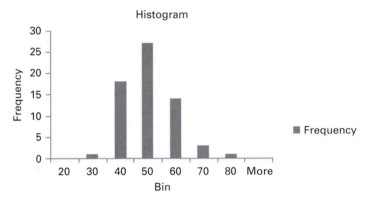

Figure 2.12. Histogram of lead content.

marks on the horizontal axis instead of coinciding with the tick marks. This problem can be eased by replacing the Bin labels that appear in the C column under the heading *Bin*. Replace 20 by 0–20, replace 30 by 20–30 and so on. Enlarge the graph by clicking on a corner and dragging until the bin labels are aligned horizontally, as shown in figure 2.13. By clicking on the x axis label, the word 'Bin' can be replaced by something more appropriate, such as 'lead content (ppb)'. The legend 'Frequency' may be removed, as it serves no useful purpose. This may be done by clicking on the legend then pressing the Delete button. The heading of the graph (referred to by Excel as the Chart Title) has been replaced in figure 2.13 with something more descriptive than the word 'Histogram' which is Excel's default heading.

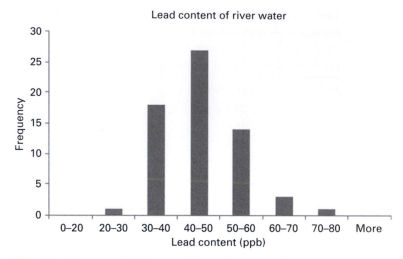

Figure 2.13. Histogram of lead content of data in sheet 2.27 with improved bin labels, and x axis label.

2.8.2 Descriptive statistics

The mean, median, mode, maximum and minimum of a list of numbers can be determined using the built in functions of Excel, as described in sections 2.5.1 to 2.5.3. However, if all of these quantities are required, then it is more efficient to use the Descriptive Statistics tool within the Analysis ToolPak. Consider data in sheet 2.30 which were gathered in an experiment to measure the density of rock specimens.

Sheet 2.30. *Density data of rock samples.*

	A	B	C	D	E	F	G	H	I	J	K
1	Density (g/cm³)	7.3	6.4	7.7	8.6	8.5	9.0	5.7	7.3	8.4	6.6

To use the Descriptive Statistics tool:

1. Click on Data>Data Analysis>Descriptive Statistics, then press OK.
2. Enter the values shown in figure 2.14 into the dialog box.
3. All the values are in row 1, so click on the option 'Grouped By Rows'.
4. Tick the Summary statistics box.
5. Click on OK. Excel returns values for the mean, median and the other statistics as shown in sheet 2.31.

Sheet 2.31. *Values returned when applying the Descriptive Statistics tool to density data.*[29]

	A	B
3	*Row1*	
4		
5	Mean	7.55
6	Standard Error	0.343592
7	Median	7.5
8	Mode	7.3
9	Standard Deviation	1.086534
10	Sample Variance	1.180556
11	Kurtosis	−1.01032
12	Skewness	−0.33133
13	Range	3.3
14	Minimum	5.7
15	Maximum	9
16	Sum	75.5
17	Count	10

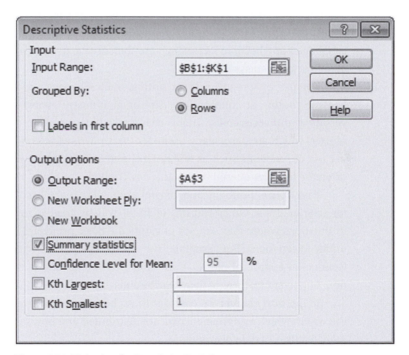

Figure 2.14. Dialog box for Descriptive Statistics.

[29] Standard error is discussed in chapter 3. Kurtosis is a statistic relating to the 'flatness' of a distribution of data. Skewness is a statistic which indicates the extent to which a distribution of data is asymmetric. For details see Spiegel and Stephens (2007).

If the data in sheet 2.30 are changed, the values returned by the Descriptive Statistics tool are <u>not</u> immediately updated. To update the values, the sequence beginning at step (1) in this section must be repeated (see figure 2.14). If the consequences need to be considered of changing one or more values, for example when a standard deviation is to be calculated, it may be better to forego use of the Descriptive Statistics tool in favour of using Excel's built in functions whose outputs are updated as soon as the contents of a cell have been changed. The trade off between ease of use offered by a tool in the Analysis ToolPak as compared to the increased flexibility and responsiveness of a 'user designed' spreadsheet needs to be considered before analysis of data using the spreadsheet begins.

2.9 Review

The analysis of experimental data may require something as straightforward as calculating a mean, to much more extensive numerical manipulation, involving the use of advanced statistical functions. The power and flexibility of computer based spreadsheets make them attractive alternatives for this purpose to, say, pocket calculators. Pocket calculators have limited analysis and presentation capabilities and become increasingly cumbersome to use when a large amount of data is to be analysed.

The use of spreadsheets for storing, analysing and presenting data has become routine at college, university or in industry owing to their power, versatility and accessibility. In this chapter we have considered some features of Excel which are particularly useful for the analysis of scientific data, including use of functions such as AVERAGE() and STDEV.S() as well as the more extensive features offered by the Analysis ToolPak.

Ease of use, flexibility and general availability make Excel a popular tool for data analysis. Other features of Excel will be treated in later chapters, in situations where they support or clarify the basic principles of data analysis under discussion. For example we will employ the Random Number Generation feature of the Analysis ToolPak to demonstrate, through simulation, important features of data distributions.

In the next chapter we consider those basic principles and in particular, the way in which experimental data are distributed and how to summarise the main features of data distributions.

End of chapter problems

(1) An equation relating the speed, v, of a wave to its wavelength, λ, for a wave travelling along the surface of a liquid is

$$v = \sqrt{\frac{\lambda g}{2\pi} + \frac{2\pi\gamma}{\rho\lambda}}, \tag{2.13}$$

where g is the acceleration due to gravity, γ is the surface tension of the liquid, and ρ is the density of the liquid. Given that

$g = 9.81\ \text{m/s}^2$,

$\gamma = 73 \times 10^{-3}\ \text{N/m}$ for water at room temperature,

$\rho = 10^3\ \text{kg/m}^3$ for water,

use Excel to calculate the speed of a water wave, v, for $\lambda = 1\ \text{cm}$ to $9\ \text{cm}$ in steps of $1\ \text{cm}$.

(2) The relationship between the mass of a body, m, and its velocity, v, can be written

$$m = \frac{m_0}{\left(1 - \dfrac{v^2}{c^2}\right)}, \tag{2.14}$$

where m_0 is the rest mass of the body and c is the velocity of light. Excel is used to calculate m, for values of v close to the speed of light, as given in column A of sheet 2.32. Note that $c = 3.00 \times 10^8\ \text{m/s}$ and we take $m_0 = 9.11 \times 10^{-31}\ \text{kg}$.

A formula has been entered into cell B2 of sheet 2.32 to calculate m. Cell D2 has been given the name, mo, and cell D3 has been given the name, c_, (see section 2.3.6 for restrictions on naming cells).

When the Enter key is pressed, the value returned is –9.34E-1. This is inconsistent with equation 2.14. Find the mistake in the formula in cell B2, correct it, and complete column B of sheet 2.32.

Sheet 2.32. *Mass of a body moving with velocity, v.*

	A	B	C	D
1	v (m/s)	m (kg)		
2	2.90E + 08	= mo/1-A2^2/c_^2	mo	9.11E-31
3	2.91E + 08		c_	3.00E + 08
4	2.92E + 08			
5	2.93E + 08			
6	2.94E + 08			
7	2.95E + 08			
8	2.96E + 08			
9	2.97E + 08			
10	2.98E + 08			
11	2.99E + 08			

(3) Sheet 2.33 contains 100 values of relative humidity (%).

Sheet 2.33. *100 relative humidity values (%).*

	A	B	C	D	E	F	G	H	I	J
1	70	62	58	76	60	55	56	60	68	59
2	69	54	61	58	62	71	68	63	73	72
3	67	68	68	63	63	64	75	65	64	67
4	68	57	72	69	61	70	73	58	63	66
5	72	63	64	65	59	65	63	63	66	58
6	73	61	72	71	69	65	58	66	77	61
7	72	66	67	65	62	70	65	67	66	72
8	64	72	61	69	70	66	64	60	61	68
9	63	70	65	62	70	75	64	79	68	62
10	63	63	72	75	70	66	72	79	78	69

- Use the built in functions on Excel to find the mean, standard deviation, maximum, minimum and range of these values.
- Use the Histogram tool to plot a histogram of the values.

(4) A diaphragm vacuum pump is used to evacuate a small chamber. Table 2.7 shows the pressure, P, in the chamber at a time, t, after the pump starts. Use Excel to:

   (i) convert the pressures in torr to pascals, given that the relationship between torr and pascals is 1 Torr = 133.3 Pa;
   (ii) plot a graph of pressure in pascals versus time in seconds. The graph should include fully labelled axes and tick marks next to the scale numbers on the axes.

Table 2.7. *Variation of pressure with time for a diaphragm pump.*

t (s)	P (torr)
1.9	750
4.2	660
6.8	570
10.7	480
16.2	390
25.2	300
49.2	210
69.0	120

(5) The vapour pressure of halothane, chloroform and trichloroethylene were measured over the temperature range 0 °C to 35 °C. The data obtained are shown in table 2.8.

Table 2.8. *Variation of vapour pressure with temperature for three volatile liquids.*

Temperature (°C)	Pressure (torr)		
	Halothane	Chloroform	Trichloroethylene
0	102	61	20
5	120	68	24
10	148	82	30
15	188	98	38
20	235	133	51
25	303	182	65
30	382	243	92
35	551	352	119

Show the variation of vapour pressure with temperature for the three liquids on an *x-y* graph.

(6) Equation 2.15 relates the electrical resistivity, ρ, of blood cells to the packed cell volume H (where H is expressed as a percentage):

$$\frac{\rho}{\rho_P} = \frac{1 + (f - 1)\frac{H}{100\%}}{1 - \frac{H}{100\%}}, \tag{2.15}$$

where ρ_P is the resistivity of the plasma surrounding the cells and f is a form factor which depends on the shape of the cells.

(i) Use Excel to determine $\frac{\rho}{\rho_P}$ for H in the range 0% to 80% in steps of 5% when $f = 4$.

(ii) Repeat part (i) for $f = 3$, 2.5, 2.0 and 1.5.

(iii) Show the $\frac{\rho}{\rho_P}$ versus H data determined in parts (i) and (ii) on the same *x-y* graph.

(7) A force, F, is applied to the string of an archer's bow, causing the string to displace a distance y. Table 2.9 contains the values of F and y.

(i) Plot a graph of displacement versus force.

(ii) Assuming a linear relationship between the displacement and the force, add a trendline and a trendline equation to the graph.

(iii) If the uncertainty in the y values is ± 0.5 cm, attach error bars to all the points on the graph.

Table 2.9. *Variation of force with displacement.*

F (N)	y (cm)
2	2.4
4	4.2
8	6.9
10	7.6
12	9.0
14	10.4
16	11.6

(8) The change in electrical characteristics of a 9 V 'extra heavy duty' battery is studied as the battery is discharged. A fresh battery is connected to various resistances, R, and the voltage, V, across each resistance is recorded. Table 2.10 shows the variation of V with R for the fresh battery.

(i) Use Excel to complete table 2.10.

(ii) Plot a graph of V against I.

(iii) Use Trendline to fit a straight line to the data. Display the equation of the straight line on the graph.

Table 2.10. *Variation of V with R for a fresh 9 V battery.*

R (Ω)	V (V)	I = V/R (A)
900	9.43	
800	9.41	
700	9.38	
600	9.35	
500	9.30	
400	9.24	
300	9.15	

The battery is allowed to discharge through a resistor for 3 h. Table 2.11 shows the variation of V with R after completion of the discharge.

(iv) Complete table 2.11.

(v) Plot V against I on the same graph as that created for part (ii) of this question.

Table 2.11. *Variation of V with R for a 9 V battery after a 3 h discharge.*

R (Ω)	V (V)	$I = V/R$ (A)
900	7.75	
800	7.73	
700	7.71	
600	7.68	
500	7.64	
400	7.58	
300	7.48	

Table 2.12. *Variation of flow rate, Q, with internal diameter, d, of a tube.*

d (mm)	Q (cm^3/s)
0.265	0.0092
0.395	0.081
0.820	0.67
1.000	1.4
1.255	3.2

    (vi) Use Trendline to fit a straight line to the data. Display the equation of the straight line on the graph.

(9) The volume flow of water, Q, through hollow tubes of varying internal diameter, d, is studied. The length of the tubes and the pressure difference between the ends of the tubes are kept constant. Table 2.12 shows the variation of Q with d.

    (i) Plot a graph of Q versus d.

    (ii) Use Trendline to fit a line represented by the equation $y = cx^b$ to the data (referred to by Excel as a 'power trendline'), where y represents the flow rate and x represents the internal diameter. c and b are constants. Show the equation on the graph.

    (iii) Change the scales on both axes from linear to logarithmic. For the x axis this can be accomplished by clicking on the graph to make it active, then click on Layout>Axes>Primary Horizontal Axis>More Primary Horizontal Axis Options. A dialog box will appear. Check the Logarithmic scale box then click Close. Follow a similar procedure to change the scale on the y axis from linear to logarithmic.

Chapter 3

Data distributions I

3.1 Introduction

We may believe that the 'laws of chance' that apply when tossing a coin or rolling dice have little to do with experiments carried out in a laboratory. Rolling dice and tossing coins are the stuff of games. Surely, well planned and executed experiments provide precise and reliable data, immune from the laws of chance. Not so. Chance, or what we refer to more formally as *probability*, has rather a large role to play in every experiment. This is true whether an experiment involves counting the number of beta particles detected by nuclear counting apparatus in one minute, measuring the time a ball takes to fall a distance through a liquid or determining the values of resistance of 100 resistors supplied by a component manufacturer. Because it is not possible to predict with certainty what value will emerge when a measurement is made of a quantity, say of the time for a ball to fall a fixed distance through liquid, we are in a similar position to a person throwing several dice, who cannot know in advance which numbers will appear 'face up'. If we are not to give up in frustration at our inability to discover the 'exact' value of a quantity experimentally, we need to find out more about probability and how it can assist rather than hamper our experimental studies.

In many situations a characteristic pattern or *distribution* emerges in data gathered when repeat measurements are made of a quantity. A distribution of values indicates that there is a probability associated with the occurrence of any particular value. Related to any distribution of 'real' data there is a *probability distribution* which allows us to calculate the probability of the occurrence of any particular value. Real probability distributions can often be approximated by a 'theoretical' probability distribution. Though it is possible to devise many theoretical probability distributions, it is the so called '*normal*' probability distribution[1] (also referred to as the Gaussian distribution) that is most widely used. This is because histograms of data obtained in many experiments have shapes that are very similar to that of the normal distribution. An attraction of the normal and other distributions is that they provide a way of describing data in a quantitative manner which complements and extends visual representations of data such as the histogram. Using the properties of the normal distribution we are usually able to summarise a whole data set, which may consist of many values, by one or two carefully chosen numbers.

Owing to the reliance placed on the normal distribution when analysing experimental data, we need to be aware in what circumstances it is inappropriate to use this distribution and what the alternatives are. In particular, we need to be able to assess how well data conform to the normal distribution.

Probability tables are often used in the calculation of probabilities and confidence intervals. Excel has several built in functions which relieve the tedium of using such tables. These functions and their application will be discussed after basic principles have been considered.

Before discussing distributions of data in detail, we introduce some of the basic rules of probability.

3.2 Probability

In science it is common to speak of 'making measurements' and accumulating 'values' which may also be referred to as 'data'. The topic of probability has its own distinctive vocabulary with which we should be familiar. If an experiment in probability requires that a die[2] is thrown 150 times, then it is usual to say that the experiment consists of 150 *trials*. The outcome of a trial is referred to as an *event*.[3] So, if on the first throw of a die a '6' occurs, we speak of the event being a '6'. We can draw equivalences between trial and measurement, and between event and value.

The record of many events constitutes the data gathered in an experiment.

Example 1

A die is thrown once, what are the possible events?

ANSWER

The possible events on a single throw of a die (i.e. a single trial) are 1 or 2 or 3 or 4 or 5 or 6.

The possible outcomes of a trial, or a number of trials, is often referred to as the *sample space*. This is written as S{}. The sample space for a single throw of a die is S{1,2,3,4,5,6}.

---

[1] Usually referred to simply as the normal distribution.

[2] The singular of dice is die, so you can throw two dice, but only one die.

[3] Also sometimes referred to as an 'outcome'.

3.2.1 Rules of probability

We will use some of the rules of probability as ideas are developed regarding the variability in data. It is useful to illustrate the rules of probability by reference to a familiar experiment which consists of rolling a die one or more times. It is emphasised that, though the outcomes of rolling a die are considered, the rules are applicable to the outcomes of many types of experiment.

Rule 1
All probabilities lie in the range 0 to 1
Notes

(i) Example of probability equal to zero: the probability of an event being a '7' on the roll of a die is zero as only whole numbers from 1 to 6 are possible.

(ii) Example of probability equal to 1: the probability of an event being from 1 to 6 (inclusive) on the roll of a die is 1, as all the possible events lie between these inclusive limits.

(iii) Owing to the symmetry of the die, it is reasonable to suppose that each number will have the same probability of occurring. Writing the probability of obtaining a 1 as $P(1)$ and the probability of obtaining a 2 as $P(2)$ and so on, it follows that,

$$P(1) + P(2) + P(3) + P(4) + P(5) + P(6) = 1.$$

If all the probabilities are equal, then $P(1) = P(2) = P(3)$ etc. $= 1/6$.

(iv) Another way to find $P(1)$ is to do an experiment consisting of n trials, where n is a large number – say 1000, and to count the number of times a 1 appears – call this number, N. The ratio N/n (sometimes called the *relative frequency*) can be taken as the probability of obtaining a 1. Assuming we use a die that has not been tampered with, it is reasonable to expect N/n to be about $1/6$. This can be looked at another way: If the probability of a particular event A is $P(A)$, then the expected number of times that event A will occur in n trials is equal to $nP(A)$.

Rule 2
When events are *mutually exclusive*, the probability of event A occurring, $P(A)$, or event B occurring, $P(B)$, is

$$P(A \text{ or } B) = P(A) + P(B).$$

Notes

(i) *Mutually exclusive events:* this means that the occurrence of one event excludes other events occurring. For example, if you roll a die and obtain a 6 then no other outcome is possible (you cannot get, for example, a 6 <u>and</u> a 5 on a single roll of a die).

(ii) Example: what is the probability of obtaining a 2 or a 6 on a single roll of a die?

$$P(2 \text{ or } 6) = P(2) + P(6) = 1/6 + 1/6 = 1/3.$$

Rule 3

When events are *independent*, the probability of event A, $P(A)$, <u>and</u> event B, $P(B)$, occurring is $P(A) \times P(B)$. This is written[4] as

$$P(A \text{ and } B) = P(A) \times P(B).$$

Notes

    (i) *Independent events:* this means that the occurrence of an event has no influence on the probability of the occurrence of succeeding events. For example, on the first roll of a die, the probability of obtaining a 6 is $1/6$. The next time the die is rolled, the probability that a 6 will occur remains $1/6$.

  (ii) Example: using the rule, the probability of throwing two sixes in succession is:

$$P(6 \text{ and } 6) = P(6) \times P(6) = 1/6 \times 1/6 = 1/36.$$

We now have sufficient probability rules for our immediate purposes.[5] Next we consider how functions which describe the probability of particular outcome occurring can be useful when describing variability observed in 'real' data. We begin by discussing probability distributions.

3.3 Probability distributions

In the previous section we considered the probability that a particular outcome or event will occur when a die is rolled. A single roll of a die has only six possible outcomes. What do we do if our data consists of many events or, as we usually prefer to call them, values – is it possible to assign a probability to each value observed and is it sensible to do this? For illustrative purposes we consider a computer based 'experiment' which consists of generating one thousand random numbers[6] with values between 0 and 1. Table 3.1 shows the first hundred random numbers.

    In contrast to the situation in which a die is rolled, we do not know the probability that a particular value appearing in table 3.1 will occur.[7] Is it possible to establish the probability of obtaining a random number, x, between,

---

[4] $P(A \text{ and } B)$ is often written as $P(AB)$.

[5] Adler and Roessler (1972) discuss other rules of probability.

[6] Random numbers between 0 and 1 can be generated using the RAN function found on many calculators such as those manufactured by CASIO. The RAND() function on Excel can also be used to generate random numbers.

[7] Unless we know the details of the algorithm that produced the random numbers.

Table 3.1. *One hundred random numbers in the interval 0 to1.*

0.632	0.328	0.696	0.166	0.665	0.157	0.010	0.391	0.454	0.396
0.322	0.454	0.087	0.540	0.603	0.138	0.021	0.203	0.272	0.763
0.055	0.095	0.410	0.422	0.109	0.713	0.834	0.029	0.577	0.984
0.575	0.932	0.772	0.043	0.464	0.112	0.234	0.062	0.657	0.839
0.600	0.894	0.421	0.186	0.213	0.676	0.504	0.028	0.916	0.809
0.798	0.841	0.927	0.335	0.505	0.549	0.352	0.430	0.984	0.853
0.803	0.302	0.389	0.814	0.175	0.309	0.607	0.198	0.569	0.177
0.711	0.445	0.279	0.091	0.469	0.572	0.719	0.901	0.993	0.034
0.571	0.277	0.345	0.119	0.688	0.512	0.437	0.141	0.903	0.453
0.048	0.597	0.532	0.864	0.936	0.040	0.553	0.129	0.077	0.706

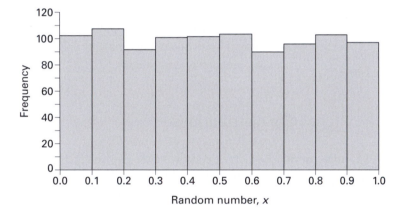

Figure 3.1. Histogram of 1000 random numbers.

say, 0.045 and 0.621, using the data that appear in table 3.1? A good starting point is to view the values in picture form to establish whether any values appear to occur more frequently than others. The histogram shown in figure 3.1 indicates the frequency of occurrence of the 1000 random numbers.

The histogram provides evidence to support the following statements.

- All values of x lie in the range 0 to 1.
- The number of values (i.e. the frequency) appearing in any given interval, say from 0.0 to 0.1, is about the same as that appearing in any other interval of the same width.

The histogram in figure 3.1 can be transformed into a probability distribution. This allows the probability of obtaining a random number in any interval to be

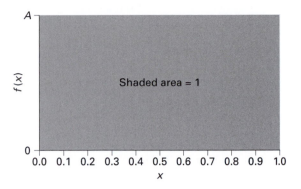

Figure 3.2. Uniform probability density function $f(x)$ versus x.

calculated. First, the bars in figure 3.1 are merged together and the frequencies scaled so that total area under the curve created by merging the bars is 1. That is, for this distribution, the probability is 1 that x lies between 0 and 1. Taking the length of each bar to be the same, the graph shown in figure 3.2 is created.

In going from figure 3.1 to 3.2 we have made quite an important step: Figure 3.1 shows the distribution of the 1000 random numbers. These numbers may be regarded as a *sample* of all the random numbers that *could* have been generated. The number of values that could have been generated is infinite in size, so the population is, in principle, infinite. By contrast, the graph in figure 3.2 allows us to calculate the probability of obtaining a value of x within a particular interval. What we have done by innocently smoothing out the 'imperfections' in the histogram shown in figure 3.1 prior to scaling, is to *infer* that the population of random numbers behaves in this manner, i.e. it consists of values that are evenly spread between 0 and 1. Successfully inferring something about a population by considering a sample consisting of values drawn from that population is a central goal in data analysis. Note that the label on the vertical axis in figure 3.2 differs from that in figure 3.1. The vertical axis is labelled $f(x)$, where $f(x)$ is referred to as the *probability density function*, or pdf for short.

As the area total under any $f(x)$ versus x curve is 1 and, in general, x may have any value between $-\infty$ to $+\infty$, we write

$$\int_{-\infty}^{\infty} f(x)dx = 1. \tag{3.1}$$

With reference to the graph in figure 3.2,

$f(x) = 0$ for $x < 0$ and $f(x) = 0$ for $x > 1$, and $f(x) = A$ for $0 \leq x \leq 1$.

Using equation 3.1 and replacing $f(x)$ by A gives

$$\int_0^1 A dx = 1,$$

therefore,

$$A[x]_0^1 = A[1 - 0] = 1.$$

It follows that $A = 1$, so $f(x) = 1$ from $x = 0$ to $x = 1$.

A word of caution here: $f(x)$ is *not* the probability of the occurrence of the value x. If that were the case then, as $f(x) = 1$ for all values of x between 0 and 1, we would conclude that the probability of observing, say, the number 0.476, is 1. This is not so. It is the *area* under the curve that is interpreted as a probability. If we want to know the probability of observing a value between x_1 and x_2, where $x_2 > x_1$, we must integrate $f(x)$ between these limits,[8] i.e.

$$P(x_1 \leq x \leq x_2) = \int_{x_1}^{x_2} f(x) dx. \tag{3.2}$$

For this probability distribution, $f(x)$ is 1 between $x = 0$ and $x = 1$, and the relationship emerges

$$P(x_1 \leq x \leq x_2) = \int_{x_1}^{x_2} 1.dx = x_2 - x_1. \tag{3.3}$$

Example 2

What is the probability of observing a random number, generated in the manner described in this section, between 0.045 and 0.621?

ANSWER

Substituting $x_2 = 0.621$ and $x_1 = 0.045$, into equation 3.3 gives

$$P(0.045 \leq x \leq 0.621) = 0.621 - 0.045 = 0.576.$$

In performing the integration given by equation 3.3, the implication is that the variable, x, represents values of a continuous quantity.[9] That is, x can take on any value between the limits x_1 and x_2. Mass, length and time are quantities regarded as continuous[10] but some values measured through experiment are discrete. An example is the counting of

---

[8] Read $P(x_1 \leq x \leq x_2)$ as 'the probability that x lies between the limits x_1 and x_2'.

[9] Here x is referred to as a *continuous random variable*, or *crv*.

[10] We can argue (with justification) that on an atomic scale, mass and length are no longer continuous, however few measurements we make in the laboratory are likely to be sensitive to the 'graininess' of matter on this scale.

particles emerging from a radioactive substance. In this case the recorded values must be a whole number of counts. The probability distributions used to describe the distribution of discrete quantities differ from those that describe continuous quantities. For the moment we focus on continuous distributions as we tend to encounter these more often than discrete distributions in the physical sciences.

The probability distribution given by figure 3.2 and represented mathematically by equation 3.3 is convenient for illustrating some of the basic features common to all probability distributions. Few scientific experiments generate data that are evenly distributed as shown in figure 3.1, meaning that the probability distribution in figure 3.2 is more of illustrative, rather than practical, value. In section 3.4 we consider data gathered in more conventional experiments and look at a probability distribution with more features in common with real data distributions.

Exercise A

(1) (a) A particular probability density function is written $f(x) = Ax$ for the range $0 \leq x \leq 4$ and $f(x) = 0$ outside this range.

      (i) Sketch a graph of $f(x)$ versus x.

      (ii) Use equation 3.1 to find A.

      (iii) Calculate the probability that x lies between $x = 3$ and $x = 4$.

      (iv) If 100000 measurements are made, how many values would you expect to lie between $x = 3$ and $x = 4$, assuming this probability function to be valid?

    (b) Another probability density function is written $f(x) = Ax^2$ for the range $-4 \leq x \leq 4$ and $f(x) = 0$ outside this range. Repeat parts (i) through to (iv) from part (a) for this new function.

(2)

The exponential probability density function[11] may be written

$$f(x) = \lambda e^{(-\lambda x)}, \quad \text{where } \lambda > 0 \text{ and } x \geq 0.$$

    (i) Show that, for $\lambda > 0$, $\int_0^\infty f(x)dx = 1$.

    (ii) If $\lambda = 0.3$, calculate the probability that x lies between $x = 0$ and $x = 2$.

    (iii) Calculate the probability that $x > 2$.

---

[11] This distribution is sometimes used to predict the lifetime of electrical and mechanical components. For details see Hayter (2006).

3.3.1 Limits in probability calculations

When calculating a probability using a probability density function, does it matter whether the limits of the interval, x_1 and x_2, are included in the calculation? In other words, does $P(x_1 \leq x \leq x_2)$ differ from $P(x_1 < x < x_2)$? The answer to this is no, and to justify this we can write

$$P(x_1 \leq x \leq x_2) = P(x_1) + P(x_1 < x < x_2) + P(x_2). \tag{3.4}$$

$P(x_1)$ is the area under the probability curve at $x = x_1$. Using equation 3.2, this probability is written

$$P(x_1) = P(x_1 \leq x \leq x_1) = \int_{x_1}^{x_1} f(x)dx. \tag{3.5}$$

Both limits of the integral are x_1 and so the right hand side of equation 3.5 must be equal to zero. Therefore, $P(x_1) = P(x_2) = P(x) = 0$, and equation 3.4 becomes,

$$P(x_1 \leq x \leq x_2) = P(x_1 < x < x_2). \tag{3.6}$$

Equation 3.6 is valid for continuous random variables.

The fact that $P(x) = 0$ can be unsettling. Let us return to table 3.1 and consider the first value, $x = 0.632$. If we are dealing with the distribution of a random variable, then $P(0.632) = 0$. However, the value 0.632 has been observed, so how are we able to reconcile this with the fact that $P(0.632) = 0$? The explanation is that $x = 0.632$ is a *rounded* value[12] and the 'actual' value could lie anywhere between $x = 0.63150$ and $x = 0.63250$. That is, though it is not obvious, we are dealing with an implied range by the way the number is written. If we now ask what is the probability that a value of x (for this distribution) lies between $x = 0.63150$ and $x = 0.63250$, we have, by using equation 3.3,

$$P(x_1 \leq x \leq x_2) = 0.63250 - 0.63150 = 0.001.$$

If 1000 measurements are made, then the number of times 0.632 is expected to occur is $1000 \times 0.001 = 1$.

3.4 Distributions of real data

We now consider some real data gathered in a diversity of laboratory based experiments with a view to identifying some general features of data distributions.

---

[12] Rounding could have been carried out by the experimenter, or the value recorded could have been limited by the precision of the instrument used to make the measurement.

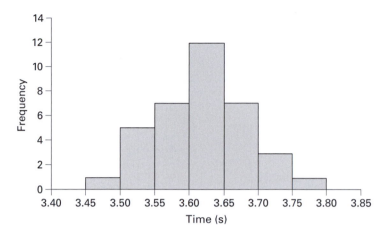

Figure 3.3. Frequency versus time of fall through oil.

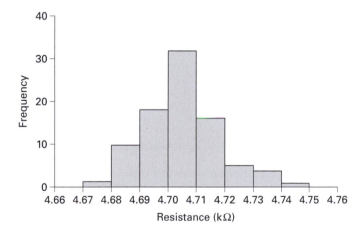

Figure 3.4. Frequency versus resistance.

Figure 3.3 shows data from an experiment to determine the viscosity of an oil. In the experiment, a small metal sphere was allowed to fall a fixed distance through the oil and the time of fall was recorded. This procedure was repeated many times. Figure 3.4 shows the resistances of a sample of 100 resistors of nominal resistance 4.7 kΩ which were measured with a high precision multi-meter. Figure 3.5 shows the number of counts detected by a scintillation counter in 100 time intervals of 10 s, when the counter was placed in close proximity to the β-particle emitter, strontium 90.

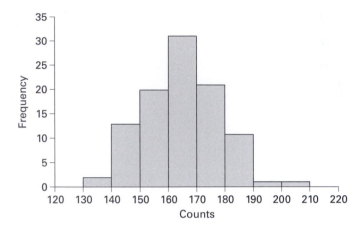

Figure 3.5. Frequency versus counts in a radioactivity experiment.

As the experiments described here are diverse and the underlying causes of variability in each experiment likely to be quite different,[13] we might anticipate the distribution of data in each experiment as shown in figures 3.3 to 3.5 to be markedly different. This is not the case, and we are struck by the similarity in the shapes of the histograms, rather than the differences between them. Note that:

- the distribution of each data set is approximately symmetric;
- there is a single central 'peak';
- most data are clustered around that peak (and, as a consequence, few data are found in the 'tails' of the distribution).

A commonly used shorthand way of expressing the main features of histograms such as those in figures 3.3 to 3.5, is to say that the data they represent follow a 'bell shaped' distribution. To illustrate this, figure 3.6 shows a bell shaped curve superposed on the time of fall data shown in figure 3.3.

We shift our attention to the mathematical function that generates the bell shaped curve in figure 3.6. This is because it appears reasonable to argue that the distribution of data in this sample of data, and more fundamentally, the distribution of the population from which the sample was drawn, can be described well by this function. A consideration of this function and its properties provides further insights into how the variability in data can be expressed concisely in numeric form.

---

[13] For example, a major source in the variability of time measured in the viscosity experiment is due to inconsistency in hand timing the duration of the fall of the ball, whereas in the resistance experiment the dominant factor is likely to be variability in the process used to manufacture the resistors.

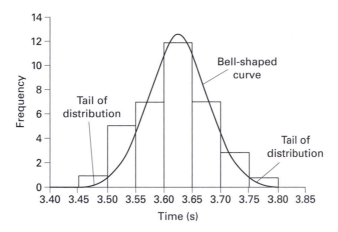

Figure 3.6. Frequency versus time of fall through liquid.

It is perhaps worth sounding a note of caution at this point. By no means do data from all experiments follow the shape given by figures 3.3 to 3.5. However, approximations to this shape are found so frequently in experimental data that it makes sense to focus attention on this distribution ahead of any other. Additionally, even if the spread of data is distinctly asymmetric, a bell shaped distribution can be constructed from the raw data[14] in a way that assists in expressing the variability or uncertainty in the *mean* of the data. This is discussed in section 3.7.

3.5 The normal distribution

The bell shaped curve appearing in figure 3.6 is generated using the probability density function

$$f(x) = \frac{1}{\sigma\sqrt{2\pi}}\,e^{-(x-\mu)^2/2\sigma^2}\,, \tag{3.7}$$

where μ and σ are the population mean and the population standard deviation respectively introduced in chapter 1 and which we used to describe the centre and spread respectively of a data set. Equation 3.7 is referred to as the *normal probability density function*. It is a 'normalised' equation which is another way of saying that when it is integrated between $-\infty \leq x \leq +\infty$, the value obtained is 1.

---

[14] The term 'raw data' refers to data that have been obtained directly through experiment or observation and have not been manipulated in any way, such as combining values to calculate a mean.

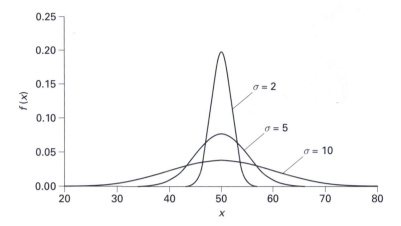

Figure 3.7. Three normal distributions with the same mean, but differing standard deviations.

Using equation 3.7 we can generate the 'bell shaped' curve associated with any combination of μ and σ and hence, by integration, find the probability of obtaining a value of x within a specified interval. Figure 3.7 shows $f(x)$ versus x for the cases in which $\mu = 50$ and $\sigma = 2$, 5 and 10.

The population mean, μ, is coincident with the centre of the symmetric distribution and the standard deviation, σ, is a measure of the spread of the data. A larger value of σ results in a broader, flatter distribution, though the total area under the curve remains equal to 1. On closer examination of figure 3.7 we see that $f(x)$ is quite small (though not zero) for x outside the interval $\mu - 2.5\sigma$ to $\mu + 2.5\sigma$.

The normal distribution may be used to calculate the probability of obtaining a value of x between the limits, x_1 and x_2. This is given by

$$P(x_1 \leq x \leq x_2) = \frac{1}{\sigma\sqrt{2\pi}} \int_{x_1}^{x_2} e^{-(x-\mu)^2/2\sigma^2} dx. \tag{3.8}$$

The steps to determine the probability are shown symbolically and pictorially in figure 3.8.

The function given by,

$$F(x \leq x_1) = P(-\infty < x < x_1) = \frac{1}{\sigma\sqrt{2\pi}} \int_{-\infty}^{x_1} e^{-(x-\mu)^2/2\sigma^2} dx \tag{3.9}$$

is usually referred to as the *cumulative distribution function*, cdf, for the normal distribution. The cdf is often used to assist in calculating probabilities, such as those shown in figure 3.8.

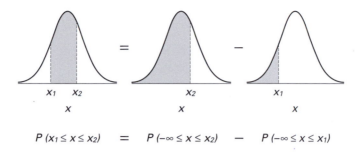

$$P\left(x_1 \leq x \leq x_2\right) \quad = \quad P\left(-\infty \leq x \leq x_2\right) \quad - \quad P\left(-\infty \leq x \leq x_1\right)$$

Figure 3.8. Finding the area under the normal curve.

As an example of the use of the normal distribution, we calculate the probability that x lies between $\pm\sigma$ of the mean, μ. Finding this probability requires that $f(x)$ in equation 3.2 be replaced by the expression in equation 3.7. The limits of the integration are $x_1 = \mu - \sigma$, and $x_2 = \mu + \sigma$. We have

$$P(\mu - \sigma \leq x \leq \mu + \sigma) = \frac{1}{\sigma\sqrt{2\pi}} \int_{\mu-\sigma}^{\mu+\sigma} e^{-(x-\mu)^2/2\sigma^2} dx. \tag{3.10}$$

To assist in evaluating the integral in equation 3.10, it is usual to change the variable from x to z, where z is given by

$$z = \frac{x - \mu}{\sigma}, \tag{3.11}$$

where z is a random variable with mean of zero and standard deviation of 1. Equation 3.7 reduces to

$$f(z) = \frac{1}{\sqrt{2\pi}} e^{-z^2/2}. \tag{3.12}$$

Equation 3.10 becomes

$$P(-1 \leq z \leq 1) = \frac{1}{\sqrt{2\pi}} \int_{-1}^{1} e^{-z^2/2} dz. \tag{3.13}$$

Figure 3.9 shows the variation of $f(z)$ with z.

The integral appearing in equation 3.13 cannot be evaluated analytically and so a numerical method for solving for the area under the curve is required. It turns out that the shaded area indicated in figure 3.9 is about 0.68. We conclude the probability that a value obtained through measurement lies within $\pm\sigma$ of the population mean is 0.68. Or, looked at another way, 68% of all values

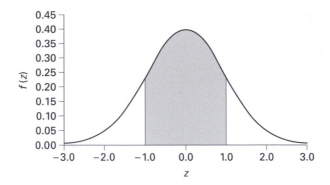

Figure 3.9. Variation of $f(z)$ with z. The shaded area is equal to the probability that z lies between –1 and +1.

obtained through experiment are expected to lie within $\pm\sigma$ of the population mean.[15]

3.5.1 Excel's NORM.DIST() function[16]

Traditionally, probability tables have been used to assist in determining probabilities associated with areas under the normal probability distribution.[17] Excel has built in functions which offer an attractive alternative for determining probabilities.

The NORM.DIST() function in Excel calculates the value of the probability density function (pdf) given by equation 3.7 and the value of the cumulative distribution function[18] (cdf) as given by equation 3.9.

The syntax of the function is as follows.

NORM.DIST(x,mean,standard deviation,cumulative)

In order to select the cdf option, the cumulative parameter in the function is set to TRUE. To choose the pdf option, the cumulative parameter is set to FALSE.

Example 3
Calculate the value of the pdf and the cdf when $x = 46$ for normally distributed data with mean = 50 and standard deviation = 4.

[15] As long as the values are normally distributed.

[16] In Excel 2010, NORM.DIST() supersedes the NORMDIST() function found in earlier versions of Excel (both functions operate identically). For backward compatibility, the NORMDIST() function is still available in Excel 2010.

[17] It remains important to know how to use such tables, as a computer will not always be 'to hand'. We consider one such table in section 3.5.2.

[18] The cumulative distribution function in Excel is referred to as the *normal cumulative distribution*.

ANSWER

The NORM.DIST() function for the pdf and cdf is shown entered in cells A1 and A2 respectively of sheet 3.1(a).

Sheet 3.1(b) shows the values returned in cells A1 and A2 after the ENTER key has been pressed.

We conclude that

$$f(x = 46) = 0.06049, \ and \ P(-\infty \le x \le 46) = 0.15866.$$

Sheet 3.1. *Using the NORM.DIST() function.*

(a) *Values entered into NORM.DIST()*

	A	B	C
1	=NORM.DIST(46,50,4,FALSE)		
2	=NORM.DIST(46,50,4,TRUE)		

(b) *Values returned by the NORM.DIST() function*

	A	B	C
1	0.060493		
2	0.158655		

Example 4

Calculate the area under the curve between $x = 46$ and $x = 51$ for normally distributed data with mean = 50 and standard deviation = 4.

ANSWER

$$P(46 \le x \le 51) = P(-\infty \le x \le 51) - P(-\infty \le x \le 46)$$

In cell A1 of sheet 3.2(a) we calculate $P(-\infty \le x \le 51)$ and in cell A2 we calculate $P(-\infty \le x \le 46)$. The two probabilities are subtracted in cell A3 to give the required result. Sheet 3.2(b) shows the values returned by Excel.

The area under the curve between $x = 46$ and $x = 51$ is 0.440051.

Sheet 3.2. *Example of the application of the NORM.DIST() function.*

(a) *Numbers entered into NORM.DIST()*

	A	B	C
1	=NORM.DIST(51,50,4,TRUE)		
2	=NORM.DIST(46,50,4,TRUE)		
3	=A1−A2		

(b) *Result of calculations*

	A	B	C
1	0.598706		
2	0.158655		
3	0.440051		

Exercise B

The masses of small metal spheres are known to be normally distributed with $\mu = 8.1$ g and $\sigma = 0.2$ g. Use the NORM.DIST() function to find the values of the cdf for values of mass given in table 3.2 to four decimal places.

Table 3.2. *Values of mass for exercise B.*

x (g)	7.2	7.4	7.6	7.8	8.0	8.2	8.4	8.6	8.8	9.0

3.5.2 The standard normal distribution

In order to assist in integrating the normal pdf in section 3.5, a mathematical 'trick' was played consisting of changing variables using equation 3.11. This has the effect of creating a new distribution with a mean of zero and a standard deviation of 1. This distribution is usually referred to as the *standard normal distribution*. It is particularly useful as it can represent any normal distribution, whatever its mean and standard deviation. So, if we wish to evaluate the area (and hence the probability) under a normal distribution curve between the two limits, x_1 and x_2, we do not need to integrate the pdf given by equation 3.7, which differs from one normal distribution to another, but equation 3.12 which is the same for all normal distributions.

To find the probability of obtaining a value of x between x_1 and x_2, first transform x values to z values, so that

$$z_1 = \frac{x_1 - \mu}{\sigma} \text{ and } z_2 = \frac{x_2 - \mu}{\sigma},$$

and find the probability that a value of z lies between the limits z_1 and z_2 by evaluating the integral

$$P(z_1 \le z \le z_2) = \frac{1}{\sqrt{2\pi}} \int_{z_1}^{z_2} e^{-z^2/2} dz. \qquad (3.14)$$

Table 3.3, as well as table 1 of appendix 1, may be used to assist in the evaluation of equation 3.14.

If we do not have access to Excel or some other spreadsheet package that is able evaluate the area under a normal curve directly, we can use 'probability' tables such as those found in appendix 1.

To explain the use of table 3.3, first consider figure 3.10. For any value of z_1, say $z_1 = 0.55$, go down the first column as far as $z_1 = 0.50$, then across to the column headed 0.05, this brings you to the entry in the table with value 0.7088. We conclude that the probability of obtaining a value of $z \le 0.55$ is

$$P(-\infty \le z \le 0.55) = \frac{1}{\sqrt{2\pi}} \int_{-\infty}^{0.55} e^{-z^2/2} dz = 0.7088.$$

Knowing that:

Table 3.3. *Normal probability table giving the area under the standard normal curve between $z = -\infty$ and $z = z_1$.*

z_1	0.00	0.01	0.02	0.03	0.04	0.05	0.06	0.07	0.08	0.09
0.00	0.5000	0.5040	0.5080	0.5120	0.5160	0.5199	0.5239	0.5279	0.5319	0.5359
0.10	0.5398	0.5438	0.5478	0.5517	0.5557	0.5596	0.5636	0.5675	0.5714	0.5753
0.20	0.5793	0.5832	0.5871	0.5910	0.5948	0.5987	0.6026	0.6064	0.6103	0.6141
0.30	0.6179	0.6217	0.6255	0.6293	0.6331	0.6368	0.6406	0.6443	0.6480	0.6517
0.40	0.6554	0.6591	0.6628	0.6664	0.6700	0.6736	0.6772	0.6808	0.6844	0.6879
0.50	0.6915	0.6950	0.6985	0.7019	0.7054	0.7088	0.7123	0.7157	0.7190	0.7224
0.60	0.7257	0.7291	0.7324	0.7357	0.7389	0.7422	0.7454	0.7486	0.7517	0.7549
0.70	0.7580	0.7611	0.7642	0.7673	0.7704	0.7734	0.7764	0.7794	0.7823	0.7852
0.80	0.7881	0.7910	0.7939	0.7967	0.7995	0.8023	0.8051	0.8078	0.8106	0.8133
0.90	0.8159	0.8186	0.8212	0.8238	0.8264	0.8289	0.8315	0.8340	0.8365	0.8389
1.00	0.8413	0.8438	0.8461	0.8485	0.8508	0.8531	0.8554	0.8577	0.8599	0.8621

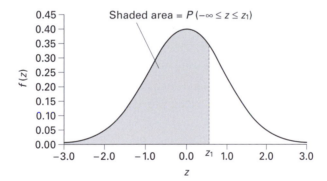

Figure 3.10. Relationship between area under curve and z_1.

- the area under the curve = 1;
- the curve is symmetric about $z = 0$,

it is possible to calculate the probabilities in many other situations.

Example 5

What is the probability that a value of z is *greater* than 0.75?

ANSWER

We have,

$P(z_1 \leq 0.75) + P(z_1 \geq 0.75) = 1$, and so (using table 3.3),
$P(z_1 \geq 0.75) = 1 - P(z_1 \leq 0.75) = 1 - 0.7734 = 0.2266.$

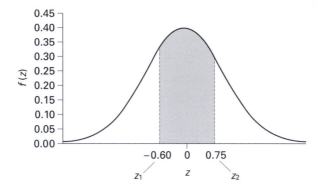

Figure 3.11. Shaded area under curve is equal to probability that z lies between –0.6 and 0.75.

Example 6

What is the probability that z lies between the limits $z_1 = -0.60$ and $z_2 = 0.75$?

ANSWER

Solving this type of problem is assisted by drawing a picture representing the area (and hence the probability) to be determined. Figure 3.11 indicates the area required.

The shaded area in figure 3.11 is equal to (area to the left of z_2) – (area to the left of z_1).

Probability tables do not usually include negative values of z, so we must make use of the symmetry of the distribution and recognise that,[19]

$$P(z \leq -|z_1|) = 1 - P(z \leq |z_1|). \tag{3.15}$$

Now,

$$P(z \leq 0.60) = 0.7257, \text{ so}$$
$$P(z \leq -0.60) = 1 - 0.7257 = 0.2743.$$

The shaded area in figure 3.11 is therefore,

$$0.7734 - 0.2743 = 0.4991 \text{ (to four significant figures)}.$$

Exercise C

With reference to figure 3.11, if $z_1 = -0.75$ and $z_2 = -0.60$, calculate the probability $P(-0.75 \leq z \leq -0.60)$.

---

[19] Table 1 in appendix 1 is slightly unusual as it gives the area under the normal curve for negative values of z.

3.5.3 Excel's NORM.S.DIST() function[20]

This function calculates the area under the standard normal curve between $z = -\infty$ and, say, $z = z_1$. This is given as

$$P(-\infty \le z \le z_1) = \frac{1}{\sqrt{2\pi}} \int_{-\infty}^{z_1} e^{-z^2/2} dz,$$

and is referred to in Excel as the *standard normal cumulative distribution function.*

The NORM.S.DIST() function is also able to calculate the value of the standard normal probability density function given by equation 3.12.

The syntax of the function is as follows.

NORM.S.DIST(z, cumulative)

To select the standard normal cumulative distribution function, the cumulative parameter in the function is set to TRUE. To choose the standard normal probability density function, the cumulative parameter is set to FALSE.

Example 7

Calculate the area under the standard normal distribution between $z = -\infty$ and $z = -1.5$.

ANSWER

Sheet 3.3 shows the NORM.S.DIST() function entered into cell A1.

Sheet 3.3(b) shows the value returned in cell A1 after the Enter key is pressed.

Sheet 3.3. *Use of NORM.S.DIST() function.*

(a) *Numbers entered into NORM.S.DIST()*

	A	B	C
1	= NORM.S.DIST(−1.5,TRUE)		
2			

(b) *Result of calculation*

	A	B	C
1	0.066807		
2			

Example 8

Calculate the area under the standard normal distribution between $z = -2.5$ and $z = -1.5$.

ANSWER

The area between $z = -2.5$ and $z = -1.5$ may be written as

---

[20] The NORM.S.DIST() in Excel 2010 function replaces the NORMSDIST() function available in earlier versions of Excel. For compatibility with earlier versions of Excel, the NORMSDIST() function remains available in Excel 2010.

$$P(-2.5 \leq z \leq -1.5) = P(-\infty \leq z \leq -1.5) - P(-\infty \leq z \leq -2.5).$$

In cell A1 of sheet 3.4(a) we calculate $P(-\infty \leq z \leq -1.5)$ and, in cell A2, $P(-\infty \leq z \leq -2.5)$. The two probabilities are subtracted in cell A3 to give the required result.

Sheet 3.4(b) shows the values returned in cells A1 to A3 after the Enter key is pressed. We conclude that $P(-2.5 \leq z \leq -1.5) = 0.060598$.

Sheet 3.4. *Example of use of NORM.S.DIST() function.*

(a) *Numbers entered into NORM.S.DIST() function*

	A	B	C
1	= NORM.S.DIST(−1.5,TRUE)		
2	= NORM.S.DIST(−2.5,TRUE)		
3	= A1−A2		

(b) *Result of calculations*

	A	B	C
1	0.066807		
2	0.00621		
3	0.060598		

Exercise D

Use the NORM.S.DIST() function to calculate the values of the standard cdf for the z values given in table 3.4. Give the values of the standard cdf to four decimal places.

3.5.4 $\bar{x}$ and s as approximations to μ and σ

Before proceeding to consider the analysis of real data using the normal distribution, we should admit once again that, in the majority of experiments, we can never know the population mean, μ, and the population standard deviation, σ, as they can only be determined if all the values that make up the population are known. If the population is infinite, as it is often in science, this is impossible. We must 'make do' with the best estimates of μ and σ, namely $\bar{x}$ and s respectively. μ and $\bar{x}$ are given by equations 1.13 and 1.6 respectively. σ and s are given by equations 1.14 and 1.16 respectively. $\bar{x}$ tends to μ and s tends to σ as the number of points, n, tends to infinity. It is fair to ask for what sample size, n, does the approximation of $\bar{x}$ for μ and s for σ become 'reasonable'?

Table 3.4. *The z values for exercise D.*

z	0.0	0.1	0.2	0.3	0.4	0.5	0.6	0.7	0.8	0.9	1.0

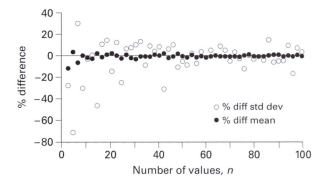

Figure 3.12. Percentage difference between population parameters and sample statistics.

Figure 3.12 shows the percentage difference between $\bar{x}$ and μ, and between σ and s for samples of size n, where n is between 2 and 100. Samples were drawn from a population consisting of simulated values known to be normally distributed.[21]

The percentage difference between the means is defined as $\left(\frac{\bar{x}-\mu}{\mu}\right) \times 100\%$ and the percentage difference between the standard deviations is defined as $\left(\frac{s-\sigma}{\sigma}\right) \times 100\%$. Figure 3.12 indicates that the magnitude of the percentage difference between μ and $\bar{x}$ for these data is small (under 5%) for $n > 10$. Figure 3.12 also shows that, as n increases, s converges more slowly to σ than $\bar{x}$ converges to μ. The percentage difference between σ and s is only consistently[22] below 20% for $n > 30$. As a rule of thumb we assume that if a sample consists of n values, where n is in excess of 30, we can take s as a reasonable approximation to σ.

3.6 From area under a normal curve to an interval

Section 3.5 assisted in answering the question 'what is the area under the normal curve between limits x_1 and x_2?'. We can also ask the closely related question 'between what limits (symmetrical about the mean) is the area under the normal curve equal to, say, $X\%$?'.

In order to determine the symmetrical limits, do the following.

(1) Choose the area under the normal curve corresponding to $X\%$ of the distribution.

---

[21] Normally distributed values used in the simulation were generated with known μ and σ using the Random Number Generation tool in Excel's Data Analysis ToolPak.

[22] However, note that one sample of $n \approx 40$ produced a percentage difference between σ and s of about –30%.

Exercise E

(1) Figure 3.6 shows a histogram of the time of fall of a small sphere through a liquid. For these data, $\bar{x} = 3.624$ s and $s = 0.068$ s.

  (i) Calculate the probability that a value lies in the interval 3.65 s to 3.70 s.

  (ii) If the data consist of 36 values, how many of these would you expect to lie in the interval 3.65 s to 3.70 s?

(2) With reference to the standard normal distribution, what is the probability that z is

  (i) less than −2?

  (ii) between −2 and +2?

  (iii) between −3 and −2?

  (iv) above +2?

  (v) Sketch the standard normal distribution for each part (i) to (iv) above, indicating clearly the probability being determined.

(3) For normally distributed data, what is the probability that a value lies further than 2.5σ from the mean?

(4) A large sample of resistors supplied by a manufacturer is tested and found to have a mean resistance of 4.70 kΩ and a standard deviation of 0.01 kΩ. Assuming the distribution of resistances is normal, what is the probability that a resistor chosen from the population will have a resistance

  (i) below 4.69 kΩ?

  (ii) between 4.71 kΩ and 4.72 kΩ?

  (iii) above 4.73 kΩ?

  (iv) If 10000 resistors are measured, how many would be expected to have resistance in excess of 4.73 kΩ?

(2) Sketch a graph of the standard normal distribution. Shade the chosen area equal to $X\%/100\%$ under the normal curve, symmetric about the mean as shown in figure 3.13.

(3) Use the normal probability tables in appendix 1 (or the NORM.S.INV() function in Excel) to determine z_1 and z_2 (note that $z_1 = -z_2$).

(4) Transform back to 'x-values' using the relationship $x = \mu + z\sigma$.

Example 9

A distribution of data consisting of many repeat measurements of the flow of water through a tube has a mean of 3.7 mL/s and a standard deviation of 0.41 mL/s. Between what limits, symmetrical about the mean, do 50% of the data lie?

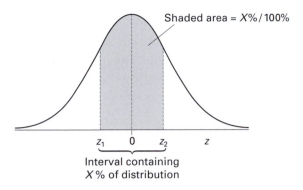

Figure 3.13. Standard normal curve showing interval containing $X\%$ of the distribution bounded by z_1 and z_2.

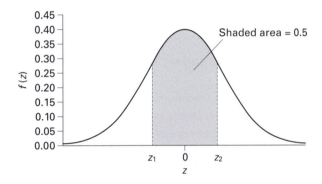

Figure 3.14. Standard normal curve with shaded area of 0.5 distributed symmetrically about $z = 0$.

ANSWER

We assume that the question provides very good estimates of the population parameters, μ and σ, so that $\mu = 3.7$ mL/s and $\sigma = 0.41$ mL/s. To find the limits it is necessary to consider the standard normal distribution and indicate the required area under the curve. The area is distributed symmetrically about $z = 0$ and is shown shaded in figure 3.14.

Table 1 in appendix 1 can be used to find z_2 so long as we can determine the area to the left of z_2, that is, the area between $-\infty \leq z \leq z_2$. As the normal distribution is symmetrical about $z = 0$, half of the total area under the curve lies between $z = -\infty$ and $z = 0$. This area is equal to 0.5, as the total area under the curve is equal to 1. As the required area is symmetrical about the mean, half the shaded area in figure 3.14 must lie between $z = 0$ and $z = z_2$. The total area under the curve between $-\infty$ and z_2 is therefore $0.5 + 0.25 = 0.75$.

The next step is to refer to the normal probability integral tables and look in the table for the probability, 0.75 (or the nearest value to it). Referring to table 1 in appendix 1, a probability of 0.75 corresponds to a z value of 0.67, so $z_2 = 0.67$. As the limits z_1 and z_2 are symmetrical about $z = 0$, it follows that $z_1 = -0.67$.

The final step is to apply equation 3.11 which is used to transform z values to x values. Rearranging equation 3.11 we obtain

$$x = \mu + z\sigma. \tag{3.16}$$

Substituting $z_1 = -0.67$ and $z_2 = 0.67$ gives $x_1 = 3.43$ mL/s and $x_2 = 3.97$ mL/s.

The interval containing 50% of the values lies between 3.43 mL/s and 3.97 mL/s. Equivalently, if a measurement is made of water flow, the probability is 0.5 that the value obtained lies between 3.43 mL/s and 3.97 mL/s.

3.6.1 Limits containing 68% and 95% of the normal distribution

As the normal distribution extends between $+\infty$ and $-\infty$, it follows that the 100% of normally distributed data lies between these limits. It is hardly very discriminating to say that all data lie between $\pm\infty$. Instead, it is sometimes useful to state an interval between which a certain percentage of the data lie. We discovered in section 3.5 that about 68% of normally distributed data lie within $\pm\sigma$ of the population mean. This interval can be written as

$$\mu - \sigma \leq x \leq \mu + \sigma. \tag{3.17}$$

In terms of the standard normal distribution variable, z, the interval containing 68% of the distribution is

$$-1 \leq z \leq 1. \tag{3.18}$$

The interval that includes 95% of the data (or equivalently, where the area under the normal curve is equal to 0.95). This is shown in figure 3.15.

As stated in the previous section, the area between $z = -\infty$ and $z = 0$ is 0.5. The area between $z = 0$ and $z = z_2$ is half the shaded area in figure 3.15. Now the $\left(\frac{\text{shaded area}}{2}\right) = \frac{0.95}{2} = 0.475$, so the area between $z = -\infty$ and $z = z_2$ is $0.5 + 0.475 = 0.975$.

The z value corresponding to a probability of 0.975, found using table 1 in appendix 1 is equal to 1.96, so that $z_2 = 1.96$. As the distribution is symmetric about $z = 0$, it follows that $z_1 = -z_2 = -1.96$. So the interval containing 95% of the data lies between $z = -1.96$ and $z = +1.96$.

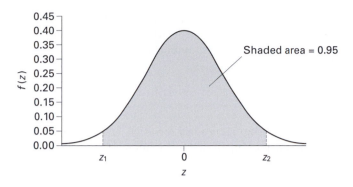

Figure 3.15. Standard normal curve indicating area of 0.95 distributed symmetrically about $z = 0$.

Example 10

The current gain of many BC107 transistors is measured.[23] The mean gain is found to be 209 with the standard deviation of 67. What are the intervals, symmetrical about the mean, which contain 68% and 95% of the distribution of the gain of the transistors?

ANSWER

Take $\mu = 209$ and $\sigma = 67$ (we assume that as 'many' transistors have been measured, 67 is a good estimate of the population standard deviation, σ).

Rearrange equation 3.11 to give

$$x = \mu + z\sigma.$$

Considering the interval containing 68% of the distribution first, the lower limit occurs for $z = -1$, giving the lower limit of x, x_1 as

$$x_1 = 209 - 67 = 142;$$

similarly, the upper limit of x, x_2 is

$$x_2 = 209 + 67 = 276.$$

We conclude that the 68% of the distribution lies between 142 and 276.

The lower limit of the interval containing 95% of the data occurs for $z = -1.96$, giving the lower limit of x, x_1 as

$$x_1 = 209 + (-1.96) \times 67 = 78.$$

The upper limit of z is 1.96, giving x_2 as,

$$x_2 = 209 + 1.96 \times 67 = 340.$$

---

[23] Current gain has no units.

We conclude that the interval containing 95% of the current gain data lies between 78 and 340. Values of z for other intervals may be calculated in a similar manner. Table 3.5 shows a summary of z values corresponding to various percentages, $X\%$, of data distributed symmetrically about the mean.

Exercise F

A water cooled heat sink is used in a thermal experiment. The variation of the temperature of the heat sink is recorded at one minute intervals over the duration of the experiment. Forty values of heat sink temperature (in $^\circ$C) are shown in table 3.6.

Assuming the data in table 3.6 are normally distributed, use the data to find the intervals, symmetric about the mean, that contain 50%, 68%, 90%, 95% and 99% of heat sink temperatures.

3.6.2 Excel's NORM.INV() function[24]

If the area is known under the normal distribution curve between $x = -\infty$ and $x = x_1$, what is the value of x_1? For a given probability, the NORM.INV() function in Excel will calculate x_1 so long as the mean and standard deviation of the normal distribution are supplied.

The syntax of the function is as follows.

NORM.INV(probability, mean, standard deviation)

Table 3.5. *Values of z corresponding to intervals containing X% of the normal distribution.*

$X\%$	50%	68%	90%	95%	99%
$z_{X\%}$	0.67	0.99	1.64	1.96	2.58

Table 3.6. *Heat sink temperature data (values are in $^\circ$C).*

18.2	17.4	19.0	18.7	17.8	19.4	18.5	18.4
18.4	18.6	18.6	17.1	17.3	19.4	18.9	17.7
19.8	17.6	18.2	19.3	18.7	18.1	18.4	18.6
18.9	18.9	18.4	18.5	19.0	18.6	18.3	19.4
17.9	18.8	18.0	18.1	18.0	18.0	18.2	17.3

[24] The NORM.INV() function in Excel 2010 supercedes the NORMINV() function in earlier versions of Excel. The NORMINV() function remains available in Excel 2010.

Example 11

Normally distributed data have a mean of 126 and a standard deviation of 18. If the area under the normal curve between $-\infty$ and x_1 for these data is 0.75, calculate x_1.

ANSWER

Sheet 3.5 shows the function required to calculate x_1 in cell A1.

Sheet 3.5(b) shows the value returned in cell A1 after the Enter key is pressed.

Sheet 3.5. *Use of the NORM.INV() function.*

(a) *Number entered into NORM.INV() function*

	A	B	C
1	= NORM.INV(0.75,126,18)		
2			

(b) *Result of calculation*

	A	B	C
1	138.1408		
2			

Example 12

Values of density of a saline solution are found to be normally distributed with a mean of $1.15\,\text{g/cm}^3$ and a standard deviation of $0.050\,\text{g/cm}^3$. Use the NORM.INV() function to find the interval, symmetric about the mean, that contains 80% of the density data.

ANSWER

It is helpful to sketch a diagram of the normal distribution and to indicate the area under the curve and the corresponding limits, x_1 and x_2. The total area to the left of x_2 in figure 3.16 is $0.8 + 0.1 = 0.9$. Now we can use the Excel function NORM.INV() to find the value of x_2.

Sheet 3.6(a) shows the NORM.INV() function entered into cell A1. Sheet 3.6(b) shows the value returned in cell A1 after the Enter key is pressed. The upper limit of the interval is $1.214\,\text{g/cm}^3$, which is more than the mean by an amount $1.214\,\text{g/cm}^3 - 1.15\,\text{g/cm}^3 = 0.064\,\text{g/cm}^3$. The lower limit is less than the mean by the same amount. It follows that the lower limit is $1.15\,\text{g/cm}^3 - 0.064\,\text{g/cm}^3 = 1.086\,\text{g/cm}^3$.

Sheet 3.6. *Use of NORM.INV() function.*

(a) *Numbers entered into NORM.INV() function*

	A	B	C
1	= NORM.INV(0.9,1.15,0.05)		
2			

(b) *Result of calculation*

	A	B	C
1	1.214078		
2			

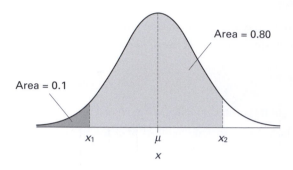

Figure 3.16. Normal curve for example 12.

3.6.3   Excel's NORM.S.INV() function[25]

If the area under the standard normal distribution between $-\infty$ and z is known, then the NORM.S.INV() function returns the value of z_1 corresponding to that area. The syntax of the function is as follows.

NORM.S.INV(probability)

Example 13

If the area under the standard normal curve between $-\infty$ and z_1 is equal to 0.85, what is the value of z_1?

ANSWER

Sheet 3.7 shows the value returned in cell A1 after the Enter key is pressed, i.e. $z_1 = 1.036433$.

Sheet 3.7. *The NORM.S.INV() function entered into cell A1.*

(a) *Use of NORM.S.INV() function*

	A	B	C
1	= NORM.S.INV(0.85)		
2			

(b) *Result of calculation*

	A	B	C
1	1.036433		
2			

Exercise G

If the area under the standard normal curve between z_1 and $+\infty$ is 0.2, use the NORM.S.INV() function to find z_1.

---

[25] The NORM.S.INV() function in Excel 2010 supercedes the NORMSINV() function in earlier versions of Excel. The NORMSINV() function remains available in Excel 2010.

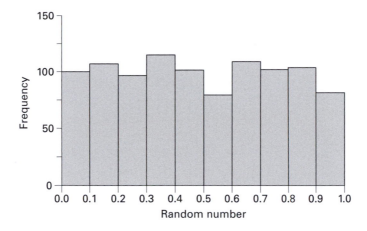

Figure 3.17. Distribution of 1000 random numbers.

3.7 Distribution of sample means

The normal distribution would be useful enough if it only described patterns observed in 'raw' data. However, the use of the normal distribution can be extended to data that are not normally distributed. Suppose, instead of drawing up a histogram of raw data, we group data which do not follow a normal distribution and take the mean of the group. We repeat this many times until we are able to plot a histogram representative of the distributions of the means. Perhaps surprisingly, the distribution of the *means* takes on the characteristics of the normal distribution. To illustrate this we revisit the situation discussed in section 3.3 in which 1000 random numbers are generated and where the underlying probability distribution is uniform, with all values falling between 0 and 1. Figure 3.17 shows a histogram of the raw data which exhibits none of the characteristics associated with the normal distribution.

Figure 3.18 is constructed by taking the data (in the order in which they were generated) as pairs of numbers, then calculating the mean of each pair. A histogram of 500 means is now constructed in the usual way. The main characteristics of the normal distribution are beginning to emerge in figure 3.18: a clearly defined centre to the distribution, symmetry about that centre and an approximately bell shaped distribution. What, if instead of plotting a histogram consisting of the means of pairs of values, we had chosen larger groupings of raw data, say containing ten values and plotted the means of these groupings? Figure 3.19 shows a histogram consisting of

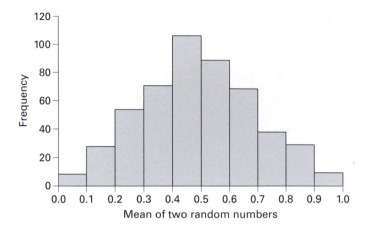

Figure 3.18. Distribution of sample means.

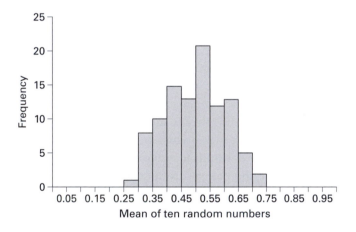

Figure 3.19. Distribution of sample means.

100 means. These means were determined using the same data as appears in figure 3.17.[26]

The distribution in figure 3.19 is narrower than those shown in figures 3.17 and 3.18 to the extent that there are no means below 0.25, or above 0.75. In the next section we quantify the 'narrowing effect' that occurs when the sample size increases.

---

[26] The distribution of means does not look as convincingly normal as say, figure 3.18. This is due to the fact that figure 3.19 consists of only 100 means.

The tendency of the distribution of sample means to follow a normal distribution, irrespective of the distribution of the raw data, is embodied in the central limit theorem.

3.8   The central limit theorem

This is an expression of what we learned in section 3.7, namely the following.

Suppose a sample consisting of n values is drawn from a population and the mean of the values is calculated. If this process is repeated many times, the distribution of means tends towards a normal distribution, irrespective of the distribution of the raw data, so long as n is large.[27]

Let the population mean of a distribution of the raw data be μ, and the population standard deviation be σ.

The population mean, $\mu_{\bar{x}}$, of the new distribution consisting of sample means is given by $\mu_{\bar{x}} = \mu$. The standard deviation of the new distribution, $\sigma_{\bar{x}}$, is given by

$$\sigma_{\bar{x}} = \frac{\sigma}{\sqrt{n}}.$$

(3.19)

The distribution of sample means can be transformed into the standard normal distribution. To do this, replace x by $\bar{x}$ and replace σ by $\sigma_{\bar{x}}$ in equation 3.11, so that z becomes

$$z = \frac{\bar{x} - \mu}{\sigma_{\bar{x}}}.$$

(3.20)

When σ is not known, the best we can do is replace σ in equation 3.19 by s so that

$$\sigma_{\bar{x}} \approx s_{\bar{x}}, \text{ where}$$
$$s_{\bar{x}} = \frac{s}{\sqrt{n}}.$$

(3.21)

The population standard deviation, σ, of the raw data is independent of sample size. By contrast, the standard deviation of the distribution of sample means, $\sigma_{\bar{x}}$, *decreases* as n increases as given by equation 3.21. This is consistent with the observation of the 'narrowing' of the distributions of the sample means shown in figures 3.18 and 3.19 for $n = 2$ and $n = 10$ respectively.

---

[27] If the underlying distribution is symmetric with a central peak, such as a triangular distribution, n of 4 or more will produce means that are normally distributed. If the underlying distribution is extremely asymmetric, n may need to be in excess of 20.

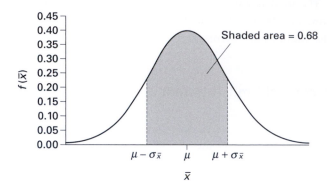

Figure 3.20. Normal distribution of sample means.

3.8.1 Standard error of the sample mean

We often refer to $\sigma_{\bar{x}}$ given in equation 3.19 as the *standard error of the mean*.[28] We can apply equation 3.19 when quoting confidence limits for the population mean. We begin by drawing the probability distribution for the sample means as shown in figure 3.20.

We discovered in section 3.6.1 that the interval, symmetric about the mean, that contains 68% of normally distributed data can be written $\mu - \sigma \le x \le \mu + \sigma$. Similarly, the interval containing 68% of the distribution of sample means is $\mu - \sigma_{\bar{x}} \le \bar{x} \le \mu + \sigma_{\bar{x}}$ as indicated in Figure 3.20.

The probability that a sample mean, $\bar{x}$, lies between $\mu - \sigma_{\bar{x}}$ and $\mu + \sigma_{\bar{x}}$ may be written as

$$P\left(\mu - \frac{\sigma}{\sqrt{n}} \le \bar{x} \le \mu + \frac{\sigma}{\sqrt{n}}\right) = 0.68. \qquad (3.22)$$

Of greater value than finding the interval containing the sample mean, $\bar{x}$, with a known probability is to find the interval, called the *confidence interval*, which contains the population mean, μ, with a known probability. By rearranging the terms within the brackets of equation 3.22, we can write

$$P\left(\bar{x} - \frac{\sigma}{\sqrt{n}} \le \mu \le \bar{x} + \frac{\sigma}{\sqrt{n}}\right) = 0.68. \qquad (3.23)$$

Indicating that there is a probability of 0.68 that μ will lie within $\pm \frac{\sigma}{\sqrt{n}}$ of $\bar{x}$.

Equation 3.23 is certainly more useful than equation 3.22, but there still is a problem: the population standard deviation, σ, appears in this equation and

---

[28] Note that $\sigma_{\bar{x}}$ is sometimes written as SE $\bar{x}$ or SEM.

we do not know its value. The best we can do is replace σ by the estimate of the population standard deviation, s. As discussed in section 3.5.4, this approximation is regarded as reasonable so long as $n \geq 30$.

Equation 3.23 becomes

$$P\left(\bar{x} - \frac{s}{\sqrt{n}} \leq \mu \leq \bar{x} + \frac{s}{\sqrt{n}}\right) = 0.68. \tag{3.24}$$

The 68% confidence interval for μ is

$$\bar{x} - \frac{s}{\sqrt{n}} \leq \mu \leq \bar{x} + \frac{s}{\sqrt{n}}. \tag{3.25}$$

We use the properties of the normal distribution to give the limits associated with other confidence intervals. The $X\%$ confidence interval is written

$$\bar{x} - z_{X\%}\frac{s}{\sqrt{n}} \leq \mu \leq \bar{x} + z_{X\%}\frac{s}{\sqrt{n}}. \tag{3.26}$$

The value of $z_{X\%}$ corresponding to the $X\%$ confidence interval is shown in table 3.5.

Example 14

Using the heat sink temperature data in table 3.6,

(i) calculate the 95% confidence interval for the population mean for these data.

(ii) How many values would be required to reduce the 95% confidence interval for the population mean to $(\bar{x} - 0.1)$ °C to $(\bar{x} + 0.1)$ °C?

ANSWER

(i) The mean, $\bar{x}$, and standard deviation, s, of the data in table 3.6, are $\bar{x} = 18.41$ °C and $s = 0.6283$ °C.[29]

Using equation 3.26, with $z_{95\%} = 1.96$,

$$z_{95\%}\frac{s}{\sqrt{n}} = 1.96 \times \frac{0.6283\,°C}{\sqrt{40}} = 0.1947\,°C.$$

The 95% confidence interval for the population mean is from $(18.41 - 0.1947)$ °C to $(18.41 + 0.1947)$ °C, i.e. from 18.22 °C to 18.60 °C.

(ii) To decrease the interval to $(\bar{x} - 0.1)$ °C to $(\bar{x} + 0.1)$ °C requires that

$$z_{95\%}\frac{s}{\sqrt{n}} = 0.1\,°C;$$

---

[29] We retain four figures for s to minimise the effect of rounding errors in later calculations.

rearranging this equation gives

$$n = \left(\frac{z_{95\%}s}{0.1^{\circ}\text{C}}\right)^2,$$

substituting $z_{95\%} = 1.96$ and $s = 0.6283\ ^{\circ}\text{C}$ gives

$$n = \left(\frac{1.96 \times 0.6283\,^{\circ}\text{C}}{0.1^{\circ}\text{C}}\right)^2 = 152.$$

Exercise H

Thirty repeat measurements are made of the density of a sample of high quality multigrade motor oil. The mean density of the oil is found to be 905 kg/m^3 and the standard deviation 25 kg/m^3. Use this information to calculate the 99% confidence interval for the population mean of the density of the motor oil.

3.8.1.1   ### Approximating $s_{\bar{x}}$

In situations in which a few values (say up to 12) are being considered, there is a rapid method by which the standard error of the mean, $s_{\bar{x}}$, can be approximated which is good enough for most purposes. $s_{\bar{x}}$ can be written as[30]

$$s_{\bar{x}} \approx \frac{\text{range}}{n}, \tag{3.27}$$

where n is number of values and range = (maximum value − minimum value). In particular, equation 3.27 is useful when the standard error is required to be within ± 20% of the value that would be obtained by using equation 3.21.

Example 15

Table 3.7 shows the input offset voltages of five operational amplifiers. Use equation 3.21 and equation 3.27 to estimate the standard error of the mean, $s_{\bar{x}}$, of these values to two significant figures.

ANSWER

Using equation 1.16 we obtain, $s = 1.795$ mV. Substituting s into equation 3.21 we obtain, $s_{\bar{x}} = \frac{s}{\sqrt{n}} = \frac{1.795\,\text{mV}}{\sqrt{6}} = 0.73$ mV. Using equation 3.27 we find

$$s_{\bar{x}} \approx \frac{\text{range}}{n} = \frac{(7.7 - 2.7)\text{mV}}{6} = 0.83\,\text{mV}.$$

---

[30] See Lyon (1980) for a discussion of equation 3.27.

Table 3.7. *Input offset voltages for six operational amplifiers.*

Input offset voltage (mV)	4.7	5.5	7.7	3.4	2.7	5.8

Table 3.8. *Values of the energy gap of germanium, as measured at room temperature.*

Energy gap (eV)	0.67	0.63	0.72	0.66	0.74	0.71	0.66	0.64

Exercise I

Table 3.8 shows values obtained from eight measurements of the energy gap of crystalline germanium at room temperature.

Use equations 3.21 and 3.27 to estimate the standard error of the mean of the values in table 3.8 to two significant figures.

3.8.2   Excel's CONFIDENCE.NORM() function[31]

The X% confidence interval of the population mean for any set of data may be found using Excel's CONFIDENCE.NORM function. The syntax of the function is as follows.

CONFIDENCE.NORM (α, standard deviation, sample size)

Note that α is sometimes referred to as the 'level of significance'. Its place in data analysis is discussed further when we consider hypothesis testing in chapter 9. For the moment we only require the relationship between α and the confidence level, X%. α is given by

$$\alpha = \frac{(100\% - X\%)}{100\%}.$$

(3.28)

As an example, a 95% confidence level corresponds to $\alpha = 0.05$.

Example 16

Fifty measurements are made of the coefficient of static friction between a wooden block and a flat metal table. The data are normally distributed with a mean of 0.340 and

---

[31] CONFIDENCE.NORM() in Excel 2010 replaces CONFIDENCE() found in earlier versions of Excel.

a standard deviation of 0.021. Use the CONFIDENCE.NORM() function to find the 99% confidence interval for the population mean.

ANSWER

Use equation 3.28 to find α:

$$\alpha = \frac{(100\% - 99\%)}{100\%} = 0.01.$$

Sheet 3.8 shows the CONFIDENCE.NORM() function entered into cell A1.

Sheet 3.8(b) shows the value returned in cell A1 after the Enter key is pressed.

The confidence interval can be written as follows.

Mean – CONFIDENCE.NORM(0.01,0.021,50) to Mean + CONFIDENCE.NORM (0.01,0.021,50)

i.e. 0.340 – 0.0077 to 0.340 + 0.0077. The 99% confidence interval for the population mean of the static friction data is therefore 0.3323 to 0.3477.

We must be cautious when using Excel's CONFIDENCE.NORM() function. This function assumes that the population standard deviation has been entered as an argument in the function. If the population standard deviation is not known, but is calculated from a sample size of less than about 30 (which would be the case in many situations), then the CONFIDENCE.NORM() function will return a value that is too small in most cases. As we are often faced with samples sizes less than 30, how do we estimate a confidence interval for the population mean? This brings us to another important probability distribution that has much in common with the normal distribution; the t distribution.[32]

Sheet 3.8. *Use of CONFIDENCE.NORM() function.*

(a) *Numbers entered into*
CONFIDENCE.NORM() function

	A	B	C
1	= CONFIDENCE(0.01,0.021,50)		
2			

(b) *Result of calculation*

	A	B	C
1	0.00765		
2			

3.9 The *t* distribution

While the shape of the normal distribution describes well the variability in the mean when sample sizes are large, it describes the variability less well when

---

[32] This is sometimes referred to as Student's *t* distribution in recognition of the author of the first paper to describe the distribution (although the author's real name was W. S. Gosset). See Student (1908).

sample sizes are small. It is important to be aware of this, as many experiments are carried out in which the number of repeat measurements is small (say ten or less). The difficulty stems from the assumption made that the estimate of the population standard deviation, s, is a good approximation to σ.

The variation in values is such that, for small data sets, s is not a good approximation to σ, and the quantity

$$\left(\frac{\bar{x} - \mu}{s/\sqrt{n}}\right),$$

where n is the size of the sample, does not follow the standard normal distribution, but another closely related distribution, referred to as the 't' distribution. If we write

$$t = \left(\frac{\bar{x} - \mu}{s/\sqrt{n}}\right), \tag{3.29}$$

we may study the distribution of t as n increases from $n = 2$ to $n = \infty$. The probability density function, $f(t)$, can be written in terms of the variable, t, as[33]

$$f(t) = K(v)\left(1 + \frac{t^2}{n-1}\right)^{-n/2}, \tag{3.30}$$

where $K(v)$ is a constant which depends on the number of degrees of freedom, v. $K(v)$ is chosen so that

$$\int_{-\infty}^{\infty} f(t)dt = 1. \tag{3.31}$$

Figure 3.21 shows the general shape of the t probability density function.

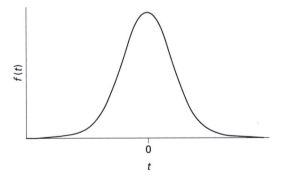

Figure 3.21. The t distribution curve.

[33] See Hoel (1984) for more information on the t probability density function.

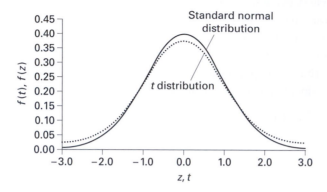

Figure 3.22. The t and z curves compared.

On the face of it, figure 3.21 has a shape indistinguishable to that of the normal distribution with a characteristic 'bell shape' in evidence. The difference becomes clearer when we compare the t and the standard normal distributions directly. Equation 3.30 predicts a different distribution for each sample size, n, with the t distribution tending to the standard normal distribution as $n \to \infty$. Figure 3.22 shows a comparison of the t distribution curve with $n = 6$ and that of the standard normal probability distribution.

An important difference between the family of t distributions and the normal distribution is the extra area in the tails of the t distributions. As an example, consider the normal distribution in which there is 5% of the area confined to the tails of the distribution. A confidence interval of 95% for the population mean is given by

$$\bar{x} - 1.96 \frac{\sigma}{\sqrt{n}} \text{ to } \bar{x} + 1.96 \frac{\sigma}{\sqrt{n}}.$$

By comparison, the 95% confidence interval for the population mean, when σ is not known, is given by

$$\bar{x} - t_{95\%,\nu} \frac{s}{\sqrt{n}} \text{ to } \bar{x} + t_{95\%,\nu} \frac{s}{\sqrt{n}}, \tag{3.32}$$

where ν is the number of degrees of freedom.

Table 3.9 gives the t values for the 68%, 90%, 95% and 99% confidence levels of the t distribution for various degrees of freedom, $\nu = n - 1$.

Table 3.9 shows $t_{95\%,\nu}$ is greater than 1.96 for $\nu \leq 99$, so that the 95% confidence interval is necessarily wider for the t distribution than the 95% confidence interval for the normal distribution that has the same standard deviation.

Table 3.9. *The t values for various confidence levels and degrees of freedom.*[34]

Number of values, n	Degrees of freedom, v	$t_{68\%,v}$	$t_{90\%,v}$	$t_{95\%,v}$	$t_{99\%,v}$
2	1	1.82	6.31	12.71	63.66
4	3	1.19	2.35	3.18	5.84
10	9	1.05	1.83	2.26	3.25
20	19	1.02	1.73	2.09	2.86
30	29	1.01	1.70	2.05	2.76
50	49	1.00	1.68	2.01	2.68
100	99	1.00	1.66	1.98	2.63
10000	9999	0.99	1.65	1.96	2.58

Example 17

The diameter of ten steel balls is shown in table 3.10. Using these data, calculate

  (i) $\bar{x}$;
 (ii) s;
(iii) the 95% and 99% confidence interval for the population mean.

ANSWER

  (i) The mean of the data in table 3.10 is 4.680 mm.
 (ii) The standard deviation of the data, calculated using equation 1.16, is 0.07149 mm.
(iii) The 95% confidence interval for the population mean is given by equation 3.32 and the 99% confidence interval by the equation

$$\bar{x} - t_{99\%,9}\frac{s}{\sqrt{n}} \text{ to } \bar{x} + t_{99\%,9}\frac{s}{\sqrt{n}}.$$

Table 3.10 gives $t_{95\%,9} = 2.26$ so the 95% confidence interval is

$$4.680\,\text{mm} - \frac{2.26 \times 0.07149\,\text{mm}}{\sqrt{10}} \text{ to } 4.680\,\text{mm} + \frac{2.26 \times 0.07149\,\text{mm}}{\sqrt{10}},$$

i.e. 4.629 mm to 4.731 mm.

Also, $t_{99\%,9} = 3.25$ so that the 99% confidence interval becomes

$$4.680\,\text{mm} - \frac{3.25 \times 0.07149\,\text{mm}}{\sqrt{10}} \text{ to } 4.680\,\text{mm} + \frac{3.25 \times 0.07149\,\text{mm}}{\sqrt{10}},$$

i.e. 4.607 mm to 4.753 mm.

[34] A more extensive table can be found in appendix 1.

Table 3.10. *Diameters of steel balls.*

Diameter (mm)				
4.75	4.65	4.60	4.80	4.70
4.70	4.60	4.65	4.75	4.60

Exercise J

Calculate the 90% confidence interval for the population mean of the data in table 3.10.

3.9.1   Excel's T.DIST.2T() and T.INV.2T() functions[35]

For a specified value of t, the T.DIST.2T() function gives the probability in the two tails of the t distribution, so long as the number of degrees of freedom, v, is specified.

The syntax for the T.DIST.2T() function is as follows.

T.DIST.2T(t, degrees of freedom)

If the tails parameter is set to 2, the function gives the area in both tails of the distribution. If the tails parameter is set to 1, then the area in one tail is given.

Example 18
  (i) Calculate the area in the tails of the t distribution when $t = 1.5$ and $v = 10$.
  (ii) Calculate the area between $t = -1.5$ and $t = 1.5$.

ANSWER
  (i) Sheet 3.9 shows the formula required to calculate the area in both tails entered into cell A1.
        Sheet 3.9(b) shows the area in the tails returned in cell A1 when the Enter key is pressed
  (ii) The area between $t = -1.5$ and $t = 1.5$ is equal to $1 - $ (area in tails) $= 1 - 0.16451 = 0.83549$.

---

[35] Versions of Excel prior to Excel 2010 included TDIST() and TINV() functions. Though TDIST() remains available in Excel2010, it has been superseded by three functions that utilise the t distribution; T.DIST(), T.DIST.2T() and T.DIST.RT(). The T.DIST() function can calculate either the value of the function given by equation 3.30, or the area in the left hand tail of the t distribution. The T.DIST.RT() function calculates the area in the right hand tail of the t distribution.

If the total area in symmetric tails of the *t* distribution is *p*, then the T.INV.2T() function[36] gives the value of *t* corresponding to that area, as long as the number of degrees of freedom, v, is specified.

The syntax for the T.INV.2T() function is as follows.

T.INV.2T(*p*, degrees of freedom)

Sheet 3.9. *Use of T.DIST.2T() function.*

(a) *Numbers entered into T.DIST.2T() function*

	A	B	C
1	= T.DIST.2T(1.5,10)		
2			

(b) *Result of calculation*

	A	B	C
1	0.164507		
2			

Example 19

If the area in the two tails of the *t* distribution is 0.6, calculate the corresponding value of *t*, assuming that the number of degrees of freedom, v, = 20.

ANSWER

Sheet 3.10 shows the T.INV.2T() function entered into cell A1.

Sheet 3.10(b) shows the value returned in cell A1 after the Enter key is pressed.

Sheet 3.10. *Use of T.INV.2T() function.*

(a) *Numbers entered into T.INV.2T() function*

	A	B	C
1	= T.INV.2T(0.6, 20)		
2			

(b) *Result of calculation*

	A	B	C
1	0.532863		
2			

3.9.2 Distribution of the standard deviation, *s*

Though generally of less importance than the distribution of sample means, the estimate of the population standard deviation, *s*, has its own distribution. It is possible to state a confidence interval which will contain the population

---

[36] The T.INV.2T() function in Excel 2010 supersedes the TINV() function in earlier versions of Excel (though TINV() remains available in Excel 2010). If the area in the left hand tail of the *t* distribution is known, then the T.INV() function will return the value of *t* that corresponds to that area.

standard deviation, σ, with a known level of confidence, $X\%$. We will not consider how the confidence interval is calculated, but make two remarks.

(1) Unlike the distribution of sample means, the distribution of s for small sample sizes is not symmetrically distributed (say for sample sizes less than 25). The consequence of this is that the confidence interval containing σ is not symmetric[37] about the peak of the distribution.

(2) The standard deviation of the distribution of s, for a large number of samples which can be written as σ_s can be approximated by[38]

$$\sigma_s \approx 0.71 \frac{\sigma}{\sqrt{n}}, \tag{3.33}$$

where n is the sample size, and σ is the population standard deviation. In general, we cannot know σ, so σ in equation 3.33 must be replaced by s.

3.9.3   Excel's CONFIDENCE.T() function

Excel's CONFIDENCE.T() function gives the confidence interval for the population mean using the t distribution. The syntax of the function is

CONFIDENCE.T(α, standard deviation, sample size)

where α is given by equation 3.28.

Example 20

Ten repeat measurements are made of the fluid flow rate of water from a narrow tube. The mean flow rate is found to be 0.670 cm^3/s and a standard deviation of 0.032 cm^3/s. Use the CONFIDENCE.T() function to find the 95% confidence interval for the population mean.

ANSWER
Use equation 3.28 to find α:

$$\alpha = \frac{(100\% - 95\%)}{100\%} = 0.05.$$

Sheet 3.11 shows the CONFIDENCE.T() function entered into cell A1.

Sheet 3.11(b) shows the value returned in cell A1 after the Enter key is pressed.

---

[37] See question 14 at the end of this chapter for an exercise that demonstrates that s is not distributed symmetrically.

[38] For more information about the sampling distribution of s see Patel and Read (1996).

The confidence interval for the population mean can be written as

Mean – CONFIDENCE.T(0.05,0.032,10) to Mean + CONFIDENCE.T(0.05,0.032,10),

i.e. (0.670 – 0.0229) cm^3/s to (0.670 + 0.0229) cm^3/s. The 95% confidence interval for the population mean of the fluid flow data is therefore 0.647 cm^3/s to 0.693 cm^3/s.

Sheet 3.11. *Use of CONFIDENCE.T() function.*

(a) *Numbers entered into the CONFIDENCE.T() function*

	A	B	C
1	= CONFIDENCE.T(0.05,0.032,10)		
2			

(b) *Result of calculation*

	A	B	C
1	0.022891		
2			

3.10 The log-normal distribution

While the normal distribution plays a central role in the analysis of experimental data in the physical sciences, not all data are distributed normally. It is important to recognise this as, for example, data rejection techniques can be used (such as that described in section 5.11.2) which assume that a particular distribution (usually the normal) applies to data. When it does not, data may be rejected which legitimately belong to the 'true' distribution. One distribution that describes the data obtained in many experiments in the physical sciences (and is closely related to the normal distribution) is the so called 'log-normal' distribution.

Suppose the continuous random variable, x, follows a log-normal distribution. If $y = \ln(x)$, then the random variable, y, is normally distributed. Examples of the probability density function for the log-normal distribution are shown in figure 3.23. The population mean of the transformed x values is 1.7 for both curves (i.e. μ_y is 1.7). Standard deviations of the transformed x values are 0.2 and 0.5 as indicated in figure 3.23.

Occurrences of log-normally distributed data are quite common in the physical sciences.[39] As an example, consider the histogram shown in figure 3.24

---

[39] Metals such as copper in river sediments, sulphate in rainwater and iodine in groundwater specimens have all been found to be approximately log-normally distributed. Crow and Shimizu (1988) offer a detailed discussion of the log-normal distribution. Limpert *et al.* (2001) also give an account of the log-normal distribution and cite examples of its applicability in several areas including geology, medicine, microbiology and ecology.

the experiment) will be very similar to the theoretical quantile if the data follow the theoretical distribution closely. Put another way, if we plot the experimental data against the theoretical quantile we would expect the points to lie along a straight line. A quantile plot may be constructed as follows.[42]

(1) Order the experimental data, x_i, from smallest to largest.
(2) Calculate the fraction of values, f_i, less than or equal to the ith value using the relationship

$$f_i = \frac{i - \frac{3}{8}}{n + \frac{1}{4}}, \tag{3.34}$$

where n is the number of values in the sample.
(3) Calculate the quantile, $q(f_i)$ for all the fractions, f_i, ($i = 1$ to n) for the standard normal distribution using

$$q(f_i) = 4.91 \left[f_i^{0.14} - (1 - f_i)^{0.14} \right]. \tag{3.35}$$

(4) Plot a graph of x_i versus $q(f_i)$.

Example 21

Table 3.11 shows the diameter (in μm) of 24 small spherical particles deposited on the surface of a film of titanium.

Construct a normal quantile plot for the data in table 3.11.

ANSWER

Table 3.12 shows f_i and $q(f_i)$. The values of particle size are shown ordered from smallest to largest (in the interests of brevity, the table shows the first 10 values in the sample only).

Figure 3.26 shows x_i versus $q(f_i)$ for all the data in table 3.11. The extreme non-linearity for $q(f_i) > 1$, indicates that the data are not adequately described by the normal distribution.

Table 3.11. *Diameter of small particles on a film of titanium (in μm).*

0.075	0.110	0.037	0.065	0.147	0.106	0.158	0.163	0.149	0.131	0.136	0.106
0.068	0.206	0.037	0.097	0.123	0.968	0.110	0.147	0.081	0.062	0.421	0.188

[42] This approach follows that of Walpole, Myers and Myers (1998).

Table 3.12. *Values of f_i and $q(f_i)$ for the data shown in table 3.11.*

i	x_i (ordered)	f_i	$q(f_i)$
1	0.037	0.025773	−1.95007
2	0.037	0.06701	−1.49944
3	0.062	0.108247	−1.23521
4	0.065	0.149485	−1.03703
5	0.068	0.190722	−0.8732
6	0.075	0.231959	−0.73025
7	0.081	0.273196	−0.6011
8	0.097	0.314433	−0.48148
9	0.106	0.35567	−0.36854
10	0.106	0.396907	−0.26024

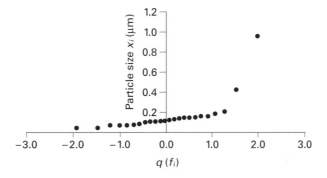

Figure 3.26. Normal quantile plot for data in table 3.11.

Exercise K

Transform the data in table 3.11 by taking the natural logarithms of the values in the table.

(i) Construct a normal quantile plot of the transformed data.
(ii) What can be inferred from the shape of this plot?

3.12 Population mean and continuous distributions

When n repeat measurements are made of a single quantity, we express the mean, $\bar{x}$, as

$$\bar{x} = \frac{\sum x_i}{n}. \tag{3.36}$$

For large n, $\bar{x} \to \mu$, where μ is the population mean. We have shown in section 3.5 that the probability of obtaining a value of x can be obtained with the assistance of the probability density function, $f(x)$. Now we go one step further to show how this function is related to the population mean, μ. We begin by considering another equation, equivalent to equation 3.36 which allows us to calculate $\bar{x}$. We can write

$$\bar{x} = \frac{\sum f_i x_i}{\sum f_i} = \frac{f_1 x_1}{\sum f_i} + \frac{f_2 x_2}{\sum f_i} + \frac{f_3 x_3}{\sum f_i} + \cdots + \frac{f_n x_n}{\sum f_i}, \tag{3.37}$$

where f_i is the frequency of occurrence of the value x_i.

As $n \to \infty$, $\bar{x} \to \mu$, and $\frac{f_i}{\sum f_i}$ becomes the *probability*, p_i, of observing the value, x_i. As $n \to \infty$, equation 3.37 becomes

$$\bar{x} \to \mu = p_1 x_1 + p_2 x_2 + p_3 x_3 + \cdots + p_n x_n. \tag{3.38}$$

This may be written more compactly as

$$\mu = \sum p_i x_i. \tag{3.39}$$

Equation 3.39 is most easily applied when dealing with quantities that take on discrete values, such as occurs when throwing dice or counting particles emitted from a radioactive source. When dealing with continuously varying quantities such as length or mass we can write the probability, p, of observing the value x in the interval, Δx, as

$$p = f(x)\Delta x, \tag{3.40}$$

where $f(x)$ is the probability density function introduced in section 3.3.

If we allow Δx to become very small, the equation for the population mean for continuous variables (equivalent to equation 3.39 for discrete variables) may be written[43]

$$\mu = \int_{-\infty}^{\infty} x f(x) dx. \tag{3.41}$$

---

[43] Here we have shown the limits of the integral extending from $+\infty$ to $-\infty$, indicating that any value of x is possible. In general, the limits of the integral may differ from $+\infty$ to $-\infty$. Most importantly, if the limits are written as a and b, then $\int_a^b f(x)dx = 1$.

Example 22

A probability density function, $f(x)$, is written

$f(x) = \frac{4}{3} - x^2$ for $0 \leq x \leq 1$, and $f(x) = 0$ for other values of x.

Use this information to determine the population mean for x.

ANSWER

Using equation 3.41 we write (with appropriate change of limits in the integration)

$$\mu = \int_0^1 x\left(\frac{4}{3} - x^2\right) dx = \int_0^1 \left(\frac{4x}{3} - x^3\right) dx.$$

Performing the integration, we have

$$\mu = \left[\frac{2}{3}x^2 - \frac{x^4}{4}\right]_0^1 = [0.6\dot{6} - 0.25] = 0.41\dot{6}.$$

Exercise L

Given that a distribution is described by the probability density function

$$f(x) = 0.2e^{(-0.2x)}, \text{ where } 0 \leq x \leq \infty,$$

determine the population mean of the distribution.

3.13 Population mean and expectation value

The mean value of x where the probability distribution governing the distribution of x is known, is also referred to as the *expectation value* of x and is sometimes written $\langle x \rangle$. For continuous quantities $\langle x \rangle$ is written

$$\langle x \rangle = \int_{-\infty}^{\infty} xf(x)dx, \tag{3.42}$$

where $f(x)$ is the probability density function.

A comparison of equations 3.42 and 3.41 indicates that $\langle x \rangle \equiv \mu$. The idea of expectation value can be extended beyond finding the mean of any function of x. For example, the expectation value of x^2 is given by

$$\langle x^2 \rangle = \int_{-\infty}^{\infty} x^2 f(x)dx. \tag{3.43}$$

In general, if the expectation value of a function $g(x)$ is required, we write

$$\langle g(x) \rangle = \int_{-\infty}^{\infty} g(x)f(x)dx. \tag{3.44}$$

From the point of view of data analysis, a useful expectation value is that of the square of the deviation of values from the mean, as this is the variance, σ^2, of values. Writing $g(x) = (x - \mu)^2$ and using equation 3.44 we have,

$$\sigma^2 = \langle (x - \mu)^2 \rangle = \int_{-\infty}^{\infty} (x - \mu)^2 f(x)dx \tag{3.45}$$

3.14 Review

Data gathered in experiments routinely display variability. This could be due to inherent variations in the quantity being measured, the influence of external factors such as electrical noise or even the care, or lack of it, shown by the experimenter in the course of carrying out the experiment. Accepting that variability is a 'fact of life', we need a formal way of expressing variability in data. In this chapter we considered how some of the basic ideas in probability can be applied to data gathered in experiments. Probability, and its application to the study of the distribution of experimental data, gives a concise, numerical way of expressing the variability in quantities. Specifically, with reference to the normal and t distributions, we are able to quote a confidence interval for the mean of a set of data in situations where data sets are large or small. In this chapter we have shown that the calculation of probabilities and confidence intervals can be eased considerably by the use of Excel's built in statistical functions.

We will apply what we know of the normal and t distributions in chapter 5 when we consider the general topic of uncertainty in experiments and how uncertainties combine when measurements are made of many physical quantities. In the next chapter we consider discrete quantities and probability distributions used to describe the variability in such quantities.

End of chapter problems

(1) A probability density function is written

$f(x) = A - x$ for $0 \leq x \leq 1$, and
$f(x) = 0$ for other values of x.

(i) Sketch the graph of $f(x)$ versus x.
(ii) Calculate A by using the relationship, $\int_0^1 f(x)dx = 1$.

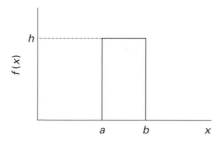

Figure 3.27. Rectangular probability density function.

     (iii) What is the probability that x lies between
        (a)   0.0 and 0.1?
        (b)   0.9 and 1.0?

(2) Consider the rectangular distribution shown in figure 3.27.
    Given that

$$f(x) = h \text{ for } a \leq x \leq b \text{ and } f(x) = 0 \text{ for } x < a \text{ or } x > b, \text{ then}$$

     (i) using $\int_a^b f(x)dx = 1$, show that $h = \frac{1}{b-a}$;
     (ii) show that the population mean, μ, of the distribution is $\mu = \frac{a+b}{2}$;
     (iii) show that the variance, σ^2, of the distribution is $\sigma^2 = \frac{(b-a)^2}{12}$.

(3) Show that points of inflexion occur in the curve of $f(z)$ versus z at $z = \pm 1$, where $f(z)$ is given by equation 3.12. (Use the result that, at the point of inflexion, $\frac{d^2 f(z)}{dz^2} = 0$.)

(4) An experiment is performed to test the breaking strength of thermoplastic fibres. Out of 1000 fibres tested, 78 fibres have breaking strength in excess of 945 MPa and 41 fibres have breaking strength of less than 882 MPa. Assuming the normal distribution to be valid for the data, calculate the mean and standard deviation of the fibre strength.

(5) A constant current is supplied to sixteen silicon diodes held at constant temperature. The voltage across each diode is recorded. Table 3.13 shows the voltage data.

Table 3.13. *Voltage data for silicon diodes.*

Voltage across diodes (V)							
0.6380	0.6421	0.6458	0.6395	0.6389	0.6364	0.6411	0.6395
0.6390	0.6464	0.6420	0.6428	0.6385	0.6401	0.6432	0.6405

Use the data to estimate the mean and standard deviation of the population from which the diodes were taken. If 200 diodes are tested, how many would you expect to have voltages is excess of 0.6400 V? (Assume the normal distribution to be valid.)

(6) The focal lengths of many lenses are measured and found to have a mean of 15.2 cm and a standard deviation of 1.2 cm.

(i) Assuming the focal lengths are normally distributed:
(a)   use the NORM.DIST() function on Excel to calculate the cumulative distribution function (cdf) for focal lengths in the range 13.0 cm to 17.0 cm in steps of 0.2 cm;
(b)   plot a graph of cdf versus focal length.
(ii) Using your cdf values found in part (i), find the probability that a lens chosen at random has a focal length in the range,
(a)   13.0 cm to 15.0 cm;
(b)   16.0 cm to 17.0 cm.

(7) Use the Excel function NORM.S.DIST() to draw up a table of the standard normal cumulative function for values of z ranging from $z = -2.00$ to $z = 2.00$ in steps of 0.01. Plot a graph of the standard normal cumulative function versus z for z in the range -2.00 to 2.00.

(8) The resonant frequency in an a.c. circuit is measured five times. The mean of the data is found to be 2106 Hz with a standard deviation of 88 Hz. Assuming the data to be drawn from a normal distribution, use the CONFIDENCE.T() function in Excel to calculate the 90% confidence interval for the population mean.

(9) The pressure in a vacuum chamber is measured each day for ten days. The data obtained are given in table 3.14 (units μPa).

Using the t distribution, calculate the 95% confidence interval for the mean of the population from which these data are drawn.

Table 3.14. *Pressure data (units μPa).*

147	128	135	137	145
153	129	125	139	142

(10) (i) Calculate the area in one tail of the t distribution between $t = 2$ and $t = \infty$ for $v = 1$.
(ii) Calculate the area in one tail of the standard normal distribution between $z = 2$ and $z = \infty$.
(iii) Repeat part (i) for $v = 2$ to 100 and plot a graph of the area in the tail versus v.

(iv) For what value of v is the area in the tail between $t = 2$ and $t = \infty$ equal to the area in the tail of the standard normal distribution between $z = 2$ and $z = \infty$ when both areas are expressed to one significant figure?

(11) In a fluid flow experiment, water is collected as it flows from a small bore hollow tube. Table 3.15 shows the volume of water gathered over 12 consecutive time periods of 20 s.

(i) Estimate the mean and standard deviation of the population of the distribution from which the data in table 3.15 are drawn.

(ii) Using the t distribution, calculate 99% confidence interval for the population mean.

(iii) Calculate the standard deviation of the estimate of the population standard deviation, s.

Table 3.15. *Amount of water collected (cm³).*

4.85	5.07	5.00	5.03	4.89	4.54
4.63	4.87	5.04	4.95	5.03	4.82

(12) The amount of TiO_2 in specimens of basement rocks drawn from a geothermal field is shown in table 3.16.

(i) Use a normal quantile plot to help decide whether the values in table 3.16 follow a normal distribution.

(ii) Transform the data in table 3.16 by taking the natural logarithms of the values. Using a normal quantile plot establish whether the transformed values follow a normal distribution (i.e. do the data follow a log-normal distribution?).

Table 3.16. *Concentration of TiO_2 in basement rocks.*

Concentration of TiO_2 (wt%)									
0.35	0.37	0.59	0.14	0.55	0.74	1.99	1.81	0.15	1.58
0.63	0.46	0.19	0.79	0.80	0.99	2.34	1.76	1.82	0.82
0.44	0.45	0.20	0.55	0.57	1.96	0.82	2.14	2.22	2.40

(13) Microspheres were added to a polymer in order to study the ultrasonic properties of inhomogeneous materials. Values for the radii of 100 microspheres are shown in table 3.17.

(i) Plot a histogram of the values in table 3.17.

Table 3.17. *Radius of microspheres added to a polymer.*

Radius of microspheres (µm)									
16.9	10.3	11.3	9.0	11.3	4.0	11.4	6.3	13.2	15.7
23.5	8.9	8.4	13.1	10.9	11.0	26.8	12.4	11.3	30.4
8.9	23.0	12.1	11.8	27.3	14.6	15.2	10.8	17.1	11.8
8.4	7.0	13.5	7.5	32.1	11.4	5.5	10.3	19.7	10.1
8.8	6.2	14.3	11.2	16.2	8.9	10.2	14.1	10.3	13.5
3.5	11.8	13.8	35.1	5.7	5.6	3.8	21.9	39.9	10.0
7.4	16.0	16.9	8.0	18.3	16.9	11.0	5.6	20.2	17.8
10.2	8.6	18.8	14.5	16.1	34.6	14.0	7.0	15.5	6.6
16.2	11.3	14.1	14.6	13.0	10.6	6.2	35.2	27.1	20.6
5.8	7.0	15.9	12.6	12.1	13.2	4.8	10.7	7.6	7.4

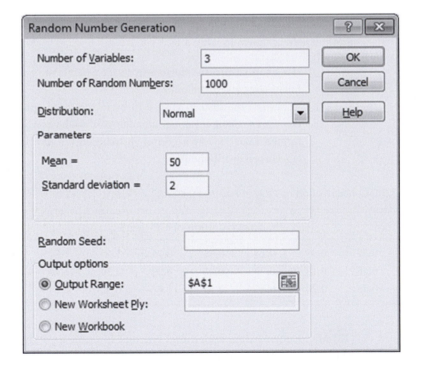

Figure 3.28. Dialog box for Random Number Generation tool for question 14.

  (ii) Take the natural logarithm of the values in table 3.17. Plot a histogram of
        the transformed values.
  (iii) Use a normal quantile plot to establish whether the transformed values
        follow a normal distribution.

(14) Open Excel's Random Number Generation tool in the Analysis ToolPak
(see section 2.8 in chapter 2 for details of the Analysis ToolPak).

  In the dialog box that appears, enter the values shown in figure 3.28.
Numbers will be generated following the normal distribution. More
specifically, columns A, B and C will each contain 1000 random numbers
drawn from a normal distribution with population mean of 50 and standard
deviation of 2.

  (i)   Calculate the standard deviation, s, for each sample of three random
        numbers in row 1 by entering =STDEV.S(A1:C1) into cell E1.
  (ii) Highlight cells E1 to E1000 then use Home>Editing>Fill>Down to calcu-
        late the standard deviations of the numbers in cells A2 to C2, etc., down.
  (iii) Use the histogram option in the Analysis ToolPak (see section 2.8.1 for
        details) to plot the distribution of the 1000 values of s. Is the distribution
        symmetric?

As the sample size increases, so the shape of the distribution of s begins to look more
normal. To demonstrate this, repeat this exercise but increase the number of random
numbers in each row to 25 by replacing 3 by 25 in the Random Number Generation
dialog box of figure 3.28. Now calculate the standard deviation of each row of 25
numbers then plot the histogram consisting of the 1000 standard deviations.

Chapter 4

Data distributions II

4.1 Introduction

In chapter 3 we considered the normal distribution largely due to its similarity to the distribution of data observed in many experiments involving repeat measurements of a quantity. In particular, the normal distribution is useful for describing the spread of values when continuous quantities such as temperature or time interval are measured.

Another important category of experiment involves counting. As examples, we may count the number of electrons scattered by a gas, the number of charge carriers thermally generated in an electronic device, or the number of beta particles emitted by a radioactive source. In these situations, distributions that describe discrete quantities must be considered. In this chapter we consider two such distributions important in science: the binomial and Poisson distributions.

4.2 The binomial distribution

One type of experiment involving discrete variables entails removing an object from a population and classifying that object in one of a finite number of ways. For example, we might test an electrical component and classify it 'within specification' or 'outside specification'. Owing to the underlying (and possibly unknown) processes causing components to fail to meet the specification, we can only give a probability that any particular component tested will satisfy the specification. When n objects are removed from a population and tested, or when a coin is tossed n times, we speak of performing n *trials*. The result of a test (e.g. 'pass') or the result of a coin toss (e.g. 'head') is referred to as an *outcome*.

Some experiments consist of trials in which the outcome of each trial can be classified as a success (S) or a failure (F). If this is the case, and if the probability of a success does not change from trial to trial, we can use the *binomial* distribution to determine the probability of a given number of successes occurring, for a given number of trials.[1] The binomial distribution is useful if we want to know the probability of, say, obtaining one or more defective light emitting diodes (LEDs) when an LED is drawn from a large population of 'identical' LEDs, or the probability of obtaining four 'heads' in six tosses of a coin. In addition, by considering a situation in which the number of trials is very large but the probability of success on a single trial is small (sometimes referred to as a 'rare' outcome), we are able to derive another discrete distribution important in science: the Poisson distribution.

How a 'success' is defined largely depends on the circumstances. For example, if integrated circuits (ICs) are tested, a successful outcome could be defined as a circuit that is fully functional. Let us write the probability of obtaining such a success as p. If a circuit is not fully functional, then it is classed as a failure. Denote the probability of a failure by q. As success or failure are the only two possibilities, we must have

$$p + q = 1. \tag{4.1}$$

As an example, suppose after testing many integrated circuits (ICs), it is found that 20% are defective. If one IC were chosen at random, the probability of a failure is 0.2 and hence the probability of a success is 0.8. If four ICs are drawn from the population, what is the probability that all four are successes? To answer this we apply one of the rules of probability discussed in section 3.2.1. So long as the trials are independent (so that removing any IC from the population does not affect the probability that the next IC drawn from the population is a success), then

$$P(4 \text{ successes}) = p \times p \times p \times p = 0.8 \times 0.8 \times 0.8 \times 0.8 = 0.4096.$$

Going a stage further in complexity, if four ICs are removed from the population, what is the probability that exactly two of them are fully functional? This is a little more tricky as, given that four ICs are removed and tested, two successes can be obtained in several ways such as two successes followed by two failures or two failures followed by two successes.

---

[1] The term *binomial distribution* derives from the fact that the probability of r successful outcomes from n trials equates to a term of the binomial expansion $(p + q)^n$, where p is the probability of a success in a single trial, and q is the probability of a failure.

The possible combinations of success (S) and failure (F) for two successes from four trials are

SSFF, FFSS, SFFS, FSSF, SFSF, FSFS.

The probability of the combination SSFF occurring,

$P(\text{SSFF}) = p \times p \times q \times q = 0.8 \times 0.8 \times 0.2 \times 0.2 = 0.0256$.

The probability that each of the other five combinations, SFFS, FFSS, etc., occurring is also 0.0256, so that the total probability of obtaining exactly two successes from four trials is

$$0.0256 + 0.0256 + 0.0256 + 0.0256 + 0.0256 + 0.0256$$
$$= 6 \times 0.0256 = 0.1536.$$

This is a tedious way to determine a probability, but if we generalise what we have done, an equation emerges which assists probability calculations in situations where the binomial distribution is valid.

4.2.1 Calculation of probabilities using the binomial distribution

What is the probability of exactly r successes in n trials? To answer this we proceed as follows.

(1) If there are r successes out of n trials, then there must be $n - r$ failures.
(2) The probability of r consecutive successes followed by $n - r$ consecutive failures is $p^r q^{n-r}$, where p is the probability of a success for a single trial and q is probability of a failure for a single trial.
(3) The number of combinations of r successes from n trials is given by[2]

$$C_{n,r} = \frac{n!}{(n-r)!r!},\tag{4.2}$$

where $n!$ is 'n factorial', given by

$$n! = n \times (n-1) \times (n-2) \times (n-3) \cdots \times 2 \times 1.\tag{4.3}$$

Multiplying the number of combinations in step 3 by the probability of any one combination in step 2, gives the probability, $P(r)$, of r successes from n trials,

$$P(r) = C_{n,r} p^r q^{n-r}.\tag{4.4}$$

---

[2] There are a variety of symbols used to represent the combinations formula including $\binom{n}{r}$ and $_nC_r$. See Meyer (1975) for more details on combinations.

Example 1

Five percent of thermocouples in a laboratory need repair. Assume the population of thermocouples to be large enough for the binomial distribution to be valid. If ten thermocouples are withdrawn from the population, what is the probability that,

    (i) two are in need of repair;

    (ii) two or fewer are in need of repair;

    (iii) more than two are in need of repair?

ANSWER

Let us designate a good thermocouple (i.e. one not in need of repair) as a success. Given that 5% of thermocouples need repair, the probability, q, of a failure is 0.05. Therefore the probability of a success, p, is 0.95.

    (i) If two thermocouples are in need of repair (failures) then the other 8 must be good (i.e. successes). We require the probability of 8 successes ($r = 8$) from 10 trials ($n = 10$). Using equation 4.2,

$$C_{n,r} = \frac{n!}{(n-r)!r!} = \frac{10!}{(10-8)!8!} = 45.$$

Equation 4.4 becomes

$$P(8) = C_{10,8}p^8 q^2 = 45 \times 0.95^8 \times 0.05^2 = 0.07463.$$

    (ii) If two or fewer thermocouples are in need of repair, then 8, 9 or 10 must be good. Therefore the probability required is

$$P(8 \leq r \leq 10) = P(8) + P(9) + P(10), \text{which may be written} \sum_{r=8}^{r=10} P(r).$$

Following the steps in part (i)

$$P(9) = C_{10,9}p^9 q^1 = 10 \times 0.95^9 \times 0.05 = 0.31512,$$

$$P(10) = C_{10,10}p^{10} q^0 = 1 \times 0.95^{10} = 0.59874,$$

$$P(8 \leq r \leq 10) = \sum_{r=8}^{r=10} P(r) = 0.07463 + 0.31512 + 0.59874 = 0.9885.$$

    (iii) If more than two thermocouples need repair, then 0, 1, 2, 3, 4, 5, 6 or 7 must be good. We require the probability, $P(0 \leq r \leq 7)$:

$$P(0 \leq r \leq 7) = P(0) + P(1) + P(2) + P(3) + P(4) + P(5) + P(6) + P(7),$$

which may be more concisely written, $P(0 \leq r \leq 7) = \sum_{r=0}^{r=7} P(r).$

We can use the result that $\sum_{r=0}^{r=10} P(r) = 1$ (i.e. the summation of probability of all possible outcomes is equal to (1), which may be written as $\sum_{r=0}^{r=7} P(r) + \sum_{r=8}^{r=10} P(r) = 1$.

Therefore $\sum_{r=0}^{r=7} P(r) = 1 - \sum_{r=8}^{r=10} P(r)$.

From part (ii) of this question $\sum_{r=8}^{r=10} P(r) = 0.9885$, so that $\sum_{r=0}^{r=7} P(r) = 1 - 0.9885 = 0.0115$.

Exercise A

The assembly of a hybrid circuit requires the soldering of 58 electrical connections. If 0.2% of electrical connections are faulty, what is the probability that an assembled circuit will have:

   (i)   no faulty connections;
  (ii)   one faulty connection;
 (iii)   more than one faulty connection?

4.2.2   Probability of a success, *p*

How do we know the probability of a success, p, in any particular circumstance? Some situations are so familiar and well defined, such as tossing a coin or rolling a dies that we regard the probability of a particular outcome occurring as known before any trials are made. For example, owing to the symmetry of a die, there are six equally possible outcomes on a single throw. The probability of a particular outcome, say a six, is therefore 1/6. In other situations, such as those requiring a knowledge of the probability of finding a defective device, we must rely on past experience where the proportion of defective devices from a large sample has been determined. Unlike the situation with the die, in which the probability of a particular outcome is fixed (so long as the die has not been tampered with) the probability of obtaining a defective device may change, especially if the manufacturing process has been modified.

4.2.3   Excel's BINOM.DIST() function

Calculating the probability of *r* successful outcomes in *n* trials using equation 4.4 is tedious if more than a few outcomes are to be considered. For example, if we want to know the probability that 10 or less outcomes are successful from 20 trials, we require the *cumulative* probability given by

$$P(r \leq 10) = \sum_{r=0}^{r=10} P(r), \tag{4.5}$$

where $P(r)$ is given by equation 4.4. Excel's BINOM.DIST() function allows for the calculation of the cumulative probability, such as that given by equation 4.5, as well as the probability of exactly r successes from n trials as given by equation 4.4. The syntax of the function is as follows.[3]

BINOM.DIST(number of successes, number of trials, probability of a success, cumulative)

Cumulative is a parameter which is set to TRUE if the cumulative probability is required and to FALSE if the probability of exactly r successful outcomes is required.

For example, if the probability of a success on a single trial is 0.45 and we require the probability of 10 or less successes from 20 trials, we would enter into a cell, =BINOM.DIST(10,20,0.45,TRUE) as shown in sheet 4.1.

Sheet 4.1. *Use of BINOM.DIST() to calculate cumulative probability.*

	A	B	C
1	= BINOM.DIST(10,20,0.45,TRUE)		
2			

After pressing the ENTER key, the number 0.750711 is returned in cell A1.

If we require the probability of <u>exactly</u> 10 successes from 20 trials, we enter into a cell , =BINOM.DIST(10,20,0.45,FALSE) as shown in sheet 4.2.

Sheet 4.2. *Use of BINOM.DIST() to calculate probability of r successful outcomes.*

	A	B	C
1	= BINOM.DIST(10,20,0.45,FALSE)		
2			

After pressing the ENTER key, the number 0.159349 is returned in cell A1.

---

[3] The BINOM.DIST() function in Excel 2010 supercedes the BINOMDIST() function available in earlier versions of Excel.

Exercise B

Given the number of trials, $n = 60$, and the success on a single trial, $p = 0.3$, use the BINOM.DIST() function in Excel to determine:

(i) $P(r = 20)$;
(ii) $P(r \leq 20)$;
(iii) $P(r < 20)$;
(iv) $P(r > 20)$;
(v) $P(r \geq 20)$.

4.2.4   Mean and standard deviation of binomially distributed data

If the probability of a success is p, and an experiment is performed which consists of n trials, then we expect np outcomes to be successful. It is possible in a real experiment for the actual number of successful outcomes from n trials to vary from 0 to n, but if the experiment is carried out many times, then the mean number of successes will tend to np. If we designate the mean number of successes over many experiments as $\bar{r}$ then, as the number of experiments becomes very large, $\bar{r}$ tends to the population mean, μ.

As an example, consider an experiment in which 10 transistors are tested. Previous experience has shown that the probability of obtaining a good transistor on a single trial is 0.7. It follows that the population mean, μ, is

$$\mu = 10 \times 0.7 = 7.$$

One hundred samples each consisting of ten transistors are removed from a large population and tested. The number of good transistors in each sample is shown in table 4.1.

If we calculate the mean of the values in table 4.1 we find $\bar{r} = 7.17$, which is close to the population mean[4] of 7.

Another important parameter representative of a population is its standard deviation. For an experiment consisting of n trials, where the probability of success is p and the probability of failure is q, the population standard deviation, σ, is given by[5]

---

[4] We cannot really know the population mean except in simple situations such as the population mean for the number of heads that would occur in ten tosses of a coin. In this situation we would make the assumption that the probability of a head is 0.5 so that in ten tosses we would expect five heads to occur.

[5] See Meyer (1975) for a proof of equation 4.6.

Table 4.1. *Number of good transistors in samples each of size n = 10, where the probability of removing a good transistor from a large population in 0.7.*

7	9	7	8	8	7	6	8	6	9
7	8	8	8	7	7	9	5	7	6
8	7	9	7	5	4	7	7	8	8
5	5	8	6	2	8	7	9	7	7
7	6	7	9	5	5	6	6	7	8
6	8	7	7	8	7	7	8	10	10
9	6	6	7	6	8	6	9	9	5
8	8	7	10	9	6	6	7	6	8
8	6	10	8	6	7	6	7	9	5
7	6	8	8	7	8	9	8	6	7

$$\sigma = \sqrt{npq} \tag{4.6}$$

For the data in table 4.1, $n = 10$, $p = 0.7$ and $q = 0.3$. Using equation 4.6, $\sigma = 1.449$.

The value for σ can be compared to the estimate of the population standard deviation, s, calculated for the data in table 4.1 by using equation 1.16. We find that

$$s = 1.39.$$

Exercise C

Two percent of electronic components are known to be defective. If these components are packed in containers each containing 400 components, determine:

  (i)   the population mean of defective items in each container;
 (ii)   the population standard deviation of defective items in each container;
(iii)   the probability that a container has two or more defective components.

4.2.5 Normal distribution as an approximation to the binomial distribution

The probability distribution for any number, n, and probability of success, p can be displayed graphically. For example, the probability of r successes, $P(r)$ from four trials, when $p = 0.2$ is shown in figure 4.1.

The probability distribution shown in figure 4.1 has little in common with the normal and t distributions considered in chapter 3. Specifically:

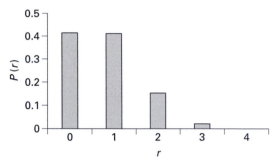

Figure 4.1. Binomial distribution for $n = 4$ and $p = 0.2$.

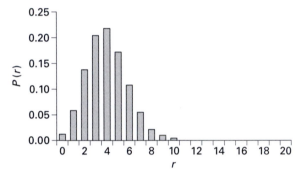

Figure 4.2. Binomial distribution with $n = 20$ and $p = 0.2$.

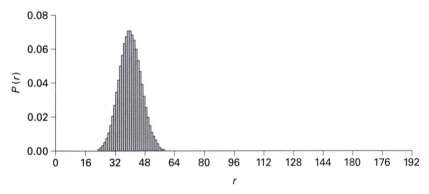

Figure 4.3. Binomial distribution with $n = 200$ and $p = 0.2$.

- the distribution is asymmetric;
- there is no well defined peak.

However, if the number of trials increases (and therefore the number of possible successes increases) there is a gradual change in the shape of the probability distribution as indicated in figures 4.2 and 4.3.

Figure 4.2 and figure 4.3 bear a striking resemblance to the normal distribution. The similarity is so close in fact, that the normal distribution with population mean, $\mu = np$ and standard deviation, $\sigma = \sqrt{npq}$ is often used as an approximation to the binomial distribution for the purpose of calculating probabilities. This is due to the fact that probabilities may be calculated more easily using the normal rather than the binomial distribution when a cumulative probability is required and the number of trials, n, is large.

As an example of the use of the normal distribution as an approximation to the binomial distribution, suppose we require the probability of three successful outcomes from 10 trials when $p = 0.4$. Using the normal distribution we find $P(3) = 0$, as finite probabilities can only be determined if an *interval* is defined. If we consider the number '3' as a rounded value from a continuous distribution with continuous random variable, x, then we would take $P(3)$ as $P(2.5 \leq x \leq 3.5)$. Comparing probabilities calculated using the binomial and normal distributions we find the following.

Binomial distribution calculation:

$P(3) = C_{10,3}0.4^3 0.6^7 = 0.2150.$

Normal distribution calculation:

Normal distribution approximation with mean, $\mu = 10 \times 0.4 = 4$, and standard deviation, $\sigma = \sqrt{10 \times 0.4 \times 0.6} = 1.549.$

$$P(3) \approx P(2.5 \leq x \leq 3.5) = P(-\infty \leq x \leq 3.5) - P(-\infty \leq x \leq 2.5). \quad (4.7)$$

The terms on the right hand side of equation 4.7 may be most easily determined using the NORM.DIST() function described in section 3.5.1, Specifically, entering =NORM.DIST(3.5,4,1.549,TRUE) into a cell in an Excel spreadsheet returns the number 0.3734. Similarly, entering =NORM.DIST(2.5,4,1.549,TRUE) into a cell returns the number 0.1664. It follows that

$$P(2.5 \leq x \leq 3.5) = 0.3734 - 0.1664 = 0.2070.$$

The approximate probability as determined by the normal distribution is about 4% less than that as given by the binomial distribution which is good enough for most purposes. In this example there is nothing to be gained in terms of efficiency using the normal distribution as an approximation to the binomial distribution. By contrast, in situations in which the number of trials, n, is large (say more than 20) and the calculation of cumulative probabilities is required, probability calculations using the normal distribution become attractive. When n is very large and r is not too close to n or zero, (say, $n > 2000$, and $1800 > (n - r) > 200$), it is not possible to calculate $C_{n,r}$ as given by equation 4.2 as an 'overflow' error occurs when using most computer based data analysis packages such as Excel.[6] In this situation there is little alternative but to use the normal distribution as an approximation to the binomial distribution in order to calculate probabilities.

Example 2

Given the number of trials, $n = 100$, and the probability of a success on a single trial, $p = 0.4$, determine the probability of between 38 and 52 successes (inclusive) occurring using the:

(i) binomial distribution;
(ii) normal distribution.

ANSWER

(i) Using the binomial distribution we require the probability, P, given by

$$P(38 \leq r \leq 52) = P(r \leq 52) - P(r < 38). \qquad (4.8)$$

The cumulative probabilities on the right hand side of equation 4.8 are most conveniently found using the BINOM.DIST() function on Excel. Entering the formula, =BINOM.DIST(52,100,0.40,TRUE) into a cell in an Excel spreadsheet returns the number 0.9942.

Entering =BINOM.DIST(37,100,0.40,TRUE) into a cell[7] returns the number 0.3068. It follows that

$$P(38 \leq r \leq 52) = P(r \leq 52) - P(r < 38) = 0.9942 - 0.3068 = 0.6874.$$

(ii) Using the normal distribution to solve this problem requires we calculate the area under a normal distribution curve between 37.5 and 52.5. Using x to represent the random variable, we write

$$P(37.5 \leq x \leq 52.5) = P(-\infty \leq x \leq 52.5) - P(-\infty \leq x \leq 37.5).$$

The population mean, $\mu = np = 100 \times 0.4 = 40$, and the population standard deviation, $\sigma = \sqrt{npq} = \sqrt{100 \times 0.4 \times 0.6} = 4.899$.

$P(37.5 \leq r \leq 52.5)$ may be found using Excel's NORM.DIST() function.[8] Specifically, entering =NORM.DIST(52.5,40,4.899,TRUE) into a cell of a spreadsheet, returns (to four significant figures) the number 0.9946. Similarly, entering =NORM.DIST(37.5,40,4.899, TRUE) into a cell returns the number 0.3049. It follows that

$P(-\infty \leq x \leq 52.5) = 0.9946$ and $P(-\infty \leq x \leq 37.5) = 0.3049$

so that, $P(-\infty \leq x \leq 52.5) - P(-\infty \leq x \leq 37.5) = 0.9946 - 0.3049 = 0.6897$.

---

[6] If overflow is to be avoided, the range of values for n and $(n - r)$ is even more restricted when using a pocket calculator to determine $C_{n,r}$.

[7] Note that $P(r \leq 37) = P(r < 38)$.

[8] The NORM.DIST() function is discussed in section 3.5.1.

Exercise D

If the probability of a success, $p = 0.3$ and the number of trials, $n = 1000$, use the binomial distribution and normal approximation to the binomial distribution to determine the probability of

  (i) between 290 and 320 successes (inclusive) occurring;

 (ii) more than 320 successes occurring.

4.3 The Poisson distribution

When introducing the binomial distribution we used the term 'trial' to describe the process of sampling a population. There are important physical processes in which outcomes may be counted, but where the term 'trial' is less appropriate. An example is the process of radioactive decay. A radioactive source emits particles due to the instability of the nuclei within the source. The emission process is random so it is not possible to say when a particular nucleus is going to decay through the emission of a particle. Nevertheless, even for a small radioactive source, there are so many nuclei that in a given time interval (say 1 minute) it is very likely that one or more nuclei will decay, ejecting a particle in the process. A system incorporating a detector and some electronics can be used to count particles emitted from the radioactive source. Characteristics of the radioactive counting experiment are as follows.

- An interval of time is chosen and the number of events (in this example an 'event' is the detection of a particle) occurring in that time interval is counted.
- The probability of an event occurring in the interval of time does not change.
- The mean number of events in the chosen time interval is a constant for the system being considered.
- The occurrence of an event can be counted but the non-occurrence cannot. So, for example, we might count 5 particles emitted from a source in 1 minute, but the question 'how many particles were *not* emitted in that minute?' has no meaning.

If we regard the occurrence of an event, such as detecting a high energy particle, as a 'success' then the binomial distribution can be used as the starting point for the derivation of an equation for the probability of an event occurring in a specified interval of time.

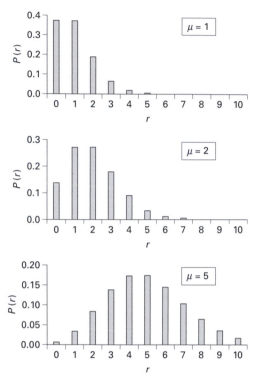

Figure 4.4. Poisson probability distribution for $\mu = 1$, 2 and 5.

If the probability of success on a single trial, p, is very small, but the number of trials, n, is very large, then the expression for the probability of r successes, $P(r)$, as given by equation 4.4 may be approximated by[9]

$$P(r) = \frac{\mu^r e^{-\mu}}{r!},$$
(4.9)

where μ is the mean number of successes in the interval chosen.

Equation 4.9 may be regarded as a new discrete probability distribution, called the Poisson distribution. The distribution is valid when n is large and p is small and p does not vary from one interval to the next. Processes (such as radioactive decay) that satisfy these conditions are sometimes referred to as 'Poisson processes'.[10]

Figure 4.4 shows how the probability given by equation 4.9 depends on the value of μ. For $\mu = 1$ the probability distribution is clearly asymmetric, but as

---

[9] See Scheaffer, Mulekar and McClave (2010) for a discussion of equation 4.9.

[10] Another example of a Poisson processes is the generation of charge carriers in semiconductor devices which leads to 'shot noise' affecting measurements of current through the devices. For further detail on Poisson processes and examples see Kingman (1993).

μ increases we again find that the characteristic shape becomes more and more 'normal like'. In situations where $\mu \geq 5$, the normal distribution can be used as an excellent approximation to the Poisson distribution and this facilitates computation of probabilities which would be very difficult to determine using equation 4.9.

Example 3

Given that $\mu = 0.8$, calculate $P(r)$ when:

(i) $r = 0$;

(ii) $r = 1$;

(iii) $r \geq 2$.

ANSWER

(i) Using equation 4.9, $P(0) = \frac{0.8^0 e^{-0.8}}{0!} = 0.4493$ (note that $0! = 1$).

(ii) Similarly, $P(1) = \frac{0.8^1 e^{-0.8}}{1!} = 0.3595$.

(iii) $P(r \geq 2) = P(2) + P(3) + P(4) \cdots + P(\infty)$. A convenient way to calculate this sum is to recognise that the sum of the probabilities for all possible outcomes $= 1$, i.e. $P(0) + P(1) + P(2) + P(3) + P(4) + \cdots + P(\infty) = 1$. It follows that

$$P(2) + P(3) + P(4) + \cdots + P(\infty) = 1 - (P(0) + P(1))$$
$$= 1 - (0.4493 + 0.3595) = 0.1912.$$

Exercise E

Given that $\mu = 0.5$, use equation 4.9 to determine

(i) $P(r = 0)$;

(ii) $P(r \leq 3)$;

(iii) $P(2 \leq r \leq 4)$.

4.3.1 Applications of the Poisson distribution

As well as nuclear counting experiments, there are other situations in which the Poisson distribution is valid. These include counting the number of:

- X-rays scattered by air molecules;
- dust particles detected per unit volume in a clean room;
- electrons scattered as they travel towards a specimen in an electron microscope;

- pinholes per unit area in a thermally evaporated thin metal film;
- particles emitted from a radioactive source.

4.3.2   Standard deviation of the Poisson distribution

To determine the population standard deviation for the Poisson distribution, we begin with the binomial distribution and determine the consequences of allowing the number of trials, n, to become very large while at the same time letting the probability of success on a single trial, p, become very small.

The standard deviation, σ, for the binomial distribution is written

$$\sigma = \sqrt{npq} = \sqrt{np(1-p)}.$$

If p is very small then $1 - p \approx 1$, so that

$$\sigma \approx \sqrt{np},$$

where $np = \mu$, so that,

$$\sigma \approx \sqrt{\mu}. \tag{4.10}$$

A useful indicator of whether data are 'Poisson distributed' is to calculate the mean and standard deviation of the number of events counted in a particular time interval or interval of area, volume, etc. If the condition $\sigma \approx \sqrt{\mu}$ does not hold then the data are not Poisson distributed. As usual, we are not able to know σ and μ, and so we must be satisfied with their best estimates, s and $\bar{x}$ respectively.

Example 4

Table 4.2 shows the number of cosmic rays detected in time intervals of 1 minute near to the surface of the Earth.

   Use the data to estimate:

  (i)   the population mean;
  (ii) the population standard deviation.

Calculate the probability of detecting in any one minute time interval:

  (iii) 0 cosmic rays;
  (iv)   more than 2 cosmic rays.

The experiment is continued so that the number of counts occurring in 1200 successive time intervals of 1 minute is recorded.

  (v)   In how many intervals would you expect the number of counts to exceed 2?

ANSWER

(i) The mean of the values in table 4.2 is 6.48. This is the best estimate of the population mean.

(ii) Using equation 4.10, the population standard deviation $\sigma \approx \sqrt{6.48} = 2.55$.

(iii) Using equation 4.9, $P(0) = \frac{6.48^0 e^{-6.48}}{0!} = 1.534 \times 10^{-3}$.

(iv) $P(r>2) = P(3) + P(4) + P(5)\cdots + P(\infty) = 1 - (P(0) + P(1) + P(2))$.

Now $P(1) = 9.939 \times 10^{-3}$ and $P(2) = 3.220 \times 10^{-2}$ so that, $P(3) + P(4) + P(5)\cdots + P(\infty)$
$$= 1 - (1.534 \times 10^{-3} + 9.939 \times 10^{-3} + 3.220 \times 10^{-2}) = 0.9563.$$

(v) The expected number of 1 minute intervals in which more than 2 counts occur $= NP(r > 2)$, where N is the total number of intervals:

$$NP(r>2) = 1200 \times 0.9563 = 1148 \text{ intervals}.$$

Table 4.2. *Number of cosmic rays detected at the Earth's surface.*

5	3	2	7	10	4	5	7	7	7
10	7	6	10	9	5	9	7	9	6
8	5	9	9	3	6	6	8	2	11
7	6	4	9	4	5	1	4	6	6
7	4	10	3	9	6	9	5	10	7

Exercise F

Small spots of contamination form on the surface of a ceramic conductor when it is exposed to a humid atmosphere. The contamination degrades the quality of electrical contacts made to the surface of the ceramic conductor. Table 4.3 shows the number of spots identified in 50 non-overlapping regions of area $100 \ \mu m^2$ at the surface of a particular sample exposed to high humidity.

Use the data in table 4.3 to determine the mean number of spots per $100 \ \mu m^2$. A silver electrode of area $100 \ \mu m^2$ is deposited on the surface of the conductor. Assuming the positioning of the electrode is random; calculate the probability that the silver electrode will

(i) cover no spots of contamination;

(ii) cover 1 spot;

(iii) cover more than 2 spots.

Table 4.3. *Number of contamination spots on the surface of a ceramic conductor.*

2	1	1	0	0	1	0	2	2	2
1	2	1	0	1	0	1	0	0	1
0	1	0	0	3	1	1	2	4	0
1	2	1	0	0	0	0	2	2	1
0	1	0	3	2	1	2	2	0	2

4.3.3   Excel's POISSON.DIST() function

Excel's POISSON.DIST() function[11] allows for the determination of probability based on equation 4.9, as well as the cumulative probability given by $P(r \leq R)$, where

$$P(r \leq R) = \sum_{r=0}^{r=R} P(r). \tag{4.11}$$

The syntax of the function is POISSON.DIST(number of events, mean, cumulative). Cumulative is a parameter which is set to TRUE for a cumulative probability given by equation 4.11. The cumulative parameter is set to FALSE if the probability of exactly r successful outcomes is required.

As an example, if the mean number of cosmic rays detected in a time interval of 1 minute is 8, and we require the probability of detecting 4 or less cosmic rays in any 1 minute interval, we would type =POISSON.DIST(4,8, TRUE) into a cell of an Excel worksheet, as shown in sheet 4.3.

Sheet 4.3. *Using Excel's POISSON.DIST() function to calculate cumulative probability.*

	A	B	C
1	= POISSON.DIST(4,8,TRUE)		
2			

After pressing the Enter key, the number 0.099632 is returned in cell A1. On the other hand, if we require the probability of *exactly* 4 cosmic rays being detected in a one minute time interval we would type =POISSON.DIST(4,8, FALSE) into a cell in Excel as shown in sheet 4.4.

After pressing the Enter key, the number 0.057252 is returned in cell A1.

---

[11] The POISSON.DIST() function in Excel 2010 replaces the POISSON() function available in earlier versions of Excel. For compatibility with earlier versions of Excel, the POISSON( ) function remains available in Excel 2010.

Sheet 4.4. *Using Excel's POISSON.DIST() function to calculate the probability of r events occurring.*

	A	B	C
1	= POISSON.DIST(4,8,FALSE)		
2			

Table 4.4. *Data from X-ray counting experiment.*

1	1	1	1	1	0	0	6	1	2
0	4	1	1	2	3	1	0	3	3
4	1	1	0	0	2	4	1	3	5
6	0	1	1	4	6	0	0	0	1
1	1	2	2	1	1	0	2	3	1
1	4	0	2	0	0	3	3	2	4
0	2	2	1	1	2	2	0	0	1
1	1	3	2	2	0	0	2	0	1
1	1	2	2	1	0	1	4	0	1
1	1	0	1	0	2	2	0	0	3

Exercise G

Table 4.4 shows the number of X-rays detected in 100 time intervals where each interval was of 1 second duration.

Use the data in table 4.4 to determine the mean number of counts per second. Assuming the Poisson distribution to be applicable, use Excel to determine the probability of observing in a 1 second time interval:

(i) 0 counts;
(ii) 1 count;
(iii) 3 counts;
(iv) between 2 and 4 counts (inclusive);
(v) more than 6 counts.

4.3.4 Normal distribution as an approximation to the Poisson distribution

Figure 4.4 indicates that when the mean, μ, equals 5, the shape of the Poisson distribution is similar to that of the normal distribution. When μ equals or exceeds 5, the normal distribution is preferred to the Poisson distribution when the calculation of probabilities is required. This is due to the fact that summing a large series is tedious, unless computational aids are available. Even with aids like Excel, some summations cannot be determined. When r, μ^r and $r!$ are large, the result of a calculation using equation 4.9 can exceed the numerical range of the computer causing an 'overflow' error to occur.

Example 5

In an X-ray counting experiment, the mean number of counts in a period of 1 minute is 200. Use the Poisson and normal distributions to calculate the probability that, in a 1 minute period, the number of counts occurring would be exactly 200.

ANSWER

To find the probability using the Poisson distribution, we use equation 4.9 with $\mu = r = 200$, i.e.

$$P(r) = \frac{\mu^r e^{-\mu}}{r!}, \text{ so that}$$

$$P(200) = \frac{200^{200} e^{-200}}{200!}.$$

This is where we must stop, as few calculators or computer programs can cope with numbers as large as 200! or 200^{200}.

Using the normal distribution we use the approximation that $P(r = 200) \cong P(199.5 \le x \le 200.5)$, where x is a continuous random variable.

Take the population standard deviation of the normal distribution to be $\sqrt{\mu} = \sqrt{200} = 14.142$.

Now

$$P(199.5 \le x \le 200.5) = P(-\infty \le x \le 200.5) - P(-\infty \le x \le 199.5). \tag{4.12}$$

The two terms on the right hand side of equation 4.12 may be determined by using the NORM.DIST() function in Excel.

Entering =NORM.DIST(200.5,200,14.142,TRUE) into a cell in Excel returns the number 0.5141 and entering =NORM.DIST(199.5,200,14.142,TRUE) into another cell returns the number 0.4859.

It follows that $P(r = 200) = 0.5141 - 0.4859 = 0.0282$.

Exercise H

Using the data in example 5, determine the probability that in any 1 minute time period, the number of counts lies between the inclusive limits of 180 and 230.

4.4 Review

When random processes, such as coin tossing, radioactive decay, or the scattering of particles, lead to outcomes which may be counted, we turn to discrete probability distributions to assist in determining the probability that certain outcomes will occur. The starting point for discussing discrete distributions is to consider the binomial distribution. In particular, we found that the binomial distribution is useful for determining probabilities when there are only two possible outcomes from a single trial and the probability of those outcomes is fixed from trial to trial. This requires that we know (or are able to determine experimentally) the probability of a successful outcome occurring in a single trial. In addition, the binomial distribution is helpful in deriving another important distribution, the Poisson distribution. This distribution is appropriate when the frequency of occurrence of 'rare' events is required, such as the decay of a radioactive nucleus. Both the binomial and the Poisson distributions become more 'normal-like' as the number of trials increases. Owing to the comparative ease with which calculations may be performed using the normal distribution, it is used extensively as an approximation to the binomial and Poisson distributions when determining probabilities or frequencies.

Whether an experiment involves discrete or continuous quantities, the outcome of a measurement or a trial cannot be predicted with complete certainty. We must therefore acknowledge that 'uncertainty' is an inherent and very important feature of all experimental work. With the results obtained in chapters 3 and 4, we are in a position to discuss the important topic of uncertainty in measurement.

End of chapter problems

(1) Determine

$$C_{10,5}, C_{15,2}, C_{42,24}, C_{580,290}$$

(suggestion: use the COMBIN() function in Excel).

(2) The screen for a laptop computer is controlled by 480000 transistors. If the probability of a faulty transistor occurring during manufacture of the screen is 2×10^{-7}, determine the probability that a screen will have

   (i) one faulty transistor;
   (ii) more than one faulty transistor.

If 5000 screens are supplied to a computer manufacturer, how many screens would you expect to have one or more faulty transistors?

(3) Ten percent of ammeters in a laboratory have blown fuses and so are unable to measure electric currents. An electronics experiment requires the use of 3 ammeters. If 25 students are each given 3 ammeters, how many students would you expect to have:

(i) 3 functioning ammeters;
(ii) 2 functioning ammeters;
(iii) less than 2 functioning ammeters?

(4) An electroencephalograph (EEG) is an instrument which uses small electrodes pressed against the scalp to detect the electrical activity of the human brain. When an electrode is pressed against the scalp there is a probability of 0.89 that it will make good enough electrical contact to the skin to allow faithful recording of brain activity. In a study, 24 electrodes are used and of those, at least 20 must make good electrical contact with the scalp for an acceptable assessment of brain activity to be made.

(i) What is the probability that at least 20 electrodes will make good contact to the scalp?

A new type of electrode is trialled which has a probability of 0.72 of making good contact with the scalp.

(ii) How many electrodes must be pressed against the scalp so that the probability that at least 20 of them will make good contact is the same as in part (i)?

(5) When a specimen is bombarded with high energy electrons (such as occurs in an electron microscope) the specimen emits X-rays. In an experiment, the number of 14.4 keV X-rays emerging from a specimen was counted in 10 successive 100 s time intervals. The data obtained are shown in Table 4.5. Use the data to estimate:

(i) the population mean for the number of counts in 100 s;
(ii) the population standard deviation.

(6) A thin tape of superconductor is inspected for flaws that would affect its capacity to carry an electrical current. Sixty strips of tape, each of length 1 m, are examined and the number of flaws in each strip is recorded. These are shown in table 4.6.

(i) Calculate the mean number of flaws per metre.
(ii) Assuming the Poisson distribution to be valid, determine the expected frequency for 0, 1, 2, 3, 4, and 5 flaws (hint: first calculate $P(0)$, $P(1)$, etc., using equation 4.9).

(7) An environmental scanning electron microscope (ESEM) is used to image specimens in a 'poor' vacuum environment. Owing to the high concentration of air molecules in the microscope, electrons emitted from the electron gun in an

Table 4.5. *Number of X-rays counted in 10 periods each of duration 100 s.*

118	131	136	119	131
122	119	129	124	98

Table 4.6. *Number of flaws in superconducting tapes.*

Number of flaws, N	Number of 1 m strips of tape with N flaws
0	18
1	19
2	12
3	9
4	1
5	1

ESEM are scattered by the molecules as they travel towards a specimen. An acceptable image of the specimen is produced as long as 20% or more of the electrons reach the specimen unscattered.

Assuming that the Poisson distribution may be used to describe the probability of electron scattering in an ESEM and that 20% of the electrons are unscattered, determine:

(i) the number of times, on average, that an electron is scattered as it travels from electron gun to specimen;

(ii) the probability that an electron will undergo more than 4 scattering 'events';

(iii) the probability that an electron will undergo two or less scattering 'events'.

(8) The mean number of cyclones to strike a region in Western Australia per year is 0.2. What is the probability that over a period of twenty years, one or more cyclones will strike that region?

Chapter 5

Measurement, error and uncertainty

5.1   Introduction

Chemists, physicists and other physical scientists are proud of the quantitative nature of their disciplines. By subjecting nature to ever closer examination, new relationships between quantities are discovered, and established relationships are pushed to the limits of their applicability. When 'numbers' emerge from an experiment, they can be subjected to quantitative analysis, compared to the 'numbers' obtained by other experimenters and be expressed in a clear and concise manner using tables and graphs. If an unfamiliar experiment is planned, an experimenter will often carry out a pilot experiment. The purpose of such an experiment might be to assess the effectiveness of the experimental methods being used, or to offer a preliminary evaluation of a theoretical prediction. It is also possible that the experimenter is acting on instinct or intuition. If the results of the pilot experiment are promising, the experimenter typically moves to the next stage in which a more thorough investigation is undertaken and where there is increased emphasis on the quality of the data gathered. The analysis of these data often provides crucial and defensible evidence sought by the experimenter to support (or refute) a particular theory or idea.

The goal of an experiment might be to determine an accurate value for a particular quantity such as the electrical charge carried by an electron. Experimenters are aware that influences exist, some controllable and others less so, that conspire to adversely affect the values they obtain. Despite an experimenter's best efforts, some uncertainty in an experimentally determined value remains. In the case of the charge on the electron, its value is recognised to be of such importance that considerable effort has gone into establishing an accurate value for it. Currently (2011) the best value[1] for the charge on the electron is $(1.602176487 \pm 0.000000040) \times 10^{-19}$ C. A very important part of the expression for the charge is the number following the $\pm$ sign. This is the *uncertainty* in the value for the electronic charge and, though the uncertainty is rather small compared to the size of the charge, it is not zero. In general, every value obtained through measurement has some uncertainty and though the uncertainty may be reduced by thorough planning, prudent choice of measuring instrument and careful execution of the experiment, it cannot be eliminated entirely.

While determining uncertainties and assessing their consequences pre-occupy many scientists, there are also everyday situations in which uncertainty is of importance to those who would not describe themselves as scientists; a driver of a car stopped by police might be interested to know whether the uncertainty in his speed determined using a 'laser-gun' is sufficiently large to call into question an allegation of speeding; in a stunt in which a motor cycle rider leaves a ramp at high velocity, only a small uncertainty in take-off velocity can be tolerated if the rider is to land safely; in medicine, a small uncertainty in the measurement of blood pressure decreases the risk of misdiagnosis; a manufacturer of engine parts who documents the uncertainty in the dimensions of components will have a competitive advantage when marketing those components internationally.[2]

While our focus is on uncertainties in scientific contexts, uncertainties impact in engineering, medicine and commerce, as well as in everyday situations.

In this chapter we consider uncertainties, how they arise, may be identi-fied and categorised, strategies for their minimisation and how they propagate when several quantities each with its own uncertainty are brought together in a calculation. We will adapt methods of describing variability in data considered in chapters 3 and 4 to support the calculation of uncertainty.

5.1.1 International dimension to uncertainty

Quantifying uncertainty allows the quality of data to be assessed. Failing to quantify uncertainty may affect the credibility of conclusions drawn from the data. We will apply methods for calculating and expressing uncertainty which are consistent with international guidelines adopted worldwide, as examples,

---

[1] Mohr, P. J., Taylor, B. N. and Newell, D. B. (2008).

[2] Examples of measurement and uncertainty from a range of contexts may be found in *Metrology in Short* published by Euramet and available as a free download from http://www.euramet.org/fileadmin/docs/Publications/Metrology_in_short_3rd_ed.pdf [accessed 3/11/2010].

in industrial and scientific laboratories. An influential document describing those guidelines is the *Guide to the Expression of Uncertainty in Measurement* (usually abbreviated to the 'GUM'). We will refer to the GUM in this and later chapters.[3]

The GUM incorporates some unfamiliar terms, such as *standard uncertainty* and *coverage factor*. We will pause from time to time in order to clarify those terms. The terminology we will use has largely been developed by metrologists (who are measurement specialists[4]) and though many of the terms are similar to those encountered in a study of statistics, there are some important differences. An authoritative resource that contains definitions of measurement-related terms is the *International Vocabulary of Metrology*, often referred to as the 'VIM'.[5]

5.1.2 Measurand

A quantity intended to be determined through measurement is called the *measurand*.[6] In many situations, determining a value for the measurand may require that other quantities be brought together. These are referred to as *input quantities*.

For example, suppose the purpose of an experiment is to measure the electrical resistance, R, of a tungsten wire. In this case R is the measurand. In order to determine R, we could measure the potential difference, V, across the wire as well as the current, I, through the wire. R is then calculated using

$$R = \frac{V}{I}. \tag{5.1}$$

Here V and I are the input quantities. In other situations a different quantity could be the measurand with R playing the role of an input quantity used in the determination of the value of the measurand. For example, the electrical resistivity, ρ, of a material can be determined experimentally with the aid of the relationship

---

[3] The GUM is available as a free download from http://www.bipm.org/utils/common/documents/jcgm/JCGM_100_2008_E.pdf [accessed 3/11/2010].

[4] The discipline of *metrology* is defined as the science of measurement and its application. Workers in that field are referred to as *metrologists*. The title 'metrologist' should not to be confused with 'meteorologist'. A meteorologist is an expert on the atmosphere and weather processes.

[5] The VIM is available as a free download from http://www.bipm.org/utils/common/documents/jcgm/JCGM_200_2008.pdf [accessed 3/11/2010].

[6] See the VIM (section 2.3).

$$\rho = \frac{RA}{l},$$ (5.2)

where A is the cross-sectional area of the wire made from the material, and l is the length of the wire.

In this situation the measurand is ρ, while R, A and l are input quantities.

In short, what is designated as the measurand and what is designated as an input quantity depends on the circumstances.

5.2 The process of measurement

Looking for patterns in data which may uncover a relationship between physical quantities demands confidence in the data gathered. One reason for this is that theories supported by dubious data mislead and divert attention from better theories or explanations. Reliable data are a consequence of reliable measurement and therefore it is important to identify factors that can affect data gathered during an experiment.

Measurement in science can involve something as simple as using an alcohol-in-glass thermometer to measure the air temperature in a room, to scanning probe microscopy incorporating state of the art technology which permits features to be measured that are of the order of nanometres in size. The process of measurement typically requires:

- choice of quantity or quantities to be measured;
- a sensor or device to be chosen which is responsive to changes in the quantity being measured;
- transformation of the output of the sensor to a form which allows easy display and/or recording of the output;
- recording or logging the transformed sensor output.

Factors that might affect the accuracy of experimental data may be illustrated by example. We consider temperature measurement, as many experiments involve the study of the variation of a quantity with temperature. That quantity could be the pressure of a gas, the electrical resistivity of an alloy, or the rate of chemical reaction.

Figure 5.1 shows the schematic diagram of a system used to measure the temperature of a copper block (the copper block might be situated, for example, in an oven or a cryostat).

The process of temperature measurement indicated in figure 5.1 involves the following.

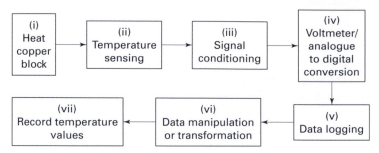

Figure 5.1. Typical system for determining the temperature of a copper block.

(i) Heat transfer from the copper block to the temperature sensor as the temperature of the block rises above that of the sensor.

(ii) A change of a physical attribute of the temperature sensor (such as its electrical resistance) in response to the temperature change of the sensor.

(iii) If the output of the sensor takes the form of an electrical signal, then 'signal conditioning' allows that signal to be modified. For example, the signal may be amplified, or perhaps a filter used to attenuate a.c. electrical noise at 50 Hz or 60 Hz.

(iv) A voltmeter or analogue to digital converter on a data acquisition card within a computer measures the output of the signal conditioner.

(v) The voltage displayed by the voltmeter is recorded 'by hand' in a notebook, or logged automatically by computer.

(vi) A calibration curve or 'look-up' table[7] is used to transform values of voltage to values of temperature.[8]

(vii) Values of temperature are recorded in a notebook or a computer.

Though we have considered temperature measurement in this example, a system that measures pressure, pH, flow rate, strain, light intensity or any other physical quantity would contain similar elements.

Steps (i) to (vii) described above, each influence the values of temperature recorded as part of an experiment. A thorough analysis of all the influences is challenging, but table 5.1 describes some of the factors that affect the value(s) of temperature recorded, along with a brief explanation.[9]

---

[7] If a simple mathematical relationship between the output of a transducer and the measured quantity cannot be found, a look-up table can be used in which each output value from a transducer is associated with a corresponding input value of the measured quantity.

[8] If a resistance thermometer is used, it is likely that voltages will be first transformed into resistances before a temperature versus resistance calibration curve is used to determine the value of temperature.

[9] Table 5.1 is not exhaustive. For a fuller discussion of the measurement of temperature and its challenges, see Nicholas and White (2001).

Table 5.1. *Factors influencing values of temperature obtained through measurement.*

Factor	Explanation
Thermal resistance between temperature sensor and copper block	No matter how intimate the contact between temperature sensor and copper block, there is some resistance to the flow of heat between the two. This affects the response time of the sensor so that, in dynamic experiments where temperature is changing with time, the sensor temperature lags the temperature of the block.
Size and constitution of temperature sensor	A temperature sensor with a large heat capacity responds slowly to heat being transferred to or from the sensor. If the temperature of the copper block is higher than that of the sensor then when the sensor is brought in contact with the block it will lower the block's temperature.
Lead wires to temperature sensor	In the case of thermocouples, thermistors and resistance thermometers, the leads attached to these sensors form pathways for heat transfer by conduction. If the temperature of the sensor is below the surroundings, heat is conducted to the sensor along the leads thus raising its temperature. Conversely, if the temperature of the sensor is higher than that of the surroundings, heat is conducted away from the sensor thus reducing its temperature.
Gain stability of signal conditioning amplifier	Changing electrical properties of the signal conditioning system can adversely affect the measurement of temperature. For example, the temperature dependence of active components within the amplifier may cause the amplifier's gain to alter in response to changes to ambient temperature.
Sampling time and voltmeter resolution	The time interval to complete the measurement of voltage (i.e. the sampling time) using a digital voltmeter can vary typically from 1 ms to 1 s. Any variation in the voltage that occurs over the time interval of the measurement cannot be known as the voltmeter displays an average value. (A typical hand held digital voltmeter makes 2 to 3 measurements per second). The resolution of a digital voltmeter is the smallest change in voltage that the voltmeter can indicate. For 3½ digit voltmeters on the 2 V range, the resolution is 1 mV.
Accuracy of calibration curve	A calibration curve is used to transform one quantity (say resistance or voltage) into temperature. A generic curve may be provided by the sensor manufacturer which relates the sensor output to temperature. Deviations from the curve for a particular sensor due to manufacturing variability is a source of error.

The factors described in table 5.1 are possible sources of experimental *error*. By error we are not referring to a mistake such as recording incorrectly a value that appears on the display of a meter, but a deviation from the 'true value' of a quantity being sought through measurement. Such a deviation might be caused, for example, by the limit of resolution of an instrument or the finite response time of a sensor. Experimental errors prevent us from being able to quote an exact value for any quantity determined through experiment. By careful consideration of all sources of error in an experiment we can quote an *uncertainty* in the measured value. A detailed consideration of the sources of uncertainties in an experiment might suggest ways by which some of those uncertainties can be minimised. While identifying and quantifying measurement uncertainties are important, it is equally important to use any insight gained to reduce the uncertainties, if possible, through better experimental design.

Quoting uncertainties allows for values to be compared. For example, values for the density of a particular alloy obtained by two experimenters working independently might be $6255 \, \text{kg/m}^3$ and $6138.2 \, \text{kg/m}^3$, respectively. If we have no knowledge of the uncertainty that can be ascribed to each value, we cannot know if the values are consistent with each other. By contrast, if we are told that the density values are $(6255 \pm 10) \, \text{kg/m}^3$ and $(6138.2 \pm 2.9) \, \text{kg/m}^3$ we are in a much stronger position to argue that the values are *not* mutually consistent and that further investigation is warranted.[10] When quoting results we are obliged to communicate all relevant details including the 'best' value of a quantity and the uncertainty in this value.

5.3   True value and error

It is often helpful to imagine that a quantity, such as the mass of a body, has a 'true' value and it is the true value that we seek through measurement. If the quantity to be measured is completely defined and instruments are perfect (for example, they are capable of infinitely fine resolution) and no outside influences conspire to affect the quantity undergoing the process of measurement, then we should be able to determine the true value of a quantity. Recognising that instruments are not perfect, and that outside influences though they can be minimised, can never be completely eliminated, the best we can do is obtain a 'best estimate' of the true value. If we *could* know the true value of a quantity

---

[10] However, we *do* need to know how the uncertainties $10 \, \text{kg/m}^3$ and $2.9 \, \text{kg/m}^3$ are calculated in order to interpret their meaning.

then, in an experiment to determine the temperature, t, of a body we could perhaps write

$$t = 45.1783128212°\text{C}.$$

The reality is that, as examples, offset drift in the measuring instrument, electrical noise and resolution limits cause measured values to vary from the true value by such an amount as to make the majority of the figures after the decimal point in t above quite meaningless in most situations. The difference between the measured value and the true value is referred to as the experimental *error*. Representing the true value by the symbol, μ, the error in the ith measured value, δx_i, is given by

$$\delta x_i = x_i - \mu, \tag{5.3}$$

where x_i is the ith value obtained through measurement. If the errors are random then we expect values of δx_i to be positive and negative, as x_i will take on values greater and less than μ. If we could sum the errors due to many repeat measurements of the same quantity then the errors would have a cancelling effect such that

$$\sum_{i=1}^{i=n} \delta x_i \approx 0, \tag{5.4}$$

where n is the number of measurements made.

Substituting for δx_i from equation 5.3 into equation 5.4 gives[11]

$$\sum (x_i - \mu) \approx 0. \tag{5.5}$$

It follows that

$$\sum x_i - n\mu \approx 0$$

or

$$\mu \approx \frac{\sum x_i}{n} = \bar{x}. \tag{5.6}$$

Equation 5.6 supports what we learned in chapters 1 and 3, namely that the best estimate of the true value of a quantity is the sample mean and, as $n \to \infty, \bar{x} \to \mu$.

The true value of a quantity is a close relation to the population mean discussed in chapter 3. The sample mean, $\bar{x}$, tends to the population mean

---

[11] Unless stated otherwise, all summations in this and future chapters are carried out between $i = 1$ and $i = n$.

when the number of values used to calculate $\bar{x}$ tends to infinity. However, the terms 'true value' and 'population mean' are not always interchangeable. As an example, ten repeat measurements of the breakdown voltage, V_B, of a *single* zener diode might be made to estimate the true value of V_B for that diode. In this case we could replace the term 'true value' by 'population mean' to describe the value being estimated. However, if the breakdown voltage of each of ten *different* zener diodes is measured then we can estimate the mean of the breakdown voltages for the population from which the zener diodes were drawn, but the term 'true value' has little meaning in this context. In addition, suppose an instrument used to measure voltages has drifted to the extent that every value of voltage displayed by the instrument is, say, 10 mV larger than the true value (this is a *systematic error* and we consider such errors in section 5.5). Taking the mean of many repeat measurements will result in a value close to the population mean being obtained using that instrument if an infinite number of measurements were made. However, the true value of the voltage being measured in this case would differ by 10 mV from the population mean.

5.4   Precision and accuracy

Repeat measurements of a quantity reveal underlying scatter or variability in that quantity. The closer the agreement between values obtained through repeat measurement of a quantity, the better is the precision of the measurement. As an example, consider values shown in table 5.2 of the acceleration of a free falling body determined at sea level close to the equator.

The mean of the values in table 5.2 is 9.605 m/s². The values in table 5.2 are clustered quite close to the mean, suggesting that effects that would act to scatter the data are small. Representing the spread of the values by the standard deviation calculated by using equation 1.16, gives $s = 0.0307$ m/s².

In the above discussion of precision there is no mention of how close the measured values are to the true value. In fact, values obtained through repeat measurements may show little scatter but be far from the true value. A particular value is regarded as *accurate* if it is close to the true value. The difficulty with this statement is that the true value is rarely if ever known, so how is it possible to say something about the accuracy of a particular value? This is not an easy or trivial question to answer. Where

Table 5.2. *values for acceleration caused by gravity, g, measured at the equator.*

g (m/s²)	9.62	9.57	9.64	9.64	9.58	9.59	9.57	9.63

measurements are designed to establish the value of a quantity or constant of wide interest to the scientific community, such as the energy gap of a semiconductor, the boiling point of ethanol or the acceleration caused by gravity near to the Earth's surface, it is possible to compare values with those published by other experimenters, Such comparisons are especially revealing where the 'other experimenters' are experts in their field and have devoted time and expertise to the accurate determination of the quantity in question.

We can compare the values of the acceleration due to gravity in table 5.2 with a value acknowledged as being accurate for the acceleration at sea level at the equator[12] which is 9.780 m/s^2. None of the values in table 5.2 is as large as 9.780 m/s^2, suggesting that these values are **not** accurate. It appears that, although the scatter of the values is quite small, each value is less than the true value by about 0.2 m/s^2. Some influence is at play which is consistently causing all the values to be too low. Such consistent or systematic influences have no effect on the precision of measurements but they do dramatically affect the accuracy of a measurement, since the mean of many values will never tend to the true value so long as systematic effects prevail.

We can summarise precision and accuracy as follows.

- Measurement *precision* is good when the scatter in the values is small BUT this does not imply that the values are close to the true value.
- Measurement is *accurate* when values obtained are close to the true value.
- The mean of n values is accurate when the mean tends to the true value as n becomes large BUT the deviation of individual values from the mean may be quite large.

The goal in experimental science is for values obtained through experiment to be both accurate and precise (close to the true value with little scatter). Experiments in which the precision is good are sometimes shown later to have produced data of poor accuracy.

5.5 Random and systematic errors

Random errors cause values to lie above and below the true value and, due to the scatter they create, they are usually easy to recognise. So long as only random errors exist, the mean of the values obtained through measurement tends towards the true value as the number of repeat measurements increases.

---

[12] For a detailed discussion on the factors affecting gravity near to the Earth's surface, see Telford, Geldart and Sheriff (2010).

Another type of error is that which causes the measured values to be consistently above or consistently below the true value. This is termed a *systematic* error, also sometimes referred to as a *bias* error. An example of this is the zero offset of a spring balance. Suppose, with no mass attached to the spring, the balance indicates 0.01 N. We can 'correct' for the zero offset

- if the facility is available, by adjusting the balance to indicate zero or,
- by subtracting 0.01 N from every value obtained using the balance.

No value, including an offset, can be known exactly so the correction applied to values will have its own error. As a consequence we cannot claim that an adjustment designed to correct for an offset has completely eliminated the systematic error – only that the error has been reduced.

Some systematic errors are difficult to identify and have more to do with the way a measurement is made rather than any obvious inadequacy or limitation in the measuring instrument. Systematic errors have no effect on the precision of values, but act to impair measurement accuracy.

5.6 Random errors

Many effects can cause values obtained through experiment to be scattered above and below the true value. Those effects may be traced to limitations in a measuring instrument or perhaps be related to a human trait, such as reaction time. In some experiments, values may show scatter due to the influence of physical processes that are inherently random. Table 5.3 describes three such sources of random error.

5.6.1 Common sources of error

Classifying the sources of error in table 5.3 as random is clear-cut as the physical processes causing the error are well characterised. We next consider several other sources of error often classified as random but indicate that in some cases the same source may be responsible for both random and systematic error.

Resolution error
A 3½ digit voltmeter on its 200 mV scale indicates voltages to the nearest 0.1 mV, i.e. the resolution of the instrument on this scale is 0.1 mV. If the scale indicates 182.3 mV, the true voltage could lie anywhere between 182.25 mV and 182.35 mV. That is, the resolution of the instrument is a source of error. If we assume that sometimes the voltmeter indicates 'rounded up'

Table 5.3. *Examples of sources of random error.*

Description	Explanation
Johnson noise	A voltage that varies randomly with time appears across any component or material that has electrical resistance. The random voltage is referred to as Johnson noise. The cause of the noise is the random thermal motion of electrons in the material and so one way the noise can be reduced is to lower the temperature of the material.[13]
Radioactive decay	The nuclei of some chemical elements are unstable and become more stable by emitting a particle. The process is random and so it is not possible to predict at what instant a particular nuclei will emit a particle. As a consequence, the number of particles detected in a given time interval by a nuclear counter varies in a random manner from one interval to the next.[14]
Molecular motion	Forces are exerted on bodies due to the random motion of molecules in gases and liquids. Though the forces are very small, in highly sensitive pressure measuring systems, or those incorporating fine torsion fibres, the forces are sufficient to be detected.

values which are greater than the true value and at other times indicates 'rounded down' values that are less than the true value, then it is reasonable to regard the resolution of an instrument to be a source of random error.

While the resolution error can be reduced by selecting another (probably more expensive) instrument with higher resolution, it is often the case that the error due to the limited resolution of an instrument is a small component of the total error in a measurement.

Parallax error

If you move your head when viewing the position of a pointer on an analogue stopwatch or the top of a column of alcohol in an alcohol-in-glass thermometer, the value (as read from the scale of the instrument) changes. The position of the viewer with respect to the scale introduces *parallax* error. In the case of the alcohol-in-glass thermometer, the parallax error stems from the displacement of the scale from the top of the alcohol column, as shown in figure 5.2.

The parallax error may be random if the eye moves with the top of the alcohol column such that the eye is positioned sometimes above or below the best viewing position shown by figure 5.2(c). However, it is possible to either consistently view the alcohol column with respect to the scale from a position

---

[13] See Simpson (1987) for a discussion of noise due to random thermal motion.

[14] Chapter 4 deals with the statistics of radioactive decay.

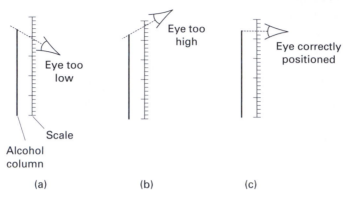

Figure 5.2. Errors introduced by parallax.

below the column (figure 5.2(a)) or above the column (figure 5.2(b)). In such situations the parallax error would be systematic, i.e. values obtained would be consistently below or above the true value.

Reaction time error

Clocks and watches are available that routinely measure the time between events to the nearest millisecond or better. If the starting and stopping of a watch is synchronised with events 'by hand' then the error introduced is often considerably larger than the resolution of the watch. Hand timing of events introduces errors which are typically 200 ms to 300 ms and therefore time intervals of the order of seconds are not really suited to hand timing. Though the timing error *may* be random, it is possible that the experimenter consistently starts or stops the watch prematurely or belatedly so that the systematic component of the total error is larger than the random component.

These examples indicate that it is not always evident how a particular error should be categorised, since the way an instrument is used has an important bearing on whether an error should be regarded as random or systematic.

5.6.2 Systematic error

Systematic errors, like random errors, cause measured values to differ from the true value. In contrast to random errors, which cause measured values to be scattered both above and below the true value, systematic errors introduce a bias so that measured values are consistently above or consistently below the true value. Random errors 'betray' themselves by apparent variability in measured values but systematic errors leave no such trace and this makes them very difficult to detect (or, put another way, very easy to overlook). Experienced experimenters are always on the look-out for sources of systematic error and devote time to identifying and correcting for, as far as possible, such errors. It is not uncommon for systematic errors to be larger than random errors and this is why we should consider carefully sources of systematic error and where possible devise methods by which those errors might be revealed. For example, inconsistency in values obtained for the same quantity which have been determined by two quite different methods might suggest the existence of sources of systematic error influencing values obtained using one or both methods.

5.6.3 Calibration errors and specifications

Measuring instruments are sources of systematic error. An expensive instrument such as a micro-voltmeter is often supplied with a calibration certificate which details the performance of the instrument as established by comparison with another, more accurate, instrument.[15] Less expensive voltmeters are supplied with a maker's specification which indicates the limits of accuracy of an instrument.

Checking that an instrument is behaving 'sensibly' is important before data are gathered, as all values obtained using the instrument may be viewed with some suspicion until there is satisfactory evidence that the values are trustworthy. Gross errors are usually easy to identify. For example, if the voltage between the terminals of a 9 V battery is measured with a voltmeter and the voltmeter indicates 15 V then it is likely that something is amiss. It is possible that the meter has malfunctioned and requires attention. Smaller discrepancies may be more difficult to detect. There may not be the time or facilities to fully assess the calibration of an instrument. A useful check that can be done quite quickly is to replace an instrument with another and assess how well the two instruments 'agree'. Though this approach is no substitute for proper instrument calibration, it can reveal inconsistencies which must be investigated further.

5.6.4 Offset and gain errors

Offset errors are common systematic errors and, in some cases, may be reduced. If the error remains fixed over time then the measured value can be

---

[15] Eventually the calibration should be traceable back to a standard held by a national measurement laboratory.

corrected by an amount so as to compensate for the offset error. As an example, consider a micrometer which is capable of measuring lengths in the range 0 to 25.40 mm. The resolution of the instrument (i.e. the separation of the smallest scale division) is 0.01 mm. Suppose that when the jaws of a particular micrometer are brought into contact, the scale of the micrometer indicates 0.01 mm. We conclude that any measurement made with this instrument will produce a value which is too large by an amount 0.01 mm. To minimise the systematic error we would subtract 0.01 mm from any value obtained using the instrument. If we do this then we are making the assumption that the offset error will remain fixed and that may or may not be a reasonable assumption. Also, we must acknowledge that any correction we apply will not be known exactly, so we cannot hope to completely eliminate the offset error.

Another important characteristic of a measuring system is its gain. For example, a system used to measure light intensity might consist of a light detector[16] and an amplifier. Focussing for a moment upon the amplifier, its purpose is to increase the small voltage generated by the light detector. If the gain of the amplifier decreases (for example, due to a temperature rise in the amplifier) then the output voltage of the amplifier for a given input voltage, will similarly decrease. This will cause a systematic error in the measurement of the light intensity. Let us consider this in more detail. For a particular instrumentation amplifier[17] the gain is fixed using an external 'gain setting' resistor, R_g. The voltage gain, G, is given by

$$G = 1 + \frac{40000 \, \Omega}{R_g}. \tag{5.7}$$

Note that R_g is temperature dependent and so G must also be temperature dependent. For many applications a small change in G with temperature will be unimportant. However, for precision applications, the systematic variation of G with temperature must be included in the assessment of the uncertainty in values obtained using the amplifier.[18]

---

[16] A typical light detector would be a photodiode.

[17] Instrumentation amplifiers are used in precision applications when small signals, often from sensors, are amplified. Usual voltage gains for such amplifiers are in the range 10 to 1000. In the example discussed here we consider the INA101HP instrumentation amplifier manufactured by Texas Instruments.

[18] There are other parameters within the amplifier that are influenced by temperature variations, such as the voltage offset of the device. Here we focus solely on the influence of temperature on the gain setting resistor and ignore other sources of error.

5.6.5 Loading errors

Some instruments affect the quantities they are designed to measure. Take as an example a Pitot tube used to measure the speed of a fluid.[19] By introducing the tube into the stream there is a slight effect on the flow of the fluid. Similarly, an ammeter measuring the electrical current in a circuit influences the value obtained because the measurement process requires that a reference resistor be placed in the path of the current. Placing a reference resistor into the circuit reduces the current slightly. The error introduced by the instrument in these circumstances is referred to as a *loading* error.

Another example of a loading error can be seen when the temperature of a body is measured using a thermometer. As the thermometer is brought into contact with the body, heat is transferred between the body and the thermometer unless both are at the same temperature. If the temperature of the thermometer is less than the body it is contact with, then the thermometer will act to lower the temperature of the body. The effect of this type of loading can be reduced by choosing a thermometer with a thermal mass much less than the thermal mass of the body whose temperature is to be measured.

Though it is often difficult to determine the effect of loading errors, there are situations in which the loading error can be quantified, especially when electrical measurements are made. Consider the situation in which a voltage across a resistor is measured using a voltmeter. Figure 5.3 shows the basic set up.

With the values of resistance shown in figure 5.3, we would expect a little more than half the voltage between the terminals of the battery to appear across

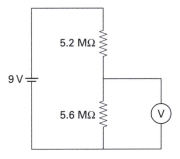

Figure 5.3. Voltmeter used to measure the voltage across a 5.6 MΩ resistor.

---

[19] See Khazan (1994) for a discussion of transducers used to measure fluid flow.

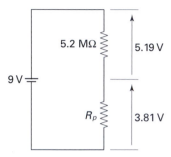

Figure 5.4. Circuit with resistor and voltmeter replaced by an equivalent resistor, R_p.

the 5.6 MΩ resistor. Calculation predicts that a 3½ digit voltmeter should indicate 4.67 V when placed across the 5.6 MΩ. In practice, when a particular voltmeter is used, it is found to indicate a value of 3.81 V. While the discrepancy could be due to the two resistors, which may have resistances far from their nominal values, the actual cause of the 'discrepancy' is the voltmeter. Voltmeters in general have a large, but not infinite, internal resistance, R_{int}. As R_{int} is in parallel with the 5.6 MΩ resistor, some of the current that would have gone through the 5.6 MΩ resistor is diverted through the voltmeter. The loading effect of the voltmeter causes the voltage across the 5.6 MΩ resistor to be reduced in comparison to the value it would take if the voltmeter were removed. We can calculate the value of the internal resistance of the voltmeter by first redrawing figure 5.3 to include the loading effect of the voltmeter. This is shown in figure 5.4 where R_p is the effective resistance of the 5.6 MΩ in parallel with the internal resistance of the voltmeter.

If the voltage across R_p is 3.81 V, then the voltage across the 5.2 MΩ resistor must be (9 – 3.81) V = 5.19 V. The current through the 5.2 MΩ resistor and R_p must be the same as they are electrically configured in series, so that

$$\frac{5.19\,\text{V}}{5.2\,\text{M}\Omega} = \frac{3.81\,\text{V}}{R_p}.$$

It follows that, $R_p = \dfrac{3.81\,\text{V} \times 5.2\,\text{M}\Omega}{5.19\,\text{V}} = 3.82\,\text{M}\Omega.$

The effective resistance, R_p, for two resistors, R and R_{int} connected in parallel is given by

$$\frac{1}{R_p} = \frac{1}{R} + \frac{1}{R_{int}}. \tag{5.8}$$

If $R_p = 3.82\,\text{M}\Omega$ and $R = 5.6\,\text{M}\Omega$, then using equation 5.8 we obtain $R_{int} = 12.0\,\text{M}\Omega$.

When R_{int} is comparable in size to (or smaller than) R, then the loading effect is considerable. To reduce the loading effect, R_{int} should be much greater than R.

Exercise A

An experimenter replaces the voltmeter in figure 5.3 by another with a higher internal resistance, R_{int}. What must be the minimum value for R_{int} if the experimenter requires the voltage indicated by the meter differ by no more than 1% from the voltage that would exist across the 5.6 MΩ in the absence of the meter? (In this problem consider only the effects on the measured values due to loading.)

Quantifying the loading error can be difficult and in many cases we rely on experience to suggest whether a loading effect can regarded as negligible.

5.6.6 Dynamic effects

In situations in which calibration, offset, gain and loading errors have been minimised or accounted for, there is still another source of error that can cause measured values to be above or below the true value and that error is due to the temporal response of the measuring system. Static, or at least slowly varying, quantities are easy to 'track' with a measuring system. However, in situations in which a quantity varies rapidly with time, the response of the measuring system influences the values obtained. Elements within a measuring system which may be time dependent include:

- the transducer used to sense the change in a physical quantity;
- the electronics used in the signal conditioning of the output from the transducer;
- the measuring device such as a voltmeter or analogue to digital converter used to measure the signal after conditioning has occurred;
- the recording component of the system, which could the experimenter or an automatic data logging system.

When dealing with the temporal response of a measuring system as a whole, it is useful to categorise systems as zero, first or second order. The order of the response relates to whether the relationship between the input and output of the system can be adequately described by a zero, first or second order ordinary differential equation. For example, a first order system may be described by the equation

$$a_1 \frac{dy}{dt} + a_0 y = bx, \tag{5.9}$$

where a_1, a_0 are constants which depend on the characteristics of the measurement system, b is a proportionality constant, y is the output signal from the measurement system and x is the input signal. Both x and y are functions of time. We will consider briefly zero and first order systems.[20]

5.6.7 Zero order system

An 'ideal' measuring system is zero order, as the output of the system follows the input of the system at all times. There is no settling time with a zero order system and the output is in phase with the input. Figure 5.5 shows the response of a zero order measuring system to a sudden or 'step' change at the input to the system at time $t = t^*$.

Though figure 5.5 shows the response of a zero order measuring system to a step input, the system would respond similarly to any variation of the input signal, i.e. the shape of the output waveform would mimic that of the input waveform. In practice, a zero order measuring system is an ideal never found as no system truly responds 'instantly' to large step changes at the input to the system.

5.6.8 First order system

While a zero order measuring system is often a good approximation to a 'real' system in situations where measured quantities vary slowly with time, we

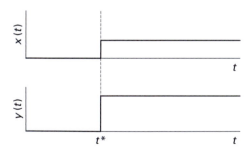

Figure 5.5. Input $x(t)$ to, and output $y(t)$ from, a zero order measuring system.

[20] Bentley (2004) considers second order measurement systems.

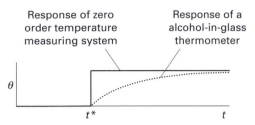

Figure 5.6. Response of a first order measuring system to a step change at the input of the system.

must accept that all measuring systems have an inherent time dependence. The time dependence can be illustrated by considering an alcohol-in-glass thermometer used to measure the temperature of a hot water bath. If the thermometer is immersed in the water at $t = t^*$ then figure 5.6 shows the response of the thermometer as well as the response that would have occurred had the thermometer responded instantly to the change in temperature.

The shape of the temperature, θ, versus time, t, graph in figure 5.6 arises because heat takes a finite time to be transferred from the water to the thermometer and only when that transference has ceased will both the thermometer and the water be at the same temperature. By examining the equation which describes the time dependence of the temperature indicated by the thermometer we are able explain the detailed form of the temperature variation with time and reveal what factors affect the response time of the thermometer. If the final temperature indicated by the thermometer is θ_f and the temperature indicated at any instant is θ, then the differential equation representing the system can be written[21]

$$\tau \frac{d\theta}{dt} + \theta = \theta_f, \tag{5.10}$$

where τ is the time constant of the system. Equation 5.10 is an example of an equation which describes the response of a first order system. A detailed analysis shows that if we can approximate the bulb of the thermometer to a sphere filled with liquid, then

$$\tau = \frac{\rho r c}{3h}, \tag{5.11}$$

[21] See Doebelin (2003) for more detail on first and higher order instruments.

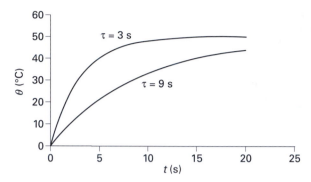

Figure 5.7. Variation of temperature with time as indicated by two thermometers with different time constants.

where ρ the density of the liquid in the thermometer, c is the specific heat of the liquid in the thermometer, h is the heat transfer coefficient for heat passing across the wall of the bulb of the thermometer and r is the radius of the thermometer bulb.

Solving equation 5.10 gives

$$\theta = \theta_f \left(1 - e^{-t/\tau} \right). \tag{5.12}$$

Figure 5.7 shows a graphical representation of equation 5.12 for $\theta_f = 50\,^{\circ}\mathrm{C}$ and two time constants, $\tau = 3$ s and $\tau = 9$ s.

Equation 5.12 and figure 5.7 indicate that, for the time response of a first order system to be improved, i.e. approach that of a first order system, τ, should be as small as possible. In the case of a alcohol-in-glass thermometer we have little control over the quantities appearing in equation 5.11 that affect the value of τ. However, if we have a number of liquid-in-glass thermometers to choose from, then for a given size of thermometer, we would expect those containing low density liquid with a small specific heat to have the smaller value of τ.

Example 1

A particular thermometer has a time constant of 5 s. If the thermometer is immersed in hot water, how much time must elapse before the value indicated by the thermometer is within 5% of the final value?

ANSWER

We begin by rearranging equation 5.12 to give

$$\frac{\theta}{\theta_f} = \left(1 - e^{-t/\tau}\right). \tag{5.13}$$

If the temperature indicated by the thermometer is within 5% of the final value then $\frac{\theta}{\theta_f} = 0.95 = \left(1 - e^{-t/\tau}\right)$. It follows that

$$e^{-t/\tau} = 0.05 \qquad \text{or} \qquad \frac{-t}{\tau} = \ln(0.05) = -2.996,$$

therefore

$$t = 2.996\tau = 2.996 \times 5 \text{ s} = 14.98 \text{ s}.$$

For a step change in temperature, the thermometer will approach within 5% of the final value after the time elapsed from the step change is greater than or equal to about 3τ. This can be generalised to any first order measuring system, i.e. after a step change at the input of the system, a time interval of at least 3τ must elapse before the output of the system is within 5% of the final value.

Exercise B

An experimenter has determined the time constant of a first order measuring system to be 2.5 s. If there is a step change at the input of this system at $t = 0$, how long must the experimenter wait before the value indicated by the system is within 0.1% of the final value (assume random errors are negligible).

To this point we have focussed on identifying sources of error, but have not attempted to quantify the effect of the errors in measurement. 'Quantifying the effect of errors in measurement' brings us to detailed consideration of uncertainty in measurement.

5.7 Uncertainty in measurement

We cannot determine the true value of a measurand, but through a consideration of the scatter of values obtained by repeat measurements (as well as other information) we are able to determine an interval within which the true value lies with a known probability. This is often referred to as the *coverage interval*.

In section 3.8.1 we considered confidence intervals. By using values obtained through repeat measurements we were able to define a confidence interval which contained the mean of the population from which the values were drawn with a known probability. When considering the factors that influence measurements more generally we discover that we need to go beyond analysing values obtained through repeat measurements to include other sources of error that affect uncertainty, such as instrument calibration. In such situations it is not strictly correct to apply the term 'confidence interval' and instead the term 'coverage interval' is preferred. From this point on, unless stated otherwise, we will adopt the term 'coverage interval' when speaking about an interval that contains the true value with a known probability.

If Y represents the true value of the measurand and y represents the *best estimate* of the true value, then the coverage interval may be expressed as

$$y - U \leq Y \leq y + U, \tag{5.14}$$

where U is the *uncertainty*[22] in y. The coverage interval $y - U$ to $y + U$ contains Y with a known probability. A more compact way to write the coverage interval is,

$$Y = y \pm U \tag{5.15}$$

What, exactly, is U and how do we calculate it?

At this point we acknowledge the 'Guide to the Expression of Uncertainty in Measurement' (GUM) which contains guidelines which have been adopted worldwide for calculating uncertainty in science, engineering and other fields in which measurement plays a pivotal role.

Before going into specific details of how uncertainty may be calculated and expressed, it is worth noting the following advice contained in the GUM.[23]

> *Although* [the *GUM] provides a framework for assessing uncertainty, it cannot substitute for critical thinking, intellectual honesty, and professional skill. The evaluation of uncertainty is neither a routine task nor a purely mathematical one; it depends on the detailed knowledge of the nature of the measurand and the measurement. The quality and utility of the uncertainty quoted for the result of a measurement therefore ultimately depend on the understanding, critical analysis, and integrity of those who contribute to the assignment of its value.*

The GUM guidelines are intended to be applicable to measurements carried out for a wide range of purposes including,

> *conducting basic research, and applied research and development, in science and engineering.*

The GUM, section 3.4.8.

A broad aim of the GUM is to encourage those communicating the results of measurements to supply full details on how uncertainties are calculated, permitting the comparison of values obtained through measurement by workers around the world.

---

[22] More correctly, U is the *expanded uncertainty*, which we discuss in section 5.9.

[23] See the GUM, p. 8.

A formal definition of uncertainty in measurement was developed for use in the GUM. It is this:

> uncertainty of measurement is a parameter, associated with the result of a measurement, that characterizes the dispersion of the values that could reasonably be attributed to the measurand.
>
> The GUM, section 2.2.3.

We now turn our attention to how the contributions to the uncertainty in measurement can be categorised so that all sources of uncertainty can be accounted for.

5.7.1 Type A and Type B evaluations of uncertainty

The GUM recommends that uncertainties be divided into two categories, referred to as Type A and Type B based on the way each is evaluated.

Type A evaluations of uncertainty apply statistical methods to values obtained from a series of measurements.

Type B evaluations of uncertainty *do not* apply statistical methods to values obtained from a series of measurements. Instead, a Type B evaluation uses other information such as the knowledge of an instrument's resolution or known tolerance limits. A Type B evaluation could also be based on information derived from data books, calibration certificates or 'historical records' of the variability in a quantity. It is a recognition that equipment, facilities, personnel or time is not available to determine the uncertainty of every measurand by a Type A evaluation that make Type B evaluations of uncertainty necessary.

The uncertainty in the value of a measurand due to a systematic error may be determined by either a Type A or Type B evaluation of uncertainty. Therefore it is not possible here (or in general) to draw an equivalence between systematic error and a Type B evaluation of uncertainty. Similarly the uncertainty in the value of a measurand due to a random error may be determined by either a Type A or Type B evaluation of uncertainty.

A worker, say, in the USA, might measure the density of a fluid (under specified conditions, for example at a fixed temperature) then employ statistical methods to evaluate the uncertainty in the density of the fluid. To the US worker, this would be a Type A evaluation of uncertainty. Others, say, in the UK, working with the same fluid (under the same conditions) may not have the time or facilities to measure its density nor establish the uncertainty in the density. They might choose to use the US worker's value for the density and the uncertainty in the density in their own work. For the UK workers who did not obtain the uncertainty by applying statistical methods to a series of measurements, the uncertainty would be categorised as a Type B evaluation. We can see that what was a Type A evaluation of uncertainty for one worker, becomes a Type B evaluation for another.

5.7.2   Evaluating uncertainty: Type A evaluation of uncertainty

In many cases the best estimate of a quantity is simply the mean of values obtained through repeat measurements of that quantity. The standard error of the mean[24] calculated using the data obtained through repeat measurements is a measure of how good the mean is as an estimate of the true value.

We now introduce a new term; the standard error of the mean now becomes the *standard uncertainty* in the best estimate of the true value obtained by a Type A evaluation.

Example 2

As part of an experiment to determine the circular cross-sectional area of a cylinder, the diameter of the cylinder is measured at positions along its length chosen at random. The values obtained for the diameter appear in table 5.4.

Use the data in table 5.4 to calculate:

(i) the best estimate of the true value for the diameter of the cylinder;
(ii) the standard uncertainty in the best estimate of the true value.

ANSWER

(i) The best estimate of the true value of the diameter is the mean of the values in table 5.4. That mean is 1.24875 mm.

(ii) We concentrate solely on the uncertainty found using the values in table 5.4 (and ignore, for example, any contribution to the overall uncertainty due to the resolution of the instrument used to measure the diameter of the cylinder).

First, use equation 1.16 to calculate the standard deviation of the values in table 5.4. This gives $s = 0.02997$ mm.

Next, calculate the standard error of the mean using equation 3.21, i.e.

$$s_{\bar{x}} = \frac{s}{\sqrt{n}} = \frac{0.02997}{\sqrt{8}} = 0.0106 \text{ mm}.$$

In this example, the standard error of the mean *is* the standard uncertainty, so the standard uncertainty in the best estimate of the diameter is 0.0106 mm.

Table 5.4. *Values of the diameter of a cylinder.*

d (mm)	1.25	1.27	1.25	1.29	1.26	1.26	1.21	1.20

---

[24] See section 3.8.1.

5.7.3 Uncertainty and significant figures

When the determination of uncertainties requires several calculations to be carried out, it is prudent not to prematurely round numbers that emerge at intermediate stages of the calculations. Doing so risks the rounding process itself being a source of error and hence the cause of an unnecessary increase in the uncertainty. Nevertheless, once the calculations are complete, it is usual to round the uncertainty to two significant figures.[25] In this text we will round uncertainties to two significant figures unless stated otherwise.[26]

Once the uncertainty has been expressed to two significant figures, then the best estimate of the true value can be given to the same of decimal places as the standard uncertainty.

Example 3

Suppose the best estimate of a magnetic field is determined as 52.2577 μT, with a standard uncertainty of 0.0672 μT. First we round the standard uncertainty to two significant figures, i.e. 0.067 μT. Next, we round the best estimate of the magnetic field to three decimal places, consistent with the standard uncertainty, so that the best estimate is expressed as 52.258 μT.

Exercise C

Re-write the best estimates and the standard uncertainties of the following quantities to the appropriate number of figures.

  (i) Best estimate of the density of an alloy = 6152.8 kg/m^3 with a standard uncertainty of 25.82 kg/m^3.
 (ii) Best estimate of a spring constant = 33.257 N/m with a standard uncertainty of 1.36 N/m.
(iii) Best estimate of the viscosity of castor oil at room temperature = 0.9855 N·s·m^{-2} with a standard uncertainty of 0.0553 N·s·m^{-2}.

5.7.4 Standard deviation, standard error and standard uncertainty

The terms 'standard deviation' and 'standard error' are employed in data analysis and statistics texts. To express the interval containing the true value

---

[25] See the GUM, section 7.2.6.

[26] It is sometimes helpful when comparing answers to exercises given in this text to give the uncertainties to more than two significant figures.

of a measurand, the GUM introduces the term, *standard uncertainty*,[27] usually represented by the symbol, u.

The introduction of the term *standard uncertainty* deserves some explanation and justification.

The standard deviation, s, given by equation 1.16, is a statistic commonly used to quantify scatter or variability in data. A statistic widely used to quantify the variability in the *mean* of samples drawn from a population, is the *standard error of the mean*, as given by equation 3.21. Both the terms *standard deviation* and *standard error* should be reserved for the statistical analysis of data obtained through experiment or observation. The term *standard uncertainty* is introduced in recognition that when determining uncertainties there are situations in which the uncertainty in the best estimate of a measurand is not determined solely by the statistical analysis of values obtained through repeat measurement, but requires the use of information from other sources (such as a calibration certificate). We may consider the term *standard uncertainty* as 'more inclusive' than *standard error* as it accounts for all sources of uncertainty, not just those determined by the statistical analysis of a series of measurements.

When the best estimate, x, of a quantity is determined solely using values obtained through repeat measurements, the standard uncertainty in x, often written as $u(x)$, is equal to the standard error of the mean, i.e.

$$u(x) = s_{\bar{x}}. \tag{5.16}$$

Exercise D

Measurements are made of the rate of flow, Q, of a fluid through a narrow tube. Values obtained are shown in table 5.5.

Determine:

(i)   the best estimate of the true value of the flow rate;
(ii) the standard deviation of the flow rate;
(iii) the standard uncertainty in the best estimate of the flow rate;
(iv) the number of degrees of freedom.

Table 5.5. *Values of rate of fluid flow through a narrow tube.*

Q (cm^3/s)	0.257	0.253	0.259	0.250	0.251	0.251	0.257	0.258	0.255	0.252

---

[27] The VIM defines standard uncertainty as 'measurement uncertainty, expressed as a standard deviation'.

5.7.5 Type B evaluation of uncertainty

Where the best estimate of an input quantity is not determined using values obtained through repeat measurements, we cannot use Type A evaluation methods to establish the standard uncertainty. In such a situation we are obliged to use information such as calibration certificates, manufacturers' specifications, data tables and knowledge of instrument resolution for both the best estimate of the input quantity and the uncertainty. An uncertainty evaluated in this manner is referred to as a 'Type B' uncertainty. In such cases care must be taken to ensure that the uncertainty, as expressed in the manufacturers' tables, calibration certificates and so on, is correctly interpreted.

For example, a calibration certificate supplied with a standard resistor may state that 'the nominal resistance at 20 $^\circ$C is R_s = 100.050 Ω, with an uncertainty of 50 mΩ at the two-standard deviation level'. In this example the best estimate of the resistance is 100.050 Ω. The standard uncertainty in this estimate, which we write as $u(R_s)$, is the same as the uncertainty at the 'one-standard deviation level'. Using the information supplied this is (50/2) mΩ, i.e. $u(R_s)$ = 25 mΩ.

The uncertainty quoted in a calibration certificate, or other document, may not be as straightforward to interpret as in this example. Often an uncertainty is quoted without additional information as to whether the uncertainty is a multiple of a standard deviation, or represents an interval with a certain 'level of confidence'.

5.7.6 Type B evaluation: estimate of true value

If a calibration certificate accompanying a standard mass, for example, gives upper and lower limits for the mass, then it is reasonable to assume that the best estimate of the true value of the mass lies midway between the limits. Denoting the upper limit as a_+ and the lower limit as a_-, then the best estimate of the true value is given by

$$\text{best estimate} = \left(\frac{a_+ + a_-}{2}\right). \tag{5.17}$$

If we are given no information regarding which values lying between the limits a_+ and a_- are likely to be more probable, we have little alternative but to assume that every value between those limits is equally probable. As a consequence, the probability distribution representing the variability of values is taken to be rectangular as shown in figure 5.8.

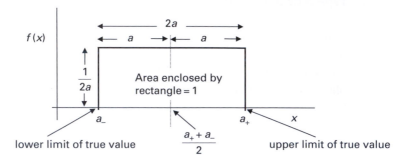

Figure 5.8. Rectangular probability distribution.

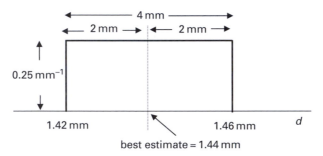

Figure 5.9. Assumed distribution of diameters when all that is known are the upper and lower bounds.

Figure 5.8 shows the situation in which it is assumed that a rectangular distribution is applicable to the distribution of values. The probability density function is $f(x)$, such that $f(x)\Delta x$ is the probability that the true value lies in an interval of width, Δx.

For example, if the maximum diameter of a wire is 1.46 mm and the minimum diameter is 1.42 mm, then the best estimate of the true value of the diameter is $\left(\frac{1.46+1.42}{2}\right) = 1.44$ mm. We can represent the distribution of diameters by redrawing figure 5.8 as shown in figure 5.9.

5.7.7 Type B evaluation: standard uncertainty in best estimate of true value when underlying distribution is rectangular

In order to determine the standard deviation in the best estimate, when the underlying distribution is rectangular, we apply the relationship for the variance, σ^2 (see equation 3.45):

$$\sigma^2 = \int_{-\infty}^{\infty} (x - \mu)^2 f(x)dx, \tag{5.18}$$

where x is a continuous random variable and μ is the population mean. In the case of the rectangular distribution shown in figure 5.8, the best estimate of the variance, s^2, is given by

$$s^2 = \frac{a^2}{3}, \tag{5.19}$$

where a is the half width of the distribution.[28]

The standard deviation (which we take to be the standard uncertainty, u, obtained through a Type B evaluation) of the distribution is given by

$$u = s = \frac{a}{\sqrt{3}}. \tag{5.20}$$

Example 4

If the true value for the diameter of a wire lies within a distribution of half width 0.02 mm, determine the standard uncertainty of the best estimate of the true value of the diameter.

ANSWER
Applying equation 5.20 gives

$$u = \frac{a}{\sqrt{3}} = \frac{0.02}{\sqrt{3}} = 0.012 \text{ mm.}$$

Exercise E

A data book indicates that the density of an alloy definitely lies between the limits $6472\,\text{kg/m}^3$ and $6522\,\text{kg/m}^3$. Using this information, determine

(i) the best estimate for the alloy density;
(ii) the Type B evaluation of standard uncertainty in the best estimate of the alloy density.

[28] A convenient way to show equation 5.19 is to consider a situation in which the expected value for x is equal to 0 (i.e. $\mu = 0$). Given that the probability density function, $f(x)$, is equal to $1/2a$ between $-a$ and $+a$ (and zero outside these limits), integrate equation 5.18 (with $\mu = 0$) between $x = +a$ to $x = -a$.

5.7.8   Type B evaluation: standard uncertainty due to limit of resolution of an instrument

Equation 5.20 is extremely useful when determining the contribution to uncertainty due to the limit of resolution of an instrument. Taking the smallest increment that can be registered by an instrument (in other words its resolution) to be δ, we assume that a values obtained using the instrument will be rounded to the nearest whole number multiple of δ.

For example, suppose a voltmeter has a resolution, δ, = 0.01 V. If the true voltage being measured is 3.1536 V (and we assume that the voltage does not change), the voltmeter would indicate 3.15 V. The half width, a, as given by equation 5.20 is equal to half the resolution, i.e.

$$a = \delta/2 = 0.005 \text{V}.$$

Applying equation 5.20 gives the standard uncertainty in the voltage, due to the limit of resolution of the voltmeter as

$$u = \frac{a}{\sqrt{3}} = \frac{0.005}{\sqrt{3}} = 0.0029 \,\text{V}.$$

Exercise F

The pH indicated by a meter is 7.5. Ignoring any other sources of error, what is the standard uncertainty in the pH as a result of the meter's resolution?

5.7.9   Type B evaluation: standard uncertainty when the underlying distribution is triangular

A rectangular distribution, in which the probability of a true value being close to the limits of the distribution is the same as the probability of the true value being close to the centre, is likely to be 'unphysical' in many situations.

An alternative to the rectangular distribution is to assume that the underlying distribution is triangular, as shown in figure 5.10.

When the distribution is triangular, the best estimate of the true value is given by

$$\text{best estimate} = \left(\frac{a_+ + a_-}{2}\right). \tag{5.21}$$

The standard uncertainty in the best estimate when the distribution is triangular (found by applying equation 5.18) is given by

$$u = s = \frac{a}{\sqrt{6}}. \tag{5.22}$$

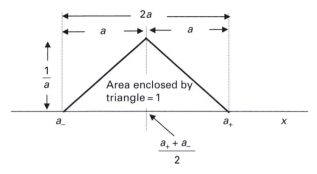

Figure 5.10. Triangular probability distribution.

Exercise G

A calibration certificate indicates that a standard mass lies between 100.052 g and 100.074 g. Assuming that a triangular distribution is appropriate, determine:

   (i) the best estimate of the true value of the weight;
  (ii) the standard uncertainty in the best estimate.

Exercise H

Show that the standard deviation of a triangular distribution of half width a (as shown in figure 5.10) is given by $s = \frac{a}{\sqrt{6}}$.

5.8 Combining uncertainties

Often the best estimates of several quantities need to be entered into an equation in order to establish the best estimate of a measurand. Assuming that the best estimate of each input quantity has some uncertainty, the value emerging from the equation (which is the best estimate of the measurand) must also have some uncertainty. This is often referred to as the 'propagation of uncertainties' and we consider it next.

5.8.1 Measurand as a function of a single input quantity

A measurand, Y, depending on a single quantity, A, may be written formally as

$$Y = f(A), \tag{5.23}$$

where A is the input quantity.

Through experiment we may establish the best estimate of A, which we write as a. This allows us to find y, which is the best estimate of Y. The relationship between y and a is written

$$y = f(a). \tag{5.24}$$

For small changes in a, we apply the approximation[29]

$$\frac{\delta y}{\delta a} \approx \frac{df}{da}, \tag{5.25}$$

where δy is a small change in y, δa is a small change in a and $\frac{df}{da}$ is the derivative of y with respect to a. Equation 5.25 can be rewritten as

$$\delta y \approx \left(\frac{df}{da}\right)\delta a. \tag{5.26}$$

The approximation indicated by equation 5.26 is sufficiently good for our purposes when δa is much less than a (which is usually the case), that we replace $\approx$ by $=$.

Note that δa may represent the error in a, in which case δy is the error in y. As the errors in a and y are unknowable, we recast equation 5.26 in terms of the quantities we *are* able to determine, namely the standard uncertainties in a and y, written as $u(a)$ and $u(y)$ respectively. Equation 5.26 now becomes

$$u(y) = \left|\frac{df}{da}\right|u(a). \tag{5.27}$$

The derivative is evaluated at the best estimate of the input quantity.

The modulus of the derivative is used to prevent $u(y)$ from assuming a negative value when $\frac{df}{da}$ is negative (uncertainties always have a positive sign). The derivative, $\frac{df}{da}$, is sometimes referred to as a 'sensitivity coefficient'.

It is conventional to use upper case symbols to represent *true values* of measurands and input quantities, and lower case symbols to represent *best estimates* of measurands and input quantities. This convention sometimes conflicts with other 'common-sense' conventions used when assigning symbols to quantities. For example, volume would be routinely represented by an upper-case V. In general we will adopt symbols most commonly used to represent physical quantities and clarify where necessary whether the symbol in question represents a true value or a best estimate.

---

[29] See Berry, J., Norcliffe, A. and Humble, S. (1989).

Example 5

The best estimate of the radius of a metal sphere is 2.10 mm with a standard uncertainty of 0.15 mm. Determine the best estimate of volume of the sphere and the standard uncertainty in the volume. Write the standard uncertainty to two significant figures.

ANSWER

The volume of a sphere, V, of radius, r, is given by

$$V = \frac{4}{3}\pi r^3. \tag{5.28}$$

Substituting $r = 2.1$ mm into equation 5.28 gives $V = 38.792$ mm^3. In order to find the uncertainty in V, we rewrite equation 5.27 in terms of V and r, i.e.

$$u(V) = \left|\frac{dV}{dr}\right| u(r). \tag{5.29}$$

If $V = \frac{4}{3}\pi r^3$ then $\frac{dV}{dr} = 4\pi r^2 = 55.418$ mm^2.

Using equation 5.29 gives, $u(V) = \left|\frac{dV}{dr}\right| u(r) = 55.418$ mm$^2 \times 0.15$ mm $= 8.312$ mm^3. Rounded to two significant figures, $u(V) = 8.3$ mm^3.

To summarise: the best estimate of the volume is 38.8 mm^3 and the standard uncertainty in the best estimate is 8.3 mm^3. We have applied the convention of expressing the best estimate to the same number of decimal places as the standard uncertainty.

Exercise I

In the following exercises write the standard uncertainties to two significant figures and round the best estimate of the quantities to the appropriate number of decimal places.

(1) An oscilloscope is used to view an electrical signal which varies sinusoidally with time. The best estimate of the period of the signal, T, is 0.552 ms and the standard uncertainty in the best estimate is 0.012 ms. Given that $f = \frac{1}{T}$, calculate the best estimate of the frequency, f, of the signal and the standard uncertainty in the best estimate.

(2) The surface area, A, of a sphere of radius, r, is given by,

$$A = 4\pi r^2.$$

Using the value for r and the standard uncertainty in r given in example 5, calculate the best estimate of A and the standard uncertainty in A.

(3) Assume k is related to n through the relationship

$$k = \ln n.$$

Given that the best estimate of n is 0.256 with a standard uncertainty of 0.013, calculate the best estimate of k and the standard uncertainty in the best estimate.

5.8.2 Measurands which are functions of several input quantities

When a measurand, Y, depends on several quantities, which have true values A, B and C, we write

$$Y = f(A, B, C). \tag{5.30}$$

Equation 5.30 can be rewritten as

$$y = f(a, b, c), \tag{5.31}$$

where y, a, b and c are best estimates of Y, A, B and C respectively.

Errors in a, b and c combine to result in an error in y. The error in y, δy, may be written as[30]

$$\delta y \approx \left(\frac{\partial f}{\partial a}\right)\delta a + \left(\frac{\partial f}{\partial b}\right)\delta b + \left(\frac{\partial f}{\partial c}\right)\delta c. \tag{5.32}$$

The errors δa, δb and δc may be positive or negative. So long as the errors in the input quantities are not correlated, the standard uncertainty in y can be found by bringing the standard uncertainties in a, b and c together.[31]

The *combined variance* $u_c^2(y)$, combines the standard uncertainties in the best estimates a, b and c, written as $u(a)$, $u(b)$ and $u(c)$ respectively, as follows[32]

$$u_c^2(y) = \left(\frac{\partial f}{\partial a}u(a)\right)^2 + \left(\frac{\partial f}{\partial b}u(b)\right)^2 + \left(\frac{\partial f}{\partial c}u(c)\right)^2, \tag{5.33}$$

where $u_c(y)$ is called the *combined standard uncertainty*. The derivatives, $\frac{\partial f}{\partial a}, \frac{\partial f}{\partial b}$ and $\frac{\partial f}{\partial c}$ are the sensitivity coefficients.

Equation 5.33 may also be written

$$u_c^2(y) = u_1^2(y) + u_2^2(y) + u_3^2(y), \tag{5.34}$$

where $u_1^2(y) = \left(\frac{\partial f}{\partial a}u(a)\right)^2$, $u_2^2(y) = \left(\frac{\partial f}{\partial b}u(b)\right)^2$ and $u_3^2(y) = \left(\frac{\partial f}{\partial c}u(c)\right)^2$.

Equation 5.34 may be written succinctly and more generally as

$$u_c^2(y) = \sum_{i=1}^{i=N} u_i^2(y), \tag{5.35}$$

where N is the number of input quantities.

---

[30] For convenience we consider three input quantities. However, this approach can be extended to any number of input quantities.

[31] For a discussion about combining uncertainties when quantities are correlated, see Kirkup and Frenkel (2006), chapter 7.

[32] A derivation of equation 5.33 can be found in appendix 2.

Example 6

The mass, m, of a wire is found to be 2.255 g with a standard uncertainty of 0.032 g. The length, l, of the wire is 0.2365 m with a standard uncertainty of 0.0035 m. The mass per unit length, μ, is given by

$$\mu = \frac{m}{l}. \tag{5.36}$$

Determine the:

(a) best estimate of μ;

(b) standard uncertainty in μ.

ANSWER

(a) Using equation 5.36, $\mu = \frac{m}{l} = \frac{2.255}{0.2365} = 9.535$ g/m.

(b) Writing equation 5.33, in terms of μ, m and l, gives

$$u_c^2(\mu) = \left(\frac{\partial\mu}{\partial m}u(m)\right)^2 + \left(\frac{\partial\mu}{\partial l}u(l)\right)^2. \tag{5.37}$$

From equation 5.36,

$$\frac{\partial\mu}{\partial m} = \frac{1}{l} = \frac{1}{0.2365} = 4.2283\,\text{m}^{-1},$$

and $\dfrac{\partial\mu}{\partial l} = -\dfrac{m}{l^2} = -\dfrac{2.255}{0.2365^2} = -40.317\,\text{g/m}^2.$

Substituting values into equation 5.37 gives,

$$u_c^2(\mu) = (4.2283 \times 0.032)^2 + (-40.317 \times 0.0035)^2$$
$$= 0.01831 + 0.01991 = 0.03822\,(\text{g/m})^2,$$

therefore,

$$u_c(\mu) = 0.1955\,\text{g/m}.$$

Exercise J

(1) Light incident at an angle, i, on a plane glass surface is refracted as it enters the glass. The refractive index of the glass, n, can be calculated by using

$$n = \frac{\sin i}{\sin r},$$

where r is the angle of refraction of the light.

The best estimate of i is $52.0°$ with a standard uncertainty of $1.0°$. The best estimate of r is $32.5°$ with a standard uncertainty of $1.5°$.

Use this information to calculate the best estimate of n. Assuming errors in i and r are not correlated, determine the combined standard uncertainty in n.

Note, for $y = \sin x$, the approximation $\frac{\delta y}{\delta x} \approx \frac{dy}{dx}$ is valid for x expressed in radians.

(2) The focal length, f, of a thin lens is related to the object distance, u, and the image distance, v, by the equation

$$\frac{1}{f} = \frac{1}{u} + \frac{1}{v}.$$

In an experiment the best estimate of u is 125.5 mm with a standard uncertainty of 1.5 mm. The best estimate of v is 628.0 mm with a standard uncertainty of 1.5 mm.

Assuming that the errors in u and v are not correlated,

  (i) calculate the best estimate of the focal length and its standard uncertainty;
  (ii) calculate the best estimate of the linear magnification, m, of the lens and the standard uncertainty in m, given that

$$m = \frac{v}{u}.$$

(3) The electrical resistivity, ρ, of a wire may be determined by using

$$\rho = \frac{RA}{l}, \tag{5.38}$$

where R is the electrical resistance of the wire measured between its ends, l is the length of the wire and A is the cross-sectional area of the wire.

For a given metal wire it is found that the:

best estimate of its length is 1.21 m with a standard uncertainty of 0.012 m;
best estimate of its diameter is 0.24 mm with a standard uncertainty of 0.015 mm;
best estimate of its resistance is 0.52 Ω with a standard uncertainty of 0.022 Ω.
Using this information determine the

  (i)   best estimate of A;
  (ii) standard uncertainty in A;
  (iii) best estimate of ρ;
  (iv) standard uncertainty in ρ.

(4) Figure 5.11 shows a flat belt wrapped around a drum. If the tension in the belt, T_1, is greater than T_2 then the drum will rotate clockwise so long as the friction between the belt and the drum is large enough. The relationship between the coefficient of static friction, μ_s between belt and drum, and the two tensions, T_1 and T_2, can be written[33]

$$\mu_s = \frac{1}{\beta} \ln\left(\frac{T_1}{T_2}\right), \tag{5.39}$$

where β is the angle defined in figure 5.11.

$T_1 = 412$ N with a standard uncertainty of 52 N;
$T_2 = 81$ N with a standard uncertainty of 13 N;

[33] See Blau (2009).

$\beta = 2.15$ rad with a standard uncertainty of 0.15 rad.

Assuming errors in T_1, T_2 and β are uncorrelated, use this information to find the best estimate of μ_s and the combined standard uncertainty in the best estimate of μ_s.

5.8.3   Combining Type A and Type B evaluations of uncertainty

Equation 5.33 may be used when we need to combine uncertainties determined by Type A and Type B evaluations.

Consider the situation in which a digital thermometer with a resolution of 0.1 °C is used to measure the temperature of air in a sealed enclosure.

Table 5.6 contains eight repeat measurements of the air temperature.

The best estimate of the true temperature can be written

$$\theta = X_\theta + Z_\theta, \tag{5.40}$$

where X_θ is the mean of the values of temperature in table 5.6. This is equal to the best estimate of the true temperature in the absence of any systematic errors. Using data in table 5.6,

$X_\theta = 32.148$ °C.

The standard deviation, s, of the value in table 5.6 is found using equation 1.16,

$s = 0.1408$ °C.

Standard uncertainty in X_θ, $u(X_\theta) = s/\sqrt{n} = 0.1408/\sqrt{8}$, i.e.

$u(X_\theta) = 0.0498$ °C.

Z_θ in equation 5.40 is a correction term which takes into account errors due to, for example, calibration error or resolution error. Here we restrict the

Table 5.6. *Values of air temperature inside a sealed container.*

Temperature (°C)	32.3	31.9	32.1	32.3	32.1	32.2	32.2	32.0

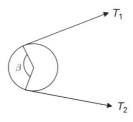

Figure 5.11. Flat belt wrapped around a drum.

Table 5.7. *Values of the mass of a steel ball.*

Mass (g)	7.054	7.053	7.053	7.054	7.053	7.053	7.052	7.054

determination of Z_θ and the standard uncertainty in Z_θ to consideration of the resolution error. The resolution error could be positive or negative. As there is no reason to favour either sign, we take $Z_\theta = 0$, so that the best estimate of the temperature is

$$\theta = (32.148 + 0) \,^{\circ}\text{C} = 32.148 \,^{\circ}\text{C}.$$

The resolution, δ, of the thermometer is $0.1\,^{\circ}\text{C}$, so the half-width $a = \delta/2 = 0.05\,^{\circ}\text{C}$. Using equation 5.20,

$$u(Z_\theta) = \frac{a}{\sqrt{3}} = \frac{0.05}{\sqrt{3}} = 0.0289 \,^{\circ}\text{C}.$$

We now apply equation 5.33, noting that $\frac{\partial \theta}{\partial X_\theta} = \frac{\partial \theta}{\partial Z_\theta} = 1$,

$$u_c^2(\theta) = u^2(X_\theta) + u^2(Z_\theta) = 0.0498^2 + 0.0289^2 = 0.00331\,^{\circ}\text{C}^2.$$

It follows that

$$u_c(\theta) = 0.058 \,^{\circ}\text{C}.$$

To summarise, the best estimate of the true value of the temperature is $32.148\,^{\circ}\text{C}$ with a combined standard uncertainty of $0.058\,^{\circ}\text{C}$.

Exercise K

A steel ball of nominal diameter 6.35 mm is weighed on a top-loading balance which has a resolution of 0.001 g. The balance is zeroed between each measurement, so that you can assume that any zero error is negligible.

The values obtained are shown in table 5.7.

Using the data in table 5.7, determine the best estimate of the true value of the mass of the steel ball, and the combined standard uncertainty in the best estimate.

5.8.4 Do we need to combine all uncertainties?

Strictly, uncertainties in all input quantities need to be established and brought together when determining a combined standard uncertainty. Careful consideration of all sources of uncertainty might lead to a redesign of an experiment to reduce the uncertainty. However there are situations, for example when the variability observed when making repeat measurements of a quantity is far larger than the resolution of the instrument used to make the measurements,

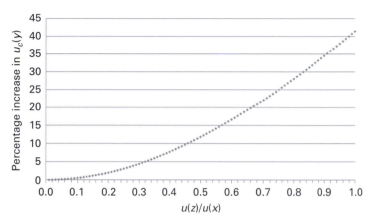

Figure 5.12. Increase in $u_c(y)$ resulting from including $u(z)$ in the calculation represented by equation 5.42.

that the contribution to the combined uncertainty due to instrument resolution can safely be ignored. Consider equation 5.41 which relates the best estimate of a quantity, y, to the best estimate of quantities x and z:

$$y = x + z. \tag{5.41}$$

Assuming uncertainties in x and z are uncorrelated, then the combined standard uncertainty can be calculated by using

$$u_c^2(y) = u^2(x) + u^2(z). \tag{5.42}$$

A consequence of combining uncertainties by summing the squares of uncertainties as given by equation 5.42 is that if one uncertainty is markedly smaller than the other, then the smaller uncertainty may be neglected.

Assuming that $u(z) < u(x)$, at what stage can we neglect $u(z)$ altogether in the calculation of $u_c(y)$? Figure 5.11 shows how much larger is $u_c(y)$, expressed as a percentage when uncertainties in both x and z are included, compared when the uncertainty in z is neglected. The horizontal axis of figure 5.11 represents the ratio $u(z)/u(x)$, which ranges from zero (when $u(z) = 0$) to 1 (when $u(z) = u(x)$).

Inspection of figure 5.12 shows that including $u(z)$ into the calculation of $u_c(y)$, will only increase $u_c(y)$ by about 12% when $u(z)$ is half of $u(x)$. Though there are no international guidelines about when to neglect an uncertainty when calculating the combined standard uncertainty (and we could argue for completeness, all uncertainties should be included), the example given here shows that for $u(z)/u(x) < 0.25$, the effect on the combined standard uncertainty of neglecting $u(z)$ is negligible.

Table 5.8. *Coverage factors in Type A evaluations for ν degrees of freedom when the level of confidence is 95%.*

ν	1	2	3	4	5	6	7	8	9	10
k	12.706	4.303	3.182	2.776	2.571	2.447	2.365	2.306	2.262	2.228

5.9   Expanded uncertainty

The combined standard uncertainty, u_c, is the basic measure of uncertainty in the GUM. However, there are situations in which it is preferable to quote a coverage interval within which there is a specified probability (often 0.95) that the true value will lie. When this probability is expressed as a percentage, it is referred to the 'level of confidence'.

Specifically, we can write

$$y - U \leq Y \leq y + U, \tag{5.43}$$

where Y is the true value of the quantity, y is the best estimate of the true value, and U is the *expanded uncertainty*. If there is a probability of, say, 95% that the true value lies in the interval given by equation 5.43, then it is referred to as the '95% coverage interval'.

Equation 5.43 may be written

$$Y = y \pm U. \tag{5.44}$$

The expanded uncertainty, U, is related to the combined standard uncertainty, u_c, by the equation:

$$U = k u_c, \tag{5.45}$$

where k is the *coverage factor*.

5.9.1   Coverage factor and Type A evaluations of uncertainty

If u_c is determined through a Type A evaluation of uncertainty, then it is usual to assume that the t distribution may be applied when determining the coverage factor, k.

When the level of confidence is 95% (i.e. the probability that the true value lies within a specified interval is 0.95), then table 5.8 gives the coverage factor for various degrees of freedom, $ν$. For values of $ν > 10$, k tends towards a value of close to 2. When $ν > 10$, experimenters often use 2 as the coverage factor when the level of confidence is 95%.

Example 7

In an experiment to calibrate a 1 mL pipette, the mass of water dispensed by the pipette was measured using an electronic balance. Ten measurements were made of the mass of water dispensed by the pipette. The mean mass of water was found to be 0.9567 g, with a standard error in the mean of 0.0035 g. Using this information, determine the:

   (i) best estimate of the true value of the mass of water dispensed by the pipette;
  (ii) standard uncertainty in the mass, u;
 (iii) coverage factor, k, for a 95% level of confidence;
 (iv) expanded uncertainty, U, for a 95% level of confidence;
  (v) 95% coverage interval for the true value of the mass of water.

ANSWER

   (i) The best estimate is taken to be the mean of values obtained through repeat measurements, which in this example is 0.9567 g.
  (ii) When Type A evaluations are carried out, the standard uncertainty is equal to the standard error in the mean, i.e. $u = 0.0035$ g.
 (iii) In this example, the number of degrees of freedom, $v = 10 - 1 = 9$. Using table 5.8, the corresponding value for k is 2.262.
 (iv) $U = ku = 2.262 \times 0.0035 = 0.0079$ g.
  (v) Using equation 5.44, the 95% coverage interval is written (0.9567 ± 0.0079) g.

Exercise L

The calcium content of five samples of powdered mineral is determined. The mean of the five values is (expressed as percent composition) 0.02725 with a standard uncertainty (obtained through a Type A evaluation) of 0.00012. Determine the expanded uncertainty at the 95% level of confidence.

5.9.2 Coverage factor and the Welch–Satterthwaite formula

The combined uncertainty in the best estimate of a quantity will, in general, be determined from Type A and Type B evaluations of uncertainty.

It is possible that the best estimate of each quantity is obtained from a different number of repeat measurements. For example, the resistivity of a metal is obtained from knowledge of the resistance, length and diameter of a wire made from the metal. Suppose five measurements are made of the resistance of the wire, seven of its diameter and four of its length. We are able to determine the combined uncertainty using equation 5.33, but how are we able to calculate the coverage factor for the expanded uncertainty? In this situation we calculate the *effective number of degrees of freedom*, v_{eff}, using the Welch–Satterthwaite formula,[34] which can be expressed:

$$v_{\text{eff}} = \frac{u_c^4(y)}{\displaystyle\sum_{i=1}^{N} \frac{u_i^4(y)}{v_i}}, \tag{5.46}$$

where v_i is number of degrees of freedom associated with each quantity contributing to the calculation of the combined standard uncertainty.

We can calculate $u_c^4(y)$ with the aid of equation 5.34. Expanding the denominator of equation 5.46 gives

$$\sum_{i=1}^{N} \frac{u_i^4(y)}{v_i} = \frac{u_1^4(y)}{v_1} + \frac{u_2^4(y)}{v_2} + \frac{u_3^4(y)}{v_3} + \cdots.$$

If y depends on quantities a and b, and c then

$$u_1^4(y) = \left(\frac{\partial f}{\partial a} u(a)\right)^4, \quad u_2^4(y) = \left(\frac{\partial f}{\partial b} u(b)\right)^4 \text{ and } u_3^4(y) = \left(\frac{\partial f}{\partial c} u(c)\right)^4.$$

To avoid overestimating v_{eff} and thereby underestimating the coverage factor, v_{eff} is rounded <u>down</u> to the nearest whole number.

Example 8

The optical density (o.d.) of a fluid is given by:

$$\text{o.d.} = \varepsilon C l, \tag{5.47}$$

where ε is the extinction coefficient, C is the concentration coefficient of the absorbing species in the liquid, and l is the path length of the light. The best estimate of each quantity, standard uncertainty and number of degrees of freedom are as follows.

$$\varepsilon = 14.9 \text{ L} \cdot \text{mol}^{-1} \cdot \text{mm}^{-1}; u(\varepsilon) = 1.2 \text{ L} \cdot \text{mol}^{-1} \cdot \text{mm}^{-1}; v_{\varepsilon} = 5$$
$$C = 0.042 \text{ mol} \cdot \text{L}^{-1}; u(C) = 0.003 \text{ mol} \cdot \text{L}^{-1}; v_c = 7$$
$$l = 1.42 \text{ mm}; u(l) = 0.21 \text{ mm}; v_l = 8$$

Determine:

(i) the best estimate of the optical density, o.d.
(ii) the combined standard uncertainty;
(iii) the effective number of degrees of freedom, v_{eff};
(iv) the coverage factor, assuming that the level of confidence is 95%.

---

[34] See Dietrich (1991).

ANSWER

(i) o.d. $= \varepsilon Cl = 14.9 \times 0.042 \times 1.42 = 0.889$ (note that optical density has no units).

(ii)

$$u_c^2(y) = \left(\frac{\partial f}{\partial \varepsilon} u(\varepsilon)\right)^2 + \left(\frac{\partial f}{\partial C} u(C)\right)^2 + \left(\frac{\partial f}{\partial l} u(l)\right)^2$$

$$\frac{\partial f}{\partial \varepsilon} = Cl, \frac{\partial f}{\partial C} = \varepsilon l, \frac{\partial f}{\partial l} = \varepsilon C.$$

Now we write

$$u_1^2(y) = \left(\frac{\partial f}{\partial \varepsilon} u(\varepsilon)\right)^2 = (Clu(\varepsilon))^2 = (0.042 \times 1.42 \times 1.2)^2 = 0.0051219,$$

$$u_2^2(y) = \left(\frac{\partial f}{\partial C} u(C)\right)^2 = (\varepsilon lu(C))^2 = (14.9 \times 1.42 \times 0.003)^2 = 0.0040289,$$

$$u_3^2(y) = \left(\frac{\partial f}{\partial l} u(l)\right)^2 = (\varepsilon Cu(l))^2 = (14.2 \times 0.042 \times 0.21)^2 = 0.017271.$$

Using equation 5.34, this gives

$$u_c^2(y) = u_1^2(y) + u_2^2(y) + u_3^2(y) = 0.0051219 + 0.0040289 + 0.017271 = 0.026422.$$

It follows that $u_c = 0.16255$.

Note that $u_1 = 0.071568$, $u_2 = 0.063474$, and $u_3 = 0.131418$.

(iii) To find the effective number of degrees of freedom, v_{eff}, we use equation 5.46, such

that[35] $v_{\text{eff}} = \dfrac{u_c^4(y)}{\sum\limits_{i=1}^{N} \dfrac{u_i^4(y)}{v_i}} = \dfrac{(0.16255)^4}{\dfrac{(0.071568)^4}{5} + \dfrac{(0.063474)^4}{7} + \dfrac{(0.131418)^4}{8}} = 15.57(\text{round to 15}).$

(iv) When the number of degrees of freedom is 15, the coverage factor for a level of confidence of 95% is 2.131 (see table 2 in appendix 1).

5.9.3 v_{eff} and Type B evaluation of uncertainties

What number of degrees of freedom do we assign to a standard uncertainty that is determined by a Type B evaluation based on an assumed probability distribution, such as the triangular or rectangular distribution? A rectangular distribution of half width a, has a standard uncertainty given by

$$u = a/\sqrt{3}. \tag{5.48}$$

The standard uncertainty given by equation 5.48 has no uncertainty, i.e. $\Delta u = 0$.

---

[35] Because we are raising numbers to the fourth power, it is prudent to keep numbers to extra figures in the calculations in order to reduce the effect of rounding errors.

It is possible to show that the number of degrees of freedom can be determined from knowledge of the standard uncertainty, u and standard deviation of the standard uncertainty, Δu (i.e. the uncertainty in the uncertainty!)

If the uncertainty in u is Δu, then the number of degrees of freedom, v, can be written[36]

$$v = \frac{1}{2}\left[\frac{\Delta u}{u}\right]^{-2} = \frac{1}{2}\left[\frac{u}{\Delta u}\right]^{2}. \tag{5.49}$$

If $\Delta u = 0$ as is the case for an assumed rectangular distribution, then using equation 5.49, it follows that $v = \infty$. The fact that $v = \infty$ does not cause a difficulty when determining the effective number of degrees of freedom using equation 5.46, as any term in the equation with an infinite number of degrees of freedom is zero.

Example 9

A 3½ digit voltmeter is used to measure the output voltage of a transistor. The mean of ten repeat measurements of voltage is found to be 5.340 V with a standard deviation of 0.11 V. Given that the voltmeter is operating on the 20 V range and the operating manual accompanying the voltmeter states that the accuracy of the voltmeter is 0.5% of the reading + 1 figure, determine:

(i)   the standard uncertainty in the voltage determined through a Type A evaluation of uncertainty;

(ii) the standard uncertainty in the voltage determined through a Type B evaluation of uncertainty;

(iii) the combined standard uncertainty;

(iv) the effective number of degrees of freedom using the Welch–Satterthwaite formula;

(v)   the coverage factor;

(vi) the expanded uncertainty at the 95% level of confidence.

ANSWER

We begin by recognising that both the variability in the values and the stated accuracy of the instrument will contribute to the combined standard uncertainty in the voltage. We begin by writing the best estimate of the true value of the voltage as

$$V = X_V + Z_V.$$

X_V is the mean of repeat measurements (and so the standard uncertainty in X_V will be determined by a Type A evaluation of uncertainty). Z_V is a correction term to account for errors due, to for example, offsets or instrument accuracy. In this example we will use

---

[36] See the GUM, section E.4.3.

data supplied by the manufacturer to determine the standard uncertainty in Z_V (and so this is a Type B evaluation of uncertainty).

(i) The standard uncertainty in X_V, $u(X_V) = \frac{s}{\sqrt{n}}$. In this example, $s = 0.11$ V and $n = 10$, so $u(X_V) = \frac{0.11V}{\sqrt{10}} = 0.03479$ V.

(ii) We assume that the accuracy of the voltmeter can be represented by a rectangular distribution with half width of 0.5% of the reading + 1 figure, 0.5% of 5.340 V is 0.0267 V, and the least significant figure for a 3½ digit voltmeter on the 20 V range is 0.01 V. The half width of the distribution, $a = 5.340 \times 0.005 + 0.01 = 0.0367$ V.

The standard uncertainty $u(Z_V)$ is found by using equation 5.20, i.e. $u(Z_V) = \frac{a}{\sqrt{3}} = \frac{0.0367}{\sqrt{3}} = 0.0212$V

(iii) The combined standard uncertainty is found using equation 5.33, i.e. $u_c^2(V) = u^2(X_V) + u^2(Z_V) = 0.03479^2 + 0.0212^2 = 0.00166 V^2$.

So $u_c(V) = 0.04073$ V.

(iv) $v_{\text{eff}} = \dfrac{u_c^4(V)}{\sum\limits_{i=1}^{N} \dfrac{u_i^4(V)}{v_i}} = \dfrac{(0.04073)^4}{\dfrac{(0.03479)^4}{9} + \dfrac{(0.0212)^4}{\infty}} = 16.9$ (round to 16).

(v) When the number of degrees of freedom is 16, then coverage factor, k, for a level of confidence of 95% is 2.120.

(vi) The expanded uncertainty, $U = ku_c(V) = 2.120 \times 0.04073 = 0.086$ V (to two significant figures).

To summarise: the best estimate of the true value of the voltage is 5.340 V and the expanded uncertainty at the 95% level of confidence is 0.086 V.

5.10 Relative standard uncertainty

In some situations we may prefer the uncertainty to be expressed as a fraction of the best estimate of the true value of the quantity. This is referred to as the *relative* uncertainty. For example, if the best estimate of a quantity is a with a standard uncertainty of $u(a)$, then the relative standard uncertainty is given by

$$\text{relative standard uncertainty} = \frac{u(a)}{|a|}. \tag{5.50}$$

The modulus of the best estimate is used to avoid the fractional uncertainty assuming a negative value.

Suppose the best estimate of the mass of a body is 45.32 g with a standard uncertainty of 0.15, then the

$$\text{relative standard uncertainty} = \frac{0.15 \text{ g}}{45.32 \text{ g}} = 3.3 \times 10^{-3}.$$

Note that the fractional uncertainty has no units, as any units appearing in the numerator and denominator of equation 5.50 'cancel'.

The uncertainty in a value can be expressed as a percentage of the best estimate of the value by multiplying the relative uncertainty by 100%, so that,

relative standard uncertainty expressed as a percentage

$$= \frac{u(a)}{|a|} \times 100\%. \tag{5.51}$$

Example 10

In a report on density measurements made of sea water, the density, ρ, is stated to be 1.050 g/cm with a standard uncertainty of 0.020 g/cm^3. Express the uncertainty as a relative standard uncertainty.

ANSWER

Using equation 5.49, the relative standard uncertainty $= \frac{u(a)}{|a|} = \frac{0.020\,\mathrm{g/cm^3}}{1.050\,\mathrm{g/cm^3}} = 0.019$.

Expressing the relative standard uncertainty as a percentage, gives $0.019 \times 100\% = 1.9\%$.

Exercise M

The temperature, t, of a furnace is given as $t = 1450\ °C$ with a percentage standard uncertainty of 5%. Express the uncertainty as both a standard uncertainty and a relative standard uncertainty.

5.11 Coping with outliers

When repeat measurements are made of a particular quantity, it is reasonable to anticipate that values will vary due to random errors. However, occasionally there are situations in which most of the values are 'well behaved' in that all but one value appears to be consistent with the others. A value far from the others is usually referred to as an 'outlier'. The question is, what should be done with an outlier? Should it be included in the analysis or should it be ignored? In the next section we deal with the issue of outliers.

5.11.1 Outliers

Identification of outliers is important as they may indicate the existence of a novel effect or that some intermittent occurrence, perhaps due to electrical interference, is adversely influencing the measurement process. Another, perhaps more common, cause of outliers is due to the incorrect recording of the measured value. Trying to identify the cause of an outlier is important, and should precede any effort to eliminate it, as systems that automatically eliminate outliers may exclude the most important data before anyone has had the opportunity to examine the cause of the outlier or its significance. Where possible, it is preferable to repeat the experiment in an effort to establish if an outlier is, in fact, a 'mistake' or an occasional, but regular occurrence.

It is not always possible to repeat an experiment due to time, cost or other reasons and we must do the best we can with the data available. However, caution is always warranted when deciding to omit an outlier based on statistical grounds only, as the process often requires the assumption that the data follow a particular distribution (such as the normal distribution). If the assumption is not valid then removing an outlier may not be justified. Where possible, the actual distribution of data should be compared with the assumed distribution (for example by assessing the normality of data as described in section 3.11).

5.11.2 Chauvenet's criterion

One approach to deciding whether an outlier should be removed from a group of values derived from repeat measurements of a quantity is to begin by asking the following question.

'Given the scatter displayed by the data as a whole, what is the probability of obtaining a value at least as far from the mean as the outlier?'

If this probability is very low then we have good statistical grounds for omitting the outlier from any further analysis. We assume that the scatter of data being considered is well described by the normal distribution.

Let us consider a specific example. Table 5.9 shows mass loss from nine samples of a ceramic when they are heated in an oxygen-free environment.

An inspection of the values in table 5.9 indicates that the value, $x = 6.6$ mg, seems to differ considerably from the others. Should this value be eliminated, and if so on what grounds? To assess the scatter in the values we first determine the mean mass loss, $\bar{x}$, and the standard deviation, s, using all the values in table 5.9. This gives

$\bar{x} = 5.711$ mg and $s = 0.3551$ mg.

Table 5.9. *Mass loss from the ceramic YBa$_2$Cu$_3$O$_7$ heated in an oxygen free environment.*

Mass loss, x, (mg)	5.6	5.4	5.7	5.5	5.6	6.6	5.8	5.7	5.5

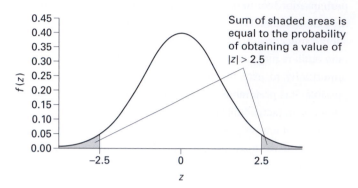

Figure 5.13. Standard normal probability distribution. The sum of the areas in the tails is equal to the probability that a value is more than 2.5 standard deviations from the mean.

The characteristic spread of the values is represented by the standard deviation, s, so it is sensible to express the 'separation' between the mean and the outlier in terms of the standard deviation. To do this we calculate z given by[37]

$$z = \frac{x_{OUT} - \bar{x}}{s}, \tag{5.52}$$

where x_{OUT} is the value of the outlier.

Inserting values from this example gives

$$z = \frac{6.6\,\mathrm{mg} - 5.711\,\mathrm{mg}}{0.3551\,\mathrm{mg}} = 2.50.$$

To find the probability that a value differs from the mean by at least 2.5 standard deviations, we assume that the values are drawn from a normal distribution. Figure 5.13 shows the standard normal distribution.

$P(-\infty \leq z \leq -2.5)$ obtained using table 1 in appendix 1 is equal to 0.00621. As the tails of the distribution are symmetric, we double this value to find the probability represented by the total shaded area in figure 5.13, i.e. the probability that a value lies at least as far from the mean as the outlier is 2×0.00621 = 0.01242. Put another way, if we collect, for say, 100 values, we would expect $100 \times 0.01242 \approx 1$ value to be at least as far from the mean as the outlier. As only nine measurements have been made, the number of values we would expect to be at least as far from the mean as the outlier is $9 \times 0.01242 \approx 0.1$.

We can use the number of values expected to be at least as far from the mean as the outlier as the basis of a criterion upon which to accept or reject an outlier. If the expected number of values at least as far from the mean as the

---

[37] Note we should really use the population standard deviation, σ, in the denominator of equation 5.52. As we cannot know the value of σ, we must replace it by the estimate of its value, s.

outlier is ≤ 0.5 then the outlier should be eliminated and the mean and standard deviation of the remaining values recalculated. This criterion is referred to as Chauvenet's criterion and it should be applied only once to a set of values, i.e. if a value is rejected and the mean and standard deviation are recalculated, the criterion should not be applied to any other value.

We can summarise the steps in applying Chauvenet's criterion as follows.

(i) Calculate the mean, $\bar{x}$, and the standard deviation, s, of the values being considered.

(ii) Identify a possible outlying value, x_{OUT}. Calculate z, where $z = \frac{x_{OUT} - \bar{x}}{s}$.

(iii) Use the normal probability tables to determine the probability, P, that a value will occur at least as far from the mean as x_{OUT}.

(iv) Calculate the number, N, expected to be at least as far from the mean as x_{OUT}, $N = nP$, where n is the number of values and P is the probability that a value is at least as far from the mean as x_{OUT}.

(v) If $N \leq 0.5$, reject the outlier.

(vi) Recalculate $\bar{x}$ and s but do not reapply the criterion to the remaining data.

Example 11

Table 5.10 shows values for the 'hand timing' of the free fall of an object through a vertical distance of 5 m.

Identify a possible outlier then use Chauvenet's criterion to decide if the possible outlier should be rejected.

ANSWER

The mean, $\bar{x}$, and standard deviation, s, of the values in table 5.10 are 1.084 s and 0.09991 s respectively.

The value furthest from the mean is 0.90 s, i.e. $x_{OUT} = 0.90$ s. Calculating z using equation 5.52 gives

$$z = \frac{x_{OUT} - \bar{x}}{s} = \frac{0.90\,\text{s} - 1.084\,\text{s}}{0.09991\text{s}} = -1.84.$$

Using table 1 in appendix 1 gives $P(-\infty \leq z \leq -1.84) = 0.03288$. The total probability that a value lies at least as far from the mean as the possible outlier is $P = 2 \times 0.03288 = 0.06576$. The number of values, N, expected to be at least as far from the mean as the outlier is $N = nP \approx 10 \times 0.066 = 0.66$. As this number is greater than 0.5, do not reject the value, $x = 0.90$ s.

Table 5.10. *Values of time for an object to fall freely through a distance of 5 m.*

Time, x, (s)	1.11	0.90	1.13	1.13	1.16	1.23	0.94	1.08	1.12	1.04

Table 5.11. *Value of time for liquid to flow from a vessel.*

t (s)	215	220	217	235	222

Exercise N

In a fluid flow experiment, the time, t, for a fixed volume of water to flow from a vessel is measured. Table 5.11 shows five values for t, obtained through repeat measurements.

   (i) Calculate the mean and standard deviation of these values.

  (ii) Identify the value in table 5.11 furthest from the mean.

 (iii) Apply Chauvenet's criterion to the outlier to decide whether the outlier should be eliminated.

 (iv) If the outlier *is* eliminated, recalculate the mean and standard deviation of the remaining values.

5.12 Weighted mean

When calculating the mean of values obtained through experiment or observation, the assumption is usually made that all values should be weighted equally. While this is reasonable in most circumstances, there are situations in which weighting should be considered. Weighting recognises that some values have smaller uncertainty than others and so the calculation of the mean is modified to account for this. Specifically, those values that have smaller uncertainty are weighted more heavily than those values that have greater uncertainty.

The standard uncertainty in each mean appears explicitly in the expression for the weighted mean. A weighted mean $\bar{x}_w$, which is found by combining several means, is written[38]

$$\bar{x}_w = \frac{\sum \frac{\bar{x}_i}{u_i^2}}{\sum \frac{1}{u_i^2}}, \tag{5.53}$$

where $\bar{x}_i$ is the ith mean and u_i is the standard uncertainty in the ith mean.

---

[38] See appendix 3 for a derivation of equation 5.53.

Table 5.12. *Thickness of thin film of platinum determined using three techniques.*

Instrument	Mean thickness (nm)	Standard uncertainty in mean (nm)
PF	325.0	7.5
SEM	330.0	5.5
AFM	329.0	4.0

Example 12

The thickness of a platinum film is measured using a profilometer (PF), a scanning electron microscope (SEM) and an atomic force microscope (AFM). Table 5.12 shows the mean of the values obtained by each method and the standard uncertainty in the mean. Calculate the weighted mean thickness of the film.

ANSWER

Using equation 5.46, we have

$$\bar{x}_w = \frac{\sum \frac{\bar{x}_i}{u_i^2}}{\sum \frac{1}{u_i^2}} = \frac{\frac{325\,\text{nm}}{(7.5\,\text{nm})^2} + \frac{330\,\text{nm}}{(5.5\,\text{nm})^2} + \frac{329\,\text{nm}}{(4.0\,\text{nm})^2}}{\frac{1}{(7.5\,\text{nm})^2} + \frac{1}{(5.5\,\text{nm})^2} + \frac{1}{(4.0\,\text{nm})^2}} = \frac{37.25\,\text{nm}}{0.1133} = 328.7\,\text{nm}.$$

Exercise O

The mean time of a ball to fall a fixed distance under gravity is found by three experimenters to be:

1.25 s with a standard uncertainty of 0.21 s;
0.98 s with a standard uncertainty of 0.15 s; and
1.32 s with a standard uncertainty of 0.32 s.

Determine the weighted mean of these times.

5.12.1 Standard uncertainty in the weighted mean

If the standard uncertainty in the means that contributed to the calculation of the weighted mean vary, the standard uncertainty in the weighted mean must take this variation into account. In appendix 4 we show that the weighted standard uncertainty, $u_{\bar{x}_w}$, may be expressed as,

$$u_{\bar{x}_w} = \left(\frac{1}{\sum \frac{1}{u_{\bar{x}_i}^2}} \right)^{1/2}.$$

(5.54)

Example 13

Calculate the standard uncertainty of the weighted mean for the mean thicknesses appearing in table 5.12.

ANSWER

Using equation 5.54,

$$u_{\bar{x}_w} = \left(\frac{1}{\sum \frac{1}{u_{\bar{x}_i}^2}} \right)^{1/2} = \left(\frac{1}{\frac{1}{(7.5\,\text{nm})^2} + \frac{1}{(5.5\,\text{nm})^2} + \frac{1}{(4.0\,\text{nm})^2}} \right)^{1/2} = \left(8.823(\text{nm})^2 \right)^{1/2}$$

$$= 3.0\,\text{nm}.$$

Exercise P

Calculate the standard uncertainty of the weighted mean for the time of fall of a ball using information given in exercise O.

5.12.2 Should means be combined?

Though we have considered how a weighted mean may be calculated, we should pause to ask whether it is reasonable to combine means if those means are considerably different. At the heart of the matter is the assumption, or hypothesis, that the mean of values obtained by whatever method is used would tend to the same value as the number of measurements increases. Suppose, for example, the velocity of sound in air is determined. Method A gives the mean velocity of 338 m/s with a standard uncertainty of 12 m/s and method B gives 344.5 m/s with a standard uncertainty of 6.2 m/s. As the difference between the two means is less than the standard uncertainty of the means it is reasonable to assert the same 'true value' underlies values obtained using both method A and method B. By contrast if the velocity of sound in air determined using method A is 338.2 m/s with a standard uncertainty of 1.2 m/s and by method B the velocity is 344.5 m/s with a standard uncertainty of 1.1 m/s, we should consider carefully before combining the means as the difference between the means is larger than the standard uncertainties. There may be many reasons for inconsistency between the means. As examples, consider the following.

- The conditions in which each experiment was performed are different (for example the ambient temperature might have changed between measurements carried out by method A and method B). In this situation, the true value of the velocity of sound is not the same in each experiment.
- One or more instruments used in method A or B may no longer be operating within specification, i.e. a large instrument based systematic error may account for the difference between values obtained.
- There could be systematic error introduced if hand timing were carried out by different experimenters as part of methods A and B.

In chapter 9 we consider more fully whether two means obtained from data are consistent with those data being drawn from the same population.

5.13 Review

It is hard to overstate the importance of careful measurement in science. Whether an experiment involves establishing the effectiveness of an anti-reflective coating on heat transfer into a car, the effect of temperature on the electrical conductive properties of a ceramic material, or the determination of the energy released during a chemical reaction, the 'values' that emerge from the experiment are subjected to extremely close scrutiny. Careful measurements can reveal an effect that has been overlooked by a less careful worker and lead to an important discovery. But no matter how carefully any experiment is performed, there is always some uncertainty in values established through measurement and so it is important we know how to determine, combine and interpret uncertainties. In this chapter we have considered measurement and some of the factors that affect the precision and accuracy of values obtained through experiment. In particular, we have seen that random and systematic errors introduce uncertainty into any measured value, how uncertainty can be quantified using internationally recognised guidelines. Our focus in the chapter has been on factors that affect repeat measurements of a quantity when the conditions in which the measurements are made are essentially unchanged. An important area we consider next is the study of the relationship between physical variables, and how the relationship can be quantified.

End of chapter problems

(1) The best estimate of the pH of a sample of river water is 5.85. If the standard uncertainty in the pH is 0.22, determine the relative standard uncertainty to two significant figures.

(2) Four repeat measurements are made of the height, h, to which a steel ball rebounds after striking a flat surface. Values obtained are shown in table 5.13.

Use the values in table 5.13 to determine:

(i) the best estimate in the rebound height;

(ii) the standard uncertainty in the best estimate of the rebound height (ignore the influence of the resolution of the instrument used to measure the rebound heights);

(iii) the coverage interval at the 95% level of confidence for the true value of the rebound height.

(iv) Assuming the resolution of the instrument used to measure the rebound height is 1 mm, recalculate the combined standard uncertainty in the best estimate of the rebound height.

Table 5.13. *Rebound heights for a steel ball.*

h (mm)	189	186	183	186

(3) A profilometer is an instrument used to measure the thickness of thin films. Table 5.14 show values of thickness of a film of aluminium on glass obtained using a profilometer. Using the values in table 5.14 find:

(i) the best estimate of the film thickness;

(ii) the standard uncertainty in the best estimate of film thickness;

(iii) the coverage interval at the 99% level of confidence for the film thickness.

Table 5.14. *Thickness of aluminium film determined using a profilometer.*

Thickness (nm)	340	310	300	360	300	360

(4) The relationship between critical angle, θ_c, and the refractive index, n, for light travelling from glass into air is

$$n = \frac{1}{\sin \theta_c}.$$

(5.55)

For a particular glass, the best estimate of θ_c is $43.2°$ with a $u(\theta_c) = 1.2°$. Use this information to determine n and $u(n)$.

(5) The porosity, r, of a material can be determined by using

$$r = \left(1 - \frac{\rho_s}{\rho_t}\right),$$

(5.56)

where ρ_s is density of a specimen of the material as determined by experiment, and ρ_t is the theoretical density of the material.

For a ceramic, $\rho_t = 6.35$ g/cm^3 (assume negligible uncertainty). The best estimate of ρ_s is 3.90 g/cm^3 with $u(\rho_s) = 0.20$ g/cm^3. Use this information to determine the best estimate of r and the standard uncertainty in the best estimate.

(6) X-rays incident on a thin film are no longer reflected when the angle of incidence exceeds a critical angle, θ_c, given by,

$$\theta_c = K\lambda\sqrt{\rho}, \tag{5.57}$$

where K is a constant for the film, λ is the wavelength of the incident X-rays and ρ is density of the film.

For a gallium arsenide film, $K = 4.44 \times 10^5$ rad·m$^{\frac{1}{2}}$·kg$^{-\frac{1}{2}}$. Given that $\lambda = 1.5406 \times 10^{-10}$ m, determine the best estimate for θ_c and the standard uncertainty in θ_c, given that the best estimate of $\rho = 6.52$ g/cm^3 with a standard uncertainty of 0.24 g/cm^3.

(7) The electrical resistance, R, of a wire at temperature, θ, is written

$$R = R_0(1+\alpha\theta+\beta\theta^2), \tag{5.58}$$

where α and β are constants with negligible uncertainty given by, $\alpha = 3.91 \times 10^{-3}\,°C^{-1}$ and $\beta = 1.73 \times 10^{-7}\,°C^{-2}$. R_0 is the resistance at 0 °C. Given that $R_0 = 100.0\ \Omega$ and that it too has negligible uncertainty, calculate the best estimate of R and the standard uncertainty in the best estimate when the best estimate of $\theta = 65.2$ °C and the standard uncertainty in the best estimate is 1.5 °C.

(8) When a ball rebounds from a flat surface, the coefficient of restitution, c, for the collision is given by

$$c = \sqrt{\frac{h_2}{h_1}}, \tag{5.59}$$

where h_1 is the initial height of the ball and h_2 is its rebound height. Given that $h_1 = 52.2$ cm with a standard uncertainty of 0.2 cm, and $h_2 = 23.5$ cm with a standard uncertainty of 0.5 cm, determine c and the combined standard uncertainty in c assuming the errors in h_1 and h_2 are not correlated.

(9) The rate of heat flow, H, across a material of thickness, l, can be written

$$H = -kA\left(\frac{t_2 - t_1}{l}\right), \tag{5.60}$$

where k is the thermal conductivity of the material, A is its cross-sectional area, t_2 and t_1 are the temperatures at opposite surfaces of the material ($t_2 > t_1$). Given that $k = 109$ W/(m·°C), $A = 0.0015$ m^2 with a standard uncertainty of 0.0001 m^2, $t_2 = 62.5$ °C with a standard uncertainty of 0.5 °C, $t_1 = 56.5$ °C with a standard uncertainty of 0.5 °C and $l = 0.15$ m with a standard uncertainty of

0.01 m, determine the best estimate of H and the combined standard uncertainty in the best estimate, assuming the errors in the quantities are not correlated.

(10) The velocity of sound may be determined by measuring the length of a resonating column of air. In an experiment, the length, l, of the resonating column is measured 10 times. The values obtained are shown in table 5.15.

  (i) Determine the mean of the values in table 5.15.
  (ii) Identify the value which lies furthest from the mean.
  (iii) Use Chauvenet's criterion to decide whether the outlier identified in part (ii) should be removed.
  (iv) If the outlier *is* removed, recalculate the mean.

Table 5.15. *Values for the length of a resonating column of air.*

l (cm)	49.2	48.6	47.8	48.5	42.7	47.7	49.0	48.8	48.3	47.7

(11) A 3½ digit voltmeter is used to measure the output of a pressure transducer. Values of voltage obtained are shown in table 5.16. The accuracy of the meter used to measure the voltage is 0.5% of the reading + 1 digit.
  Using this information determine:

  (i) the best estimate of the true voltage from the pressure transducer;
  (ii) the combined standard uncertainty in the best estimate;
  (iii) the 95% coverage interval for the true value of the voltage.

Table 5.16. *Voltages from a pressure transducer determined using a 3½ digit voltmeter.*

Voltage (mV)	167	167	165	163	168

(12) In the process of calibrating a 1 mL bulb pipette, the mass of water dispensed by the pipette was measured using an electronic balance. The process was repeated with the same pipette until 10 values were obtained. Values obtained are shown in table 5.17.
  Using the values in table 5.17 determine:

  (i) the best estimate of the mass of water dispensed;
  (ii) the standard uncertainty in the best estimate of the mass;
  (iii) the 95% coverage interval for the true value of the mass dispensed by the pipette.

Table 5.17. *Mass of water dispensed by a 1 mL bulb pipette.*

Mass (g)	0.9737	0.9752	0.9825	0.9569	0.9516	0.9617	0.9684	0.9585	0.9558	0.9718

(13) The reducing agent sodium thiosulphate is added to an analyte during a titration. The amount of sodium thiosulphate required to bring the reaction to its end point for six repeat titrations is shown in table 5.18.

    (i) Calculate the mean of the values in table 5.18.
    (ii) Calculate the standard deviation, s, of the values in table 5.18.
    (iii) Identify a possible outlier and apply Chauvenet's criterion to determine whether the outlier should be removed.
    (iv) If the outlier *is* removed, calculate the new mean and standard deviation.

Table 5.18. *Volume of sodium thiosulphate added to bring a reaction to end point.*

Volume (mL)	33.2	33.4	33.4	33.9	33.3	33.1

(14) The density of an aqueous solution is determined by two experimenters. Experimenter A obtains a density of 1.052 g/cm^3 for the solution with a standard uncertainty of 0.023 g/cm^3. Experimenter B obtains a density of 1.068 g/cm^3 for the solution with a standard uncertainty of 0.011 g/cm^3. Determine the weighted mean of these densities and the standard uncertainty in the weighted mean.

Chapter 6

Least squares I

6.1 Introduction

Establishing and understanding the relationship between quantities are principal goals in the physical sciences. As examples, we might be keen to know how the:

- size of a crystal depends on the growth time of the crystal;
- output intensity of a light emitting diode varies with the emission wavelength;
- amount of light absorbed by a chemical species depends on the species concentration;
- electrical power supplied by a solar cell varies with optical power incident on the cell;
- viscosity of an engine oil depends upon the temperature of the oil;
- rate of flow of a fluid through a hollow tube depends on the internal diameter of the tube.

Once an experiment is complete and the data presented in the form of an x–y graph, an examination of the data assists in answering important qualitative questions such as: Is there evidence of a clear trend in the data? If so, is that trend linear, and do any of the data conflict with the general trend? A qualitative analysis often suggests which quantitative methods of analysis to apply.

There are many situations in the physical sciences in which prior knowledge or experience suggests a relationship between measured quantities. Perhaps we are already aware of an equation which predicts how one quantity depends on another. Our goal in this situation might be to discover how well the equation can be made to 'fit' the data.

As an example, in an experiment to study the variation of the viscosity of engine oil between room temperature and 100 °C, we observe that the viscosity of the oil decreases with increasing temperature, but we would like to know more.

- What is the quantitative relationship between viscosity and temperature? Does viscosity decrease linearly with increasing temperature, or in some other way?
- Can we find an equation that represents the variation of viscosity with temperature? Perhaps this would allow us to predict values of viscosity at any given temperature, i.e. permit interpolation between measured temperatures.
- Is it possible to use the data to test a theory which predicts how viscosity should depend on temperature?
- Can a useful parameter be estimated from viscosity–temperature data, such as the change in viscosity for a temperature rise of 1 °C?

A powerful and widely used method for establishing a quantitative relationship between quantities is that of *least squares*, also known as *regression*, and it is this method that we will focus upon in this chapter.

The method of least squares is computationally demanding, especially if there are many data to be considered. We will use Excel to assist with least squares analysis at various points in this chapter after the basic principles have been considered.

6.2 The equation of a straight line

It is quite common for there to be a linear relationship between two quantities measured in an experiment. The data obtained through an experiment devised to study the relationship between two quantities are routinely represented as points on an x–y graph. By fitting a straight line to the data, a quantitative expression may be found that relates the two quantities.

What we would really like to do is find the equation that describes the *best* line that passes through (or at least close to) the data. We assume that the equation of the 'best line' is the closest we can come to finding the relationship between the quantities with the data available.

Figure 6.1 shows a straight line drawn on an x–y graph. The y value, y_i, of any point on the line is related to the corresponding x value, x_i, by the equation[1]

---

[1] It is common to find the equation of a straight line written in other ways, such as $y_i = mx_i + c$, $y_i = mx_i + b$ or $y_i = b_0 + b_1 x_i$.

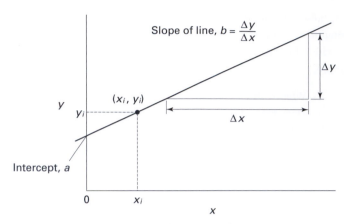

Figure 6.1. Straight line relationship between x and y.

$$y_i = a + bx_i,$$

where a is the intercept and b is the slope of the line.

If all the x–y data gathered in an experiment were to lie along a straight line, there would be no difficulty in determining a and b and our discussion would end here. We would simply use a rule to draw a line through the points. Where the line intersects the y axis at $x = 0$ gives a, and b is found by dividing Δy by Δx, as indicated in figure 6.1.

In situations in which 'real' data are considered, even if the underlying relationship between x and y is linear, it is highly unlikely that all the points will lie on a straight line, as sources of error act to scatter the data. So how do we find the best line through the points in circumstances where data are scattered?

6.2.1 The 'best' straight line through x–y data

When dealing with experimental data, we commonly plot the quantity that we are able to control on the x axis. This quantity is often referred to as the *independent* (or the *predictor*) variable. A quantity that changes in response to changes in the independent variable is referred to as the *dependent* (or the *response*) variable, and is plotted on the y axis.

As an example, consider an experiment in which the velocity of sound in air is measured at different temperatures. Here temperature is the independent variable and velocity is the dependent variable. Table 6.1 shows temperature–velocity data for sound travelling through dry air.

Table 6.1. *Temperature–velocity data for sound travelling through dry air.*

Temperature, θ (°C)	Velocity, v (m/s) (± 5 m/s)
−13	322
0	335
9	337
20	346
33	352
50	365

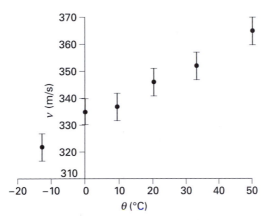

Figure 6.2. An *x–y* graph showing velocity of sound versus temperature.

The data in table 6.1 are plotted in figure 6.2. Error bars are attached to each point to indicate the standard uncertainty in the values of velocity.

From an inspection of figure 6.2, it appears reasonable to propose that there is a linear relationship between velocity, v, and temperature, θ. This relationship can be written as

$$v = A + B\theta, \tag{6.1}$$

where A and B are constants.

Finding the best straight line through the points *means* finding best estimates for A and B, as it is these two parameters that describe the relationship between v and θ. We can draw a line 'by eye' through the points using a transparent plastic rule and from that line estimate A and B. The difficulty with this approach is that, when data are scattered, it is difficult to find the position of the best line through the points.

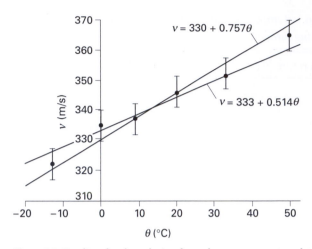

Figure 6.3. Two lines fitted to velocity of sound versus temperature data. The equation describing each line is shown.

Figure 6.3 shows two attempts at drawing a line through the velocity versus temperature data along with the equation that describes each line. Can either of the two lines be regarded as the best line through the points? If the answer is no, then how *do* we find the best line?

The guesswork associated with drawing a line by eye through data can be eliminated by applying the technique of least squares.

6.2.2   Unweighted least squares

To find the best line through x–y data, we need to decide upon a numerical measure of the 'goodness of fit' of the line to the data. One approach is to take that measure to be the 'sum of squares of residuals', which we will discuss for the case where there is a linear relationship between x and y. The least squares method discussed in this section rests on the assumptions described in table 6.2.

Figure 6.4 shows a line drawn through the x–y data. The vertical distances from each point to the line are labelled Δy_1, Δy_2, Δy_3, etc., and are referred to as the *residuals* (or *deviations*). A residual is defined as the difference between the observed y value and the y value on the line for the same x value. Referring to the ith observed y value as y_i, and the ith predicted value found using the equation of the line as $\hat{y}_i$, the residual, Δy_i, is written

Table 6.2. *Assumptions upon which the unweighted least squares method is based.*

Assumption	Comment
The y values are influenced by random errors only.	Any measurement is affected by errors introduced by such sources as noise and instrument resolution (see chapter 5). Systematic errors cannot be accounted for using least squares.
The y values measured at a particular value of x have a normal distribution.	If errors in measured values are normally distributed, then measured values will exhibit the characteristics of the normal distribution (see chapter 3).
The standard deviation of y values at all values of x are the same.	If this assumption is true then every point on an x–y graph is as reliable as any other and, in using the least squares method to fit a line to data, one point must not be favoured or 'weighted' more than any other point. This is referred to as *unweighted* least squares.
There are no errors in the x values.	This is the most troublesome assumption. In most circumstances the x quantity is controlled in an experiment. It is therefore likely to be less influenced by random errors than the y quantity. But this is not always true. We will consider one situation in which the assumption is not valid, i.e. where the errors in the x values are much larger than the errors in y values.

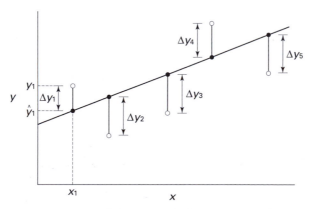

Figure 6.4. An x–y graph showing residuals.

$$\Delta y_i = y_i - \hat{y}_i, \tag{6.2}$$

where

$$\hat{y}_i = a + bx_i. \tag{6.3}$$

We propose that the underlying or 'true' relationship between x and y is given by

$$y = \alpha + \beta x, \tag{6.4}$$

where α is the true intercept and β is the true slope. We can never know α and β exactly, but best estimates of α and β, written as a and b respectively, are obtained by applying the 'Principle of Maximum Likelihood' as discussed in appendix 3. The outcome of applying this principle is that the best line is given by values of a and b which minimise the *sum of the square of the residuals, SSR*. *SSR* is given by

$$SSR = \sum (y_i - \hat{y}_i)^2, \tag{6.5}$$

where $\hat{y}_i$ is given by equation 6.3.

In principle, values of a and b that minimise *SSR* could be found by trial and error, or by a systematic numerical search using a computer. However, when a straight line is fitted to data, an equation for the best line can be found analytically.

If the assumptions in table 6.2 are valid, a and b are given by[2]

$$a = \frac{\sum x_i^2 \sum y_i - \sum x_i \sum x_i y_i}{n \sum x_i^2 - (\sum x_i)^2} \tag{6.6}$$

and

$$b = \frac{n \sum x_i y_i - \sum x_i \sum y_i}{n \sum x_i^2 - (\sum x_i)^2}, \tag{6.7}$$

where n is the number of data points and each summation is carried out between $i = 1$ and $i = n$.

Example 1

Table 6.3 contains x-y data which are shown plotted in figure 6.5.

Using the data in table 6.3:

  (i) find the value for the slope and intercept of the best line through the points;
  (ii) draw the line of best fit through the points;
 (iii) calculate the sum of squares of residuals, *SSR*.

---

[2] For derivations of equations 6.6 and 6.7 see section A3.2 in appendix 3.

ANSWER

To calculate a and b we need the sums appearing in equations 6.6 and 6.7, namely $\sum x_i$, $\sum y_i$, $\sum x_i y_i$, and $\sum x_i^2$. Many pocket calculators are able to calculate these quantities[3] (in fact, some pocket calculators are able to perform unweighted least squares fitting to give a and b directly).

A word of caution here: as there are many steps in the calculations of a and b, it is advisable *not* to round numbers in the intermediate calculations,[4] as rounding can significantly influence the values of a and b.

(i) Using the data in table 6.3 we find that $\sum x_i = 30$, $\sum y_i = 284$, $\sum x_i y_i = 1840$ and $\sum x_i^2 = 220$. Substituting these values into equations 6.6 and 6.7 (and noting that the number of points, $n = 5$) gives

$$a = \frac{220 \times 284 - 30 \times 1840}{5 \times 220 - (30)^2} = 36.4,$$

$$b = \frac{5 \times 1840 - 30 \times 284}{5 \times 220 - (30)^2} = 3.4.$$

(ii) The line of best fit through the data in table 6.3 is shown in figure 6.6.
(iii) The squares of residuals and their sum, *SSR*, are shown in table 6.4.

Exercise A

Use least squares to fit a straight line to the velocity versus temperature data in table 6.1. Calculate the intercept, a, the slope, b, of the line and the sum of squares of the residuals, *SSR*.

Table 6.3. *Linearly related x–y data.*

x	y
2	43
4	49
6	59
8	63
10	70

---

[3] If there are many data, a spreadsheet is preferred to a pocket calculator for calculating the sums.

[4] This problem is acute when a calculation involves subtracting two numbers that are almost equal, as can happen in the denominator of equations 6.6 and 6.7. Premature rounding can cause the denominator to tend to zero, resulting in very large (and very likely incorrect) values for a and b.

Table 6.4. *Calculation of sum of squares of residuals*

x_i	y_i	$\hat{y}_i = 36.4 + 3.4x_i$	$(y_i - \hat{y}_i)^2$
2	43	43.2	0.04
4	49	50.0	1.00
6	59	56.8	4.84
8	63	63.6	0.36
10	70	70.4	0.16
			$SSR = 6.4$

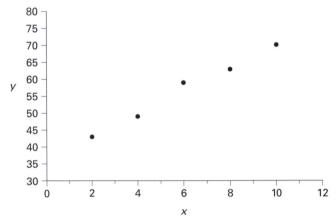

Figure 6.5. Linearly related *x*–*y* data.

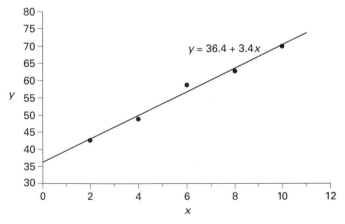

Figure 6.6. Line of best fit through data given in table 6.3.

6.2.3 Trendline in Excel

Excel may be used to add the best straight line to an x–y graph using the Add Trendline option. Excel uses equations 6.6 and 6.7 to determine the intercept and slope. Section 2.7.1 describes plotting an x–y graph using Excel and how a line of best fit can be added using Trendline.

Advantages of using Excel's Trendline include the following.

(i) It requires that data be plotted first so that we are encouraged to consider whether it is *reasonable* to draw a straight line through the points.

(ii) The best line is added automatically to the graph.

(iii) The equation of the best line can be displayed if required.

(iv) Excel is capable of drawing the best line through points for relationships between x and y other than linear, such as a logarithmic or power relationship.

(v) If data are changed then the points on the graph, the line of best and the equation of the best line are updated and displayed immediately, as the graph and the line of best fit are linked 'dynamically' to the data.

Though excellent for determining and displaying the best line through points, Trendline does not,

(i) give the standard uncertainties in a and b;

(ii) allow 'weighted fitting'. This is required when there is evidence to suggest that some x–y values are more reliable than others. In this situation the best line should be 'forced' to pass close to the more reliable points. Weighted fitting is considered in section 6.10;

(iii) plot residuals. Residuals are extremely helpful for assessing whether it is appropriate to fit a straight line to data in the first place. Residuals are considered in section 6.7.

Despite the usefulness of the Trendline option in Excel, there are often situations in which we need to extract more from the data than just a and b. In particular, the uncertainties in a and b, expressed as the standard uncertainties in a and b, are important as they are required for the calculation of the coverage intervals for the intercept and slope. We consider uncertainties in intercept and slope next.

6.2.4 Uncertainty in a and b

One of the basic assumptions made when fitting a line to data using least squares is that the dependent variable is subject to random error. It is reasonable to expect therefore that a and b are themselves influenced by the errors in the dependent variable. A preferred way of expressing the uncertainties in a and b is in terms of their respective standard uncertainties,[5] as this permits us to calculate coverage intervals for a and b. Standard uncertainties in a and b may be found using the ideas of propagation of uncertainties discussed in chapter 5. Provided the uncertainty in each y value is the same,[6] the standard uncertainties in a and b are given by s_a and s_b, where[7]

$$s_a = \frac{s\left(\sum x_i^2\right)^{1/2}}{\left[n\sum x_i^2 - \left(\sum x_i\right)^2\right]^{1/2}},$$ (6.8)

$$s_b = \frac{sn^{1/2}}{\left[n\sum x_i^2 - \left(\sum x_i\right)^2\right]^{1/2}},$$ (6.9)

and where s is the standard deviation of the observed y values about the fitted line. The calculation of s is similar to the calculation of the estimate of the population standard deviation, s, of univariate data given by equation 1.16.

Here, s is given by

$$s = \left[\frac{1}{n-2}\sum (y_i - \hat{y}_i)^2\right]^{1/2}.$$ (6.10)

Example 2

Absorbance–concentration data shown in table 6.5 were obtained during an experiment in which standard silver solutions were analysed by flame atomic absorption spectrometry.

Assuming that absorbance is linearly related to concentration of the silver solution, use unweighted least squares to find the intercept and slope of the best line through the points and the standard uncertainties in these quantities.

ANSWER

Regarding the concentration as the independent variable, x, and the absorbance as the dependent variable, y, we write

$y = a + bx.$

Using the data in table 6.5 we find, $\sum x_i = 105$ (ng/mL), $\sum y_i = 2.667$, $\sum x_i y_i = 57.585$ (ng/mL) and $\sum x_i^2 = 2275$ (ng/mL)^2.

---

[5] Many data analysis and statistics texts refer to the standard error or standard deviation of parameter estimates. Consistent with the vocabulary introduced in chapter 5, we will use the term *standard uncertainty* when describing the uncertainty in parameter estimates.

[6] If repeat measurements are made of y at a particular value of x, the standard deviation in the y values remains the same regardless of the value of x chosen.

[7] See appendix 4 for derivations of equations 6.8 and 6.9.

Table 6.5. *Data obtained in a flame atomic absorption experiment.*

Concentration (ng/mL)	Absorbance (arbitrary[8] units)
0	0.002
5	0.131
10	0.255
15	0.392
20	0.500
25	0.622
30	0.765

Using equations 6.6 and 6.7,

$$a = \frac{2275(\text{ng/mL})^2 \times 2.667 - 105(\text{ng/mL}) \times 57.585(\text{ng/mL})}{7 \times 2275 \,(\text{ng/mL})^2 - (105(\text{ng/mL}))^2} = 4.286 \times 10^{-3},$$

$$b = \frac{7 \times 57.585 \,(\text{ng/mL}) - 105(\text{ng/mL}) \times 2.667}{7 \times 2275 \,(\text{ng/mL})^2 - (105(\text{ng/mL}))^2} = 2.511 \times 10^{-2}\text{mL/ng}.$$

In this example units have been included explicitly[9] in the calculation of a and b to emphasise that, in most situations in the physical sciences, we deal with quantities that have units and these must be carried through to the 'final answers' for a and b.

In order to use equations 6.8 and 6.9, first calculate s as given by equation 6.10. Table 6.6 has been constructed to assist in the calculation of s.

Summing the values in the last column of table 6.6 gives

$$\sum (y_i - \hat{y}_i)^2 = 3.2686 \times 10^{-4}.$$

Using equation 6.10,

$$s = \left[\frac{1}{(7-2)} \times 3.2686 \times 10^{-4} \right]^{1/2} = 0.008085.$$

Now use equations 6.8 and 6.9 to find s_a and s_b:

---

[8] When values indicate the relative size of a quantity (in this case absorbance), but are not expressed in a recognised unit of measure, such as that based on the SI system, we speak of that the unit of measure on that scale as being 'arbitrary'. Arbitrary units are often found on the y-axis of a graph.

[9] In other examples in this chapter, units do not appear (for the sake of brevity) in the intermediate calculations of a and b or s_a and s_b.

Table 6.6. *Calculation of squares of residuals.*

x_i (ng/mL)	y_i (arbitrary units)	$\hat{y}_i = 4.286 \times 10^{-3} + 2.511 \times 10^{-2} x_i$	$(y_i - \hat{y}_i)^2$
0	0.002	0.004286	5.2245×10^{-6}
5	0.131	0.129857	1.3061×10^{-6}
10	0.255	0.255429	1.8367×10^{-7}
15	0.392	0.381000	1.2100×10^{-4}
20	0.500	0.506571	4.3184×10^{-5}
25	0.622	0.632143	1.0288×10^{-4}
30	0.765	0.757714	5.3082×10^{-5}

$$s_a = \frac{s\left(\sum x_i^2\right)^{1/2}}{\left[n\sum x_i^2 - \left(\sum x_i\right)^2\right]^{1/2}} = \frac{0.008085 \times (2275)^{1/2}}{\left[7 \times 2275 - (105)^2\right]^{1/2}} = 5.509 \times 10^{-3},$$

$$s_b = \frac{sn^{1/2}}{\left[n\sum x_i^2 - \left(\sum x_i\right)^2\right]^{1/2}} = \frac{0.008085 \times (7)^{1/2}}{\left[7 \times 2275 - (105)^2\right]^{1/2}} = 3.056 \times 10^{-4}\text{mL/ng}.$$

Using the properties of the normal distribution,[10] we can say that the true value for the intercept has a probability of approximately 0.7 of lying between $(4.3-5.5) \times 10^{-3}$, and $(4.3 + 5.5) \times 10^{-3}$, i.e. the 70% coverage interval for α is between approximately -1.2×10^{-3} and 9.8×10^{-3}. Similarly, the 70% coverage interval for β is between 2.480×10^{-2} mL/ng and 2.542×10^{-2} mL/ng.

We must admit to a misdemeanour in applying the normal distribution here: In this example we are dealing with a small number of values (seven only), so we should use the t distribution rather than the normal distribution when calculating coverage intervals for α and β. We discuss this further in section 6.2.6.

Exercise B

Calculate s_a and s_b for the data in table 6.3.

6.2.5 Least squares, intermediate calculations and significant figures

In this text we adopt the convention that uncertainties are rounded to two significant figures. a and b are then presented to the number of figures consistent with the magnitude of the uncertainties. Where a pocket calculator or a

---

[10] See section 3.5.

spreadsheet has been used to determine a and b, all intermediate calculations are held to the full internal precision of the calculator, rounding only occurring in the presentation of the final parameter estimates.

6.2.6   Coverage intervals for α and β

We are able to determine the coverage interval for the true intercept, α, and that of the true slope, β, of a straight line fitted to data. We adopt the 'rule of thumb' that whenever there are fewer than 30 data points, it is appropriate to use the t distribution rather than the normal distribution when quoting coverage intervals. The 95% coverage interval for α is written as

$$\alpha = a \pm t_{95\%,v} s_a, \tag{6.11}$$

where v is the number of degrees of freedom, and $t_{95\%,v}$ is the critical t value corresponding to the 95% coverage level.[11]

When fitting a straight line, the experimental data are used to calculate a and b. By using the data to estimate two parameters, the number of degrees of freedom is reduced by two, so that v is given by[12]

$$v = n - 2,$$

where n is the number of data.

Similarly, the 95% coverage interval for β is written

$$\beta = b \pm t_{95\%,v} s_b, \tag{6.12}$$

where a and b are calculated using equations 6.6 and 6.7 respectively, and s_a and s_b are calculated using equations 6.8 and 6.9.

When the X% coverage interval is required, $t_{95\%,v}$ is replaced in equations 6.11 and 6.12 by $t_{X\%,v}$. Table 2 in appendix 1 gives the values of t for various confidence levels, X%, and degrees of freedom, v.

Example 3

Using information given in example 2, calculate the 95% coverage interval for α and β.

ANSWER

Relevant information from example 2:

---

[11] It would be quite acceptable to substitute the coverage factor, k, for $t_{95\%,v}$.

[12] See Devore (2007) for more information on degrees of freedom.

$a = 4.286 \times 10^{-3}, s_a = 5.5 \times 10^{-3},$

$b = 2.511 \times 10^{-2} \text{mL/ng}, s_b = 3.1 \times 10^{-4} \text{mL/ng},$

$v = n - 2 = 7 - 2 = 5.$

Using table 2 in appendix 1, $t_{95\%,5} = 2.571$.

From equation 6.11, the 95% coverage interval for α is

$4.286 \times 10^{-3} \pm 2.571 \times 5.5 \times 10^{-3},$

i.e. $\alpha = (4 \pm 14) \times 10^{-3}$.

Using equation 6.12,

$\beta = (2.511 \times 10^{-2} \pm 2.571 \times 3.1 \times 10^{-4}) \,\text{mL/ng},$

i.e.

$\beta = (2.511 \pm 0.080) \times 10^{-2} \,\text{mL/ng}.$

Exercise C

(1) Calculate the 99% coverage intervals for α and β in example 2.
(2) The data in table 6.7 were obtained in an experiment to study the variation of the electrical resistance, R, with temperature, θ, of a tungsten wire.

Assuming the relationship between R and θ can be written $R = A + B\theta$, calculate:

(i) the values of the intercept, a, and slope, b, of the best line through the resistance–temperature data;
(ii) the standard uncertainties in a and b;
(iii) the 95% coverage intervals for A and B.

6.3 Excel's LINEST() function

Calculating a and b and their respective standard uncertainties by using equations 6.6 to 6.10 is tedious, especially if there are many x–y values to consider. It is possible to use a spreadsheet to perform the calculations and this lessens the effort considerably. An even quicker method of calculating a and b is to use the LINEST()

Table 6.7. *Variation of resistance of a tungsten wire with temperature.*

θ (°C)	1	4	10	19	23	28	34	40	47	60	66	78	82
R (Ω)	10.2	10.3	10.7	11.0	11.2	11.4	11.8	12.2	12.5	12.8	13.2	13.5	13.6

function in Excel. This function estimates the parameters of the line of best fit and returns those estimates into an array of cells in the spreadsheet. The LINEST() function is versatile and we will consider it again in chapter 7. For the moment we use it to calculate a, b, s_a and s_b. The syntax of the function is as follows.

LINEST(y values, x values, constant, statistics)

y values:	Give range of cells containing the y values.
x values:	Give range of cells containing the x values.
constant:	Set to True to fit the equation $y = a + bx$ to the data. If we want to force the line through the origin (0, 0) (i.e. fit the equation $y = bx$ to data) we set the constant to False. See problem 11 at the end of the chapter for a brief consideration of fitting the equation $y = bx$ to data.
statistics:	Set to True to calculate the standard uncertainties[13] in a and b.

Example 4

Consider the x–y data shown in sheet 6.1. Use the LINEST() function to find the parameters a, b, s_a and s_b for the best line through these data.

ANSWER

Data are entered into columns A and B of the Excel spreadsheet as shown in sheet 6.1. We require Excel to return a, b, s_a and s_b. To do this:

(1) Move the cursor to cell D3. With the left hand mouse button held down, pull down and across to cell E4. Release the mouse button. Values returned by the LINEST() function will appear in the four cells, D3 to E4.
(2) Type =**LINEST(B2:B9,A2:A9,TRUE,TRUE)**.
(3) Hold down the Crtl and Shift keys together, then press the Enter key.

Figure 6.7 shows part of the screen as it appears after the Enter key has been pressed. Labels have been added to the figure to identify a, b, s_a and s_b.

Sheet 6.1. *The x–y data for example 4.*

	A	B
1	x	y
2	1	−2.3
3	2	−8.3
4	3	−11.8
5	4	−15.7

---

[13] In fact, the LINEST() function is able to return other statistics, but we will focus on the standard uncertainties for the moment.

Sheet 6.1. (*cont.*)

	A	B
6	5	−20.8
7	6	−25.3
8	7	−34.2
9	8	−37.2

Exercise D

Consider the *x–y* data in table 6.8.

   (i) Use the LINEST() function to determine the intercept and slope of the best line through the *x–y* data, and the standard uncertainty in intercept and slope.

 (ii) Plot the data on an *x–y* graph and show the line of best fit.

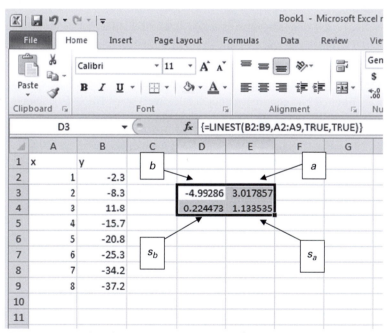

Figure 6.7. Screen shot of Excel showing use of LINEST() function.

Table 6.8. *The x–y data for exercise D.*

x	0.1	0.2	0.3	0.4	0.5	0.6	0.7	0.8	0.9
y	3.7	7.7	8.5	12.1	13.5	15.5	15.3	15.0	18.7

6.4 Using the line of best fit

There are several reasons why a straight line might be fitted to data. These include the following.

(i) A visual inspection of data presented as an x-y graph appears to indicate that there is a linear relationship between the dependent and independent variables. A line through the points helps to confirm this and may reveal anomalies such as points deviating systematically from the line.

(ii) A model or theory predicts a linear relationship between independent and the dependent variables. The best line through the points can provide strong evidence to support or refute the theory.[14]

(iii) The best line may be required for interpolation or extrapolation purposes. That is, a value of y may be found for a particular value of x, x_0, where x_0 lies within the range of x values used in the determination of the line of best fit (interpolation) or outside the range (extrapolation). The equation of the best line may also be used to determine an x value for a given y value.

We consider the use of the line of best fit next, but will return to the important matter of comparing models and data in later sections of this chapter, and again in chapter 7.

6.4.1 Comparing a 'physical' equation to $y = a + bx$

In some situations, a and b derived from fitting a straight line to experimental data can be directly related to a physical constant or parameter. In fact, it may be that something is known of the 'theoretical' relationship between variables prior to performing the experiment and that the main purpose of the experiment is to establish the value of a particular physical constant featuring in an equation derived from theory.

Take, as an example, a thermal expansion experiment in which the length of a rod is measured at different temperatures. Studies on the expansion of solids indicates that the relationship between the length of the rod, l (in metres), at temperature, θ (in degrees Celsius) may be written as

---

[14] In section 6.7 we will see how the scatter of residuals can provide convincing evidence for the appropriateness, or otherwise, of fitting a straight line to data.

$$l = l_o(1 + \alpha\theta), \tag{6.13}$$

where l_o is the length of the rod at $0\,°C$, and α is the temperature coefficient of expansion.

The right hand side of equation 6.13 can be expanded to give

$$l = l_o + l_o\alpha\theta. \tag{6.14}$$

Equation 6.14 is of the form $y = a + bx$, where $l = y$ and $\theta = x$. Comparing equation 6.14 to the equation of a straight line reveals,

$$a = l_o \text{ and } b = l_o\alpha.$$

It follows that

$$\alpha = \frac{b}{a}, \tag{6.15}$$

i.e. the ratio, b/a, gives the temperature coefficient of expansion of the material being studied. It would be usual to compare the value of α obtained through analysis of the length–temperature data with values reported by other experimenters who have studied the same, or similar, materials.

Exercise E

The pressure, P, at the bottom of a water tank is related to the depth of water, h, in the tank by the equation

$$P = \rho g h + P_A, \tag{6.16}$$

where P_A is the atmospheric pressure, g is the acceleration due to gravity, and ρ is the density of the water.

In an experiment, P is measured as the depth of water, h, in the tank increases. Let us assume equation 6.16 is valid.

(i) What would you choose to plot on each axis of an x-y graph in order to obtain a straight line?

(ii) How are the intercept and slope of that line related to ρ, g, and P_A in equation 6.16?

6.4.1.1 Uncertainties in parameters which are functions of *a* and *b*

The thermal expansion example discussed in the previous section brings up an important question: How may we establish the uncertainty in the temperature coefficient of expansion, α, given the uncertainties in a and b? α is a function of a and b, so at first sight it seems reasonable to apply the usual relationship for propagation of uncertainties (as derived in appendix 2) to find the standard uncertainty s_α in α using

$$s_\alpha^2 = \left(\frac{\partial \alpha}{\partial a}\right)^2 s_a^2 + \left(\frac{\partial \alpha}{\partial b}\right)^2 s_b^2. \tag{6.17}$$

However, equation 6.17 is only valid if there is no correlation between the errors in a and b. It turns out that the errors in a and b *are* correlated[15] and so we must consider this matter in more detail.

The line of best fit (for an unweighted fit) always passes through the point $(\bar{x}, \bar{y})$, where[16]

$$\bar{x} = \frac{\sum x_i}{n} \quad \text{and} \quad \bar{y} = \frac{\sum y_i}{n},$$

Using these relationships, we can write the intercept, a, as

$$a = \bar{y} - b\bar{x}. \tag{6.18}$$

Any equation which is a function of a and b (such as equation 6.15) can now be written as a function of $\bar{y}$ and b, by replacing a by $\bar{y} - b\bar{x}$. The advantage in this is that the errors in $\bar{y}$ and b are *not* correlated. This allows us to apply equation 5.33 to find the uncertainty in a parameter whose calculation involves both a and b.

As an example, consider the thermal expansion problem described in section 6.4.1. The temperature coefficient of expansion, α, is given by equation 6.15. Replacing a in equation 6.15 by $\bar{y} - b\bar{x}$, gives

$$\alpha = \frac{b}{\bar{y} - b\bar{x}}. \tag{6.19}$$

The standard uncertainty, s_α, is now written

$$s_\alpha^2 = \left(\frac{\partial \alpha}{\partial \bar{y}}\right)^2 s_{\bar{y}}^2 + \left(\frac{\partial \alpha}{\partial b}\right)^2 s_b^2. \tag{6.20}$$

After determining the partial derivatives in equation 6.20 and substituting for the variances in $\bar{y}$ and b we obtain[17]

$$s_\alpha = \frac{s}{a^2}\left[\frac{b^2}{n} + \frac{n\bar{y}^2}{n\sum x_i^2 - \left(\sum x_i\right)^2}\right]^{1/2}, \tag{6.21}$$

where n is the number of data points, s is given by equation 6.10, a is given by equation 6.6, and b is given by equation 6.7.

---

[15] An estimated slope, b, that is slightly larger than the true slope will consistently coincide with an estimated intercept, a, that is slightly smaller than the true intercept, and vice versa. See Weisberg (2005) for a discussion of correlation between a and b.

[16] See appendix 3.

[17] $s_{\bar{y}} = \frac{s}{\sqrt{n}}$ and s_b is given by equation 6.9.

Table 6.9. *Length–temperature data for an alumina rod.*

$T(°C)$	100	200	300	400	500	600	700	800	900	1000
$l\,(m)$	1.2019	1.2018	1.2042	1.2053	1.2061	1.2064	1.2080	1.2078	1.2102	1.2122

Exercise F

In an experiment to study thermal expansion, the length of an alumina rod is measured at various temperatures. The data are shown in table 6.9.

(i) Calculate the intercept and slope of the best straight line through the length–temperature data and the standard uncertainties in the intercept and slope.
(ii) Determine the temperature coefficient of expansion, α, for the alumina.
(iii) Calculate the standard uncertainty in α.

6.4.2 Estimating *y* for a given *x*

Once the intercept and slope of the best line through the points have been determined, we can predict the value of y, $\hat{y}_0$, at an arbitrary x value, x_0, using the relationship

$$\hat{y}_0 = a + bx_0. \qquad (6.22)$$

In the absence of systematic errors, $\hat{y}_0$ is the best estimate of the true value of the y quantity at $x = x_0$.

The true value of y at $x = x_0$ may be written $\mu_{y|x_0}$. Just as the uncertainties in the measured y values contribute to the uncertainties in a and b, so the uncertainties in a and b contribute to the uncertainty in $\hat{y}_0$. As in section 6.4.1.1, we avoid the problem of correlation of errors in a and b by replacing a by $\bar{y} - b\bar{x}$, so that equation 6.22 becomes

$$\hat{y}_0 = \bar{y} + b(x_0 - \bar{x}). \qquad (6.23)$$

As the errors in $\bar{y}$ and b are independent, the standard uncertainty in $\hat{y}_0$, written as $s_{\hat{y}_0}$, is given by

$$s_{\hat{y}_0} = s\left(\frac{1}{n} + \frac{n(x_0 - \bar{x})^2}{n\sum x_i^2 - (\sum x_i)^2}\right)^{1/2}, \qquad (6.24)$$

where s is given by equation 6.10.

We conclude that $s_{\hat{y}_0}$ and hence any coverage interval for $\mu_{y|x_0}$ depends upon the value of x at which the estimate of $\mu_{y|x_0}$ (i.e. $\hat{y}_0$) is calculated. The closer x_0 is to $\bar{x}$, the smaller is the second term inside the brackets of equation 6.24, and therefore the smaller is $s_{\hat{y}_0}$.

In general, the $X\%$ coverage interval for $\mu_{y|x_0}$ may be written as

$$\hat{y}_0 \pm t_{X\%,\nu} s_{\hat{y}_0}, \tag{6.25}$$

where $t_{X\%,\nu}$ is the critical t value corresponding to the $X\%$ level of confidence evaluated with ν degrees of freedom.[18]

Example 5

Consider the data in table 6.10.

Assuming a linear relationship between the x–y data in table 6.10, determine:

(i)   the parameters a and b of the best line through the points;
(ii) $\hat{y}_0$ for $x_0 = 12$ and $x_0 = 22.5$;
(iii) the standard uncertainty in $\hat{y}_0$ when $x_0 = 12$ and $x_0 = 22.5$.
(iv) Plot the data in table 6.10 on an x–y graph showing the line of best fit and the limits of the 95% coverage interval for $\mu_{y|x_0}$ for values of x_0 between 0 and 45.

ANSWER

(i) a and b are found using equations 6.6 and 6.7. We find $a = 42.43$ and $b = -2.527$, so that the equation for $\hat{y}_0$ can be written as

$$\hat{y}_0 = 42.43 - 2.527x_0.$$

(ii) When $x_0 = 12$, $\hat{y}_0 = 12.11$. When $x_0 = 22.5$, $\hat{y}_0 = -14.43$.

(iii) Using the data in table 6.10: $\bar{x} = 22.5, \sum x_i^2 = 5100, \sum x_i = 180, s = 4.513$ (s is calculated using equation 6.10).
Substituting values into equation 6.24 gives for $x_0 = 12$, $s_{\hat{y}_0} = 2.2$. When $x_0 = 22.5$, $s_{\hat{y}_0} = 1.6$.

(iv) When the number of degrees of freedom = 6, the critical t value for the 95% coverage interval, given by table 2 in appendix 1 is $t_{95\%,6} = 2.447$. Equation 6.25 is used to find lines which represent the 95% coverage intervals for $\mu_{y|x_0}$ for values of x_0 between 0 and 45. These are indicated on the graph in figure 6.8.

Table 6.10. *The x–y data for example 5.*

x	5	10	15	20	25	30	35	40
y	28.1	5	−0.5	−7.7	−14.8	−27.7	−48.5	−62.9

---

[18] $t_{X\%,\nu}$ may be found using table 2 in appendix 1 or the T.INV.2T() function in Excel, as discussed in section 3.9.1.

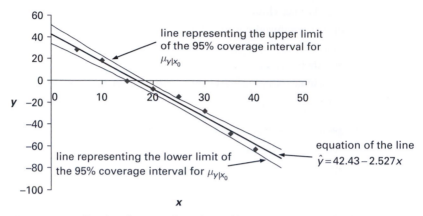

Figure 6.8. Line of best fit and upper and lower limits of the 95% coverage interval for data in table 6.10.

Exercise G

For the x–y data in example 5, calculate the 99% coverage interval for $\mu_{y|x_0}$ when $x_0 = 15$.

6.4.2.1 Uncertainty in prediction of y at a particular value of x

Let us assume that we have fitted the best straight line to a set of x–y data. If we make a measurement of y at $x = x_0$, between what limits would we expect the measured value of y to lie? This is different from considering coverage limits associated with the estimate of the population mean at $x = x_0$ because two factors must be taken into consideration:

(i) the uncertainty in the line of best fit to the data;
(ii) the uncertainty in the measurement of y made at $x = x_0$.

Using addition of uncertainties as discussed in section 5.8.2, we combine the uncertainty in the measurement of y with the uncertainty in the line of best to give[19]

$$s_{P\hat{y}_0} = s\left(1 + \frac{1}{n} + \frac{n(x_0 - \bar{x})^2}{n\sum x_i^2 - (\sum x_i)^2}\right)^{1/2}, \tag{6.26}$$

where $s_{P\hat{y}_0}$ represents that standard uncertainty in the predicted value of y at $x = x_0$.

The X% prediction interval for y at $x = x_0$ is written

$$\hat{y}_0 \pm t_{X\%,\nu} s_{P\hat{y}_0}, \tag{6.27}$$

where $\hat{y}_0$ is the best estimate of the predicted value. Note that $\hat{y}_0$ is the same as the best estimate of the population mean at $x = x_0$ and is given by equation 6.22;

---

[19] We assume that the error in the line of best fit and the error in the measurement of y made at $x = x_0$, are uncorrelated.

$t_{X\%,\nu}$ is the critical t value corresponding to the $X\%$ level of confidence evaluated with ν degrees of freedom.

Equation 6.26 is very similar to equation 6.24. However, the unity term within the brackets of equation 6.26 dominates over the other terms even when n is small and so the prediction interval for a y value at $x = x_0$ is larger than the coverage interval of the population mean $\mu_{y|x_0}$ at $x = x_0$.

Exercise H

Using information supplied in example 5, calculate the 95% prediction interval for y if a measurement of y is to be made at $x_0 = 12$.

6.4.3 Estimating x for a given y

Using the best straight line through points, we can estimate a value of x for any particular value of y. The equation of the best straight line through points is rearranged to give

$$\hat{x}_0 = \frac{\bar{y}_0 - a}{b},$$ (6.28)

where $\hat{x}_0$ is the value of x when $y = \bar{y}_0$, and where $\bar{y}_0$ is the mean of repeated measurements of the dependent variable. The question arises, as there is uncertainty in a, b and $\bar{y}_0$, what will be the uncertainty in $\hat{x}_0$? As discussed in section 6.4.1.1, the errors in a and b are correlated. Replacing a in equation 6.28 by $\bar{y} - b\bar{x}$, we have

$$\hat{x}_0 = \bar{x} + \frac{\bar{y}_0 - \bar{y}}{b}.$$ (6.29)

Assuming the uncertainties in $\bar{y}_0$, $\bar{y}$ and b to be uncorrelated (and that there is no uncertainty in $\bar{x}$), we write

$$s_{\hat{x}_0}^2 = \left(\frac{\partial \hat{x}_0}{\partial \bar{y}_0}\right)^2 s_{\bar{y}_0}^2 + \left(\frac{\partial \hat{x}_0}{\partial \bar{y}}\right)^2 s_{\bar{y}}^2 + \left(\frac{\partial \hat{x}_0}{\partial b}\right)^2 s_b^2,$$ (6.30)

where $s_{\bar{y}_0} = \frac{s}{\sqrt{m}}$ (m is the number of repeat measurements made of y; s is given by equation 6.10); $s_{\bar{y}} = \frac{s}{\sqrt{n}}$ and s_b is given by equation 6.9.

Calculating the partial derivatives in equation 6.30 and substituting the expressions for the variances in $\bar{y}_0$, $\bar{y}$ and b, gives

$$s_{\hat{x}_0} = \frac{s}{b}\left[\frac{1}{m} + \frac{1}{n} + \frac{n(\bar{y}_0 - \bar{y})^2}{b^2\left(n\sum x_i^2 - (\sum x_i)^2\right)}\right]^{1/2}.$$ (6.31)

The $X\%$ coverage interval for $\hat{x}_0$ can be written,

$$\hat{x}_0 \pm t_{X\%,\nu} s_{\hat{x}_0}, \tag{6.32}$$

where ν is the number of degrees of freedom.

Example 6

A spectrophotometer is used to measure the concentration of arsenic in solution. Table 6.11 shows calibration data of the variation of absorbance[20] with arsenic concentration.

Assuming that the absorbance is linearly related to the arsenic concentration:

(i) plot a calibration graph of absorbance versus concentration;
(ii) find the intercept and the slope of the best straight line through the data.
Three values of absorbance are obtained from repeat measurements on a sample of unknown arsenic concentration. The mean of these values is found to be 0.3520.
(iii) Calculate the concentration of arsenic corresponding to this absorbance and the standard uncertainty in the concentration.

ANSWER

(i) Figure 6.9 shows a plot of the variation of absorbance versus concentration data contained in table 6.11.
(ii) a and b are determined using equations 6.6 and 6.7:

$$a = 0.01258, b = 0.0220 \, \text{ppm}^{-1}.$$

(iii) Using the relationship, $\hat{x}_0 = \frac{\bar{y}_0 - a}{b}$,

$$\hat{x}_0 = \left(\frac{0.3520 - 0.01258}{0.02220}\right) \text{ppm} = 15.29 \, \text{ppm}.$$

We use equation 6.31 to obtain the standard uncertainty in $\hat{x}_0$. Using the information in the question and the data in table 6.11 we find $n = 5, m = 3, \bar{y}_0 = 0.3520$, $\bar{y} = 0.37842, \sum x_i^2 = 1852.363, \sum x_i = 82.483, s = 0.01153$ (s is calculated by using equation 6.10).

Substituting these values into equation 6.31 gives

$$s_{\hat{x}_0} = 0.38 \, \text{ppm}.$$

It is worth remarking that the third term in the brackets of equation 6.31 becomes large for y values far from the mean of the y values obtained during the calibration procedure. The third term is zero (and hence the standard uncertainty in $\hat{x}_0$ is smallest) when the y value of the sample under test is equal to the mean y values obtained during calibration.

---

[20] Absorbance is proportional to the amount of light absorbed by the solution as the light passes from source to detector within the spectrophotometer.

Table 6.11. *Absorbance as a function of arsenic concentration.*

Concentration (ppm)	Absorbance
2.151	0.0660
9.561	0.2108
16.878	0.3917
23.476	0.5441
30.337	0.6795

Table 6.12. *Variation of peak area with nitrite concentration.*

Concentration (ppm)	2.046	3.068	5.114	7.160	10.23
Peak area (arbitrary units)	37 752	47 658	75 847	105 499	154 750

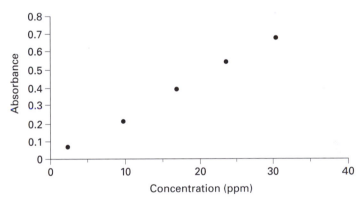

Figure 6.9. Absorbance versus concentration for a solution containing arsenic.

Exercise I

The data in table 6.12 were obtained in an experiment to determine the amount of nitrite in solution using high performance liquid chromatography (HPLC).

(i) Regarding the peak area as the dependent variable, determine the equation of the best straight line through the data.

(ii) Four repeat measurements are made on a solution with unknown nitrite concentration. The mean peak area is found to be 57 156. Use the equation of the best line to find the concentration corresponding to this mean peak area and the standard uncertainty in the concentration.

6.5 Fitting a straight line to data when random errors are confined to the x quantity

One of the assumptions we rely upon consistently in least squares analysis is that errors in x–y data are confined to the y quantity. The validity of this assumption must be considered on a case by case basis, but it is reasonable to argue that all measurements have some error so that there will be some error in the x values. It is possible to derive equations for slope, intercept and the uncertainties in these quantities in situations in which there are uncertainties in both the x and the y values.[21] If the errors in the x values are constant and errors in the y values are negligible, we can use the results already derived in this chapter to find the best line through the data. We write the equation of the line through the data as

$$x = a^* + b^* y, \tag{6.33}$$

where x is regarded as the dependent variable and y as the independent variable, a^* is the intercept (i.e. the value of x when $y = 0$) and b^* is the slope of the line. To find a^* and b^*, we must minimise the sum of squares of the residuals of the observed values of x from the predicted values based on a line drawn through the points. In essence we recreate the argument begun in section 6.2.2, but with y replacing x and x replacing y. The equation for the best line through the x–y data in this case is given when the intercept, a^*, is

$$a^* = \frac{\sum y_i^2 \sum x_i - \sum y_i \sum x_i y_i}{n \sum y_i^2 - \left(\sum y_i\right)^2}, \tag{6.34}$$

and the slope, b^*, is,

$$b^* = \frac{n \sum x_i y_i - \sum x_i \sum y_i}{n \sum y_i^2 - \left(\sum y_i\right)^2}. \tag{6.35}$$

Compare these equations with 6.6 and 6.7 for a and b when the sum of squares of residuals in the y values is minimised.

Equation 6.33 can be rewritten as

$$y = \frac{-a^*}{b^*} + \frac{x}{b^*}. \tag{6.36}$$

It is tempting to compare equation 6.36 with $y = a + bx$ and reach the conclusion that

$$a = \frac{-a^*}{b^*} \tag{6.37}$$

[21] This is beyond the scope of this text. For a review of least squares methods when both x and y variables are affected by error, see Cantrell (2008).

Table 6.13. *The x-y data.*

x	y
2.52	2
3.45	4
3.46	6
4.25	8
4.71	10
5.47	12
6.61	14

and

$$b = \frac{1}{b^*}.$$ (6.38)

However, a and b given by equations 6.37 and 6.38 are equal to a and b given by equations 6.6 and 6.7 only if both least squares fits (i.e. that which minimises the sum of the squares of the x residuals and that which minimises the sum of the squares of the y residuals) produce the same straight line through the points. The only situation in which this happens is when there are *no* errors in either the x or the y values, i.e. all the data lie exactly on a straight line!

As an example, consider the x-y data in table 6.13.

We can perform a least squares analysis assuming the following.

(i) The errors are in the x values only. Using equations 6.34 and 6.35 we find, $a^* = 1.844$ and $b^* = 0.3136$. Using equations 6.37 and 6.38 we obtain $a = -5.882$ and $b = 3.189$.

(ii) The errors are in the y values only. Using equations 6.6 and 6.7 we find, $a = -5.348$ and $b = 3.067$.

As anticipated, the parameter estimates, a and b, depend on whether minimisation occurs in the sum of the squares of the x residuals or the sum of the squares of the y residuals.

6.6 Linear correlation coefficient, *r*

When the influence of random errors on values is slight, it is usually easy to establish 'by eye' whether x and y quantities are related. However, when data are scattered it is often quite difficult to be sure of the extent to which y is dependent on x. It is useful to define a parameter directly related to the extent of

---

[22] During the experiment the current through the diode is held constant.

Table 6.14. *Voltage across a diode,*
V, as a function of temperature, θ.

θ (°C)	V (V)
2.0	0.6859
10.0	0.6619
19.0	0.6379
26.4	0.6139
40.9	0.5899
48.8	0.5659
59.7	0.5419
65.0	0.5179
80.0	0.4939
91.0	0.4699
101.3	0.4459

Exercise J

The voltage, V, across a silicon diode is measured as a function of temperature, θ, of the diode.[122] Temperature–voltage data are shown in table 6.14.

Theory suggests that the relationship between V and θ is,

$$V = k_0 + k_1\theta. \tag{6.39}$$

As V is the dependent variable and θ the independent variable, it would be usual to plot V on the y axis and θ on the x axis. However, the experimenter has evidence that the measured values of diode voltage are more precise than those of temperature.

(i) Use least squares to obtain best estimates for k_0 and k_1, where values of V are assumed to have most error.
(ii) Use least squares to obtain best estimates for k_0 and k_1, where values of θ are assumed to have most error.

the correlation between x and y. That parameter is called the linear correlation coefficient, ρ. The estimate of this parameter, obtained from data, is represented by the symbol, r.

When all points lie on the same straight line, i.e. there is perfect correlation between x and y, then equations 6.37 and 6.38 correctly relate a and b (which are determined assuming errors are confined to the y quantity) with a^* and b^* (which are determined assuming errors are confined to the x quantity). Focussing on equation 6.38 we can say that, for perfect correlation,

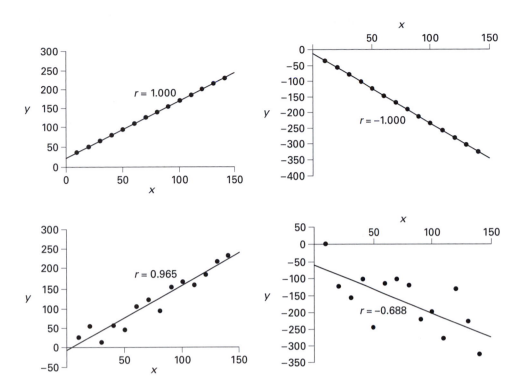

Figure 6.10. Correlation coefficients for x–y data exhibiting various amounts of scatter.

$$bb^* = 1.$$

When there is no correlation between x and y, there is no tendency for y to increase with increasing x nor x to increase with increasing y. Put another way, with no correlation we would expect both b and b^* to be close to zero, so that $bb^* \approx 0$. This suggests that it may be useful to use the product bb^* as a measure of the correlation between x and y. We define the linear correlation coefficient, r, as

$$r = \sqrt{bb^*}.$$ (6.40)

Substituting for b and b^* using equations 6.7 and 6.35 respectively, we get

$$r = \frac{n \sum x_i y_i - \sum x_i \sum y_i}{\left[n \sum x_i^2 - (\sum x_i)^2\right]^{1/2} \left[n \sum y_i^2 - (\sum y_i)^2\right]^{1/2}}.$$ (6.41)

For perfect correlation, r is either $+1$ or -1. Note that r has the same sign as that of the slope (b or b^*). Figure 6.10 shows graphs of x–y data along with the value of r for each.

As $|r|$ decreases from 1 to 0, the correlation between x and y becomes less and less convincing. Notions of 'goodness' relating to values of r can be misleading. A value of $|r|$ close to unity does indicate good correlation, however it is possible that x and y are not *linearly* related but still result in a value for $|r|$ in excess of 0.99. This is illustrated by the next example.

Example 7

Thermoelectric coolers (TECs) are devices widely used to cool electronic components, such as laser diodes and are also found in portable refrigerators. A TEC consists of a hot and cold surface with the temperature difference between the surfaces maintained by an electric current flowing through the device.

In an experiment, the temperature difference, ΔT, between the hot and the cold surface is measured as a function of the electrical current, I, passing through the TEC. Data gathered are shown in table 6.15.

(i) Calculate the value of the correlation coefficient, r.
(ii) Plot a graph of ΔT versus I and show the line of best fit through the points.

ANSWER

(i) We begin by drawing up a table containing all the values needed to calculate r using equation 6.41.

Summing the values in each column gives $\sum x_i = 4.2$, $\sum y_i = 113.2$, $\sum x_i y_i = 93.02$, $\sum x_i^2 = 3.64$, $\sum y_i^2 = 2402.22$.

Substituting the summations into equation 6.41 gives $r = \dfrac{175.7}{2.8 \times 63.2558} = 0.992$.

A fit of the equation $y = a + bx$ to the x–y data given in table 6.16 gives intercept, $a = 2.725\ °C$, and slope, $b = 22.41 °C/A$.

(ii) Figure 6.11 shows the data points and the line of best fit.

A value of $r = 0.992$ indicates a high degree of correlation between x and y. That x and y are correlated is revealed by the graph, but we might be overlooking something more important. Is the relationship between temperature difference and current *really* linear? We assumed that to be the case when calculating the line of best fit, though a close inspection of the graph above seems to indicate that a curve through the points is more appropriate than a straight line. This is a point worth emphasising: a correlation coefficient with magnitude close to unity is no guarantee that the x–y data are linearly related. We need to inspect an x–y graph of the raw data, the line of best fit and preferably a plot of the residuals (dealt with in section 6.7) in order to be satisfied that the assumption of linearity between x and y is reasonable.

Table 6.15. *Temperature difference,*
ΔT, of a TEC as a function of current, I.

I (A)	ΔT (°C)
0.0	0.8
0.2	7.9
0.4	12.5
0.6	17.1
0.8	21.7
1.0	25.1
1.2	28.1

Table 6.16. *Values needed to calculate r using equation 6.41.*

$x_i\,(=I)$	$y_i\,(=\Delta T)$	$x_i y_i$	x_i^2	y_i^2
0.0	0.8	0.0	0.0	0.64
0.2	7.9	1.58	0.04	62.41
0.4	12.5	5.0	0.16	156.25
0.6	17.1	10.26	0.36	292.41
0.8	21.7	17.36	0.64	470.89
1.0	25.1	25.1	1.0	630.01
1.2	28.1	33.72	1.44	789.61

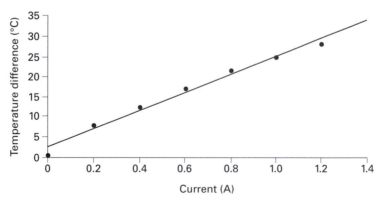

Figure 6.11. Temperature difference for a TEC versus current.

Table 6.17. *Variation of temperature of the cold surface of TEC with the volume of heat sink.*

Volume, V (cm^3)	Temperature, T (°C)
5	37.0
10	25.5
15	17.1
25	11.5
50	6.4

Exercise K

The temperature reached by the cold surface of a thermoelectric cooler depends on the size of the heat sink to which the hot surface of the cooler is attached. Table 6.17 shows the temperature, T, of the cold surface of the TEC for aluminium heat sinks of various volumes, V.

  (i) Calculate the linear correlation coefficient for these data.
 (ii) Find the equation of the line of best fit through the data assuming that a linear relationship is appropriate (plot T on the y axis and V on the x axis).
(iii) Plot a graph of temperature versus volume and indicate the line of best fit.
(iv) Do you think that the assumption of linearity between T and V is valid?

6.6.1 Calculating r using Excel

The CORREL() function in Excel calculates the correlation coefficient, r, and returns the value into a cell. The syntax of the function is as follows.

      CORREL(y values, x values)

Consider the calculation of r for the data shown in sheet 6.2.

Sheet 6.2. *The x–y data.*

	A	B
1	x	y
2	42	458
3	56	420
4	56	390
5	78	380
6	69	379
7	92	360
8	102	351
9	120	300
10		

Table 6.18. *The x–y data for exercise L.*

x	2	4	6	8	10	12	14	16	18	20	22	24
y	16.5	49.1	65.2	71.6	101.5	90.1	101.4	113.7	127.7	156.5	203.6	188.4

Table 6.19. *Correlation coefficient for ten sets of randomly generated y values correlated with the x column values.*

x	y1	y2	y3	y4	y5	y6	y7	y8	y9	y10
0.2	0.020	0.953	0.508	0.324	0.233	0.872	0.446	0.673	0.912	0.602
0.4	0.965	0.995	0.231	0.501	0.265	0.186	0.790	0.911	0.491	0.186
0.6	0.294	0.159	0.636	0.186	0.227	0.944	0.291	0.153	0.780	0.832
0.8	0.561	0.096	0.905	0.548	0.187	0.002	0.331	0.051	0.862	0.255
1.0	0.680	0.936	0.783	0.034	0.860	0.363	0.745	0.083	0.239	0.363
r	**0.400**	**−0.323**	**0.743**	**−0.392**	**0.655**	**−0.455**	**0.094**	**−0.822**	**−0.541**	**−0.242**

To calculate *r*:

(i) enter the data shown in sheet 6.2 into an Excel spreadsheet;

(ii) type =**CORREL(B2:B9,A2:A9)** into cell B10;

(iii) press the Enter key;

(iv) the value −0.9503 is returned in cell B10.

Exercise L

Use the CORREL() function to determine the correlation coefficient of the *x–y* data shown in table 6.18.

6.6.2   Is the value of *r* significant?

We have seen that we need to be cautious when using *r* to infer the extent to which there is a linear relationship between *x–y* data, as values of $|r| > 0.8$ may be obtained when the underlying relationship between *x* and *y* is not linear. There is another issue: values of $|r| > 0.8$ may be obtained when *x* and *y* are totally uncorrelated, especially when the number of *x–y* values is small. To illustrate this, consider the values in table 6.19.

The first column of table 6.19 contains five values of *x* from 0.2 to 1. The remainder of the columns contain numbers between 0 and 1 that have been randomly generated so that there is *no underlying correlation between x and y*. The bottom row shows the correlation coefficients calculated when each column of *y* is correlated in turn with the column containing *x*.

The column of values headed $y8$, when correlated with the column headed x, gives a value of $|r| > 0.8$, which might be considered in some circumstances as 'good' but in this case has just happened 'by chance'. The basic message is that, if there are only a few points (say ≤ 5), it is possible that consecutive numbers will be in ascending or descending order. If this is the case then the magnitude of the correlation coefficient can easily exceed 0.8. How, then, do you know if the correlation coefficient is significant? The question can be put as follows.

On the assumption that a set of x–y data are uncorrelated, what is the probability of obtaining the value of r at least as large as that calculated from the available data?

If that probability is small (say < 0.05) then it is highly unlikely that the x–y data are uncorrelated, i.e. the data *are* likely to be correlated. Table 6.20 shows the probability of obtaining a particular value for $|r|$ when the x–y data are uncorrelated. The probability of obtaining a particular value for $|r|$ decreases, when the x–y values are uncorrelated, as the number of values, n, increases.[23]

A correlation coefficient is considered significant if the probability of it occurring, when the data are uncorrelated,[24] is less than 0.05. Referring to table 6.20 we see that, for example, with only four data points ($n = 4$), a value of r of 0.9 is not significant as the probability that this could happen with uncorrelated data is 0.10. When the number of data values is 10 or more, then values of r of greater than about 0.6 become significant.

Table 6.20. *Probabilities of obtaining calculated r values when x–y data are uncorrelated.*

n	$|r|$ (calculated from x–y data)					
	0.5	0.6	0.7	0.8	0.9	1.0
3	0.67	0.59	0.51	0.41	0.29	0
4	0.50	0.40	0.30	0.20	0.10	0
5	0.39	0.28	0.19	0.10	0.037	0
6	0.31	0.21	0.12	0.056	0.014	0
7	0.25	0.15	0.080	0.031	0.006	0
8	0.21	0.12	0.053	0.017	0.002	0
9	0.17	0.088	0.036	0.010	0.001	0
10	0.14	0.067	0.024	0.005	<0.001	0

[23] See Bevington and Robinson (2002) for a discussion of the calculation of the probabilities in table 6.20.

[24] This value of probability is often used in tests to establish statistical significance, as discussed in chapter 9.

Table 6.21. *The x–y data for exercise M.*

x	y
2.67	1.54
1.56	1.60
0.89	1.56
0.55	1.34
−0.25	1.33

Exercise M

Consider the *x–y* data in table 6.21.

(i) Calculate the value of the correlation coefficient, *r*.

(ii) Is the value of *r* significant?

6.7 Residuals

Another indicator of 'goodness of fit', complementary to the correlation coefficient and often more revealing, is the distribution of residuals, Δy. The residual, $\Delta y = y - \hat{y}$, is plotted on the vertical axis against *x*. If we assume that a particular equation is 'correct' in that it accurately describes the data that we have collected, and only random errors exist to obscure the true value of *y* at any value of *x*, then the residuals would be expected to scattered randomly about the $\Delta y = 0$ axis. Additionally, if the residuals are plotted in the form of a histogram, they would appear to be normally distributed if the errors are normally distributed.

   If we can discern a regular pattern in the residuals then one or more factors must be at play to cause the pattern to occur. It could be:

(i) we have chosen an equation of dubious validity to fit to the data;

(ii) there is an outlier in the data which has a large influence on the line of best fit;

(iii) we should be using a weighted fit, that is, the uncertainties in the measured *y* values are not constant and we need to take this into account when applying least squares to obtain parameter estimates (section 6.10 considers weighted fitting of a straight line to data).

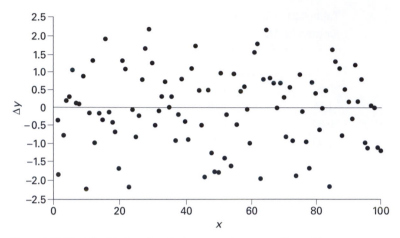

Figure 6.12. Ideal distribution of residuals – no regular pattern discernible.

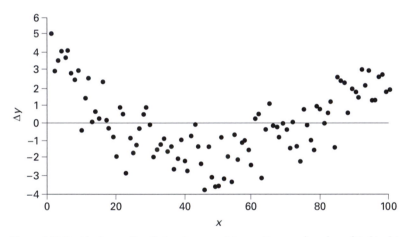

Figure 6.13. Residuals revealing that an inappropriate equation may have been fitted to data.

Typical plots of residuals[25] versus x, which illustrate patterns that can emerge, are shown in figures 6.12 to 6.15.

No discernible regular pattern is observable in the residuals in figure 6.12. This offers strong evidence to support the notion that the fit of the equation to the data is good. If the standard deviation of the residuals is constant as x

---

[25] Another way to display residuals is to plot Δy_i versus $\hat{y}_i$.

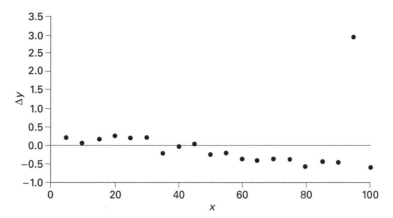

Figure 6.14. Effect of outlier on residuals.

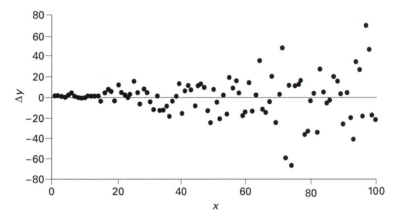

Figure 6.15. Pattern in residuals suggesting weighted fitting should be used.

increases (as in figure 6.12), those residuals are sometimes referred to as being *homoscedastic*.

In figure 6.13 the residuals reveal a systematic variation as x increases. This should lead us to question the appropriateness of the equation we have fitted to the data.

Figure 6.14 shows an extreme example of an outlier existing in x–y data. If there are many data, the intercept and slope, a and b, are little affected by such an outlier. However, the situation is quite different if there are far fewer data points (say 10 or so). In such a situation an outlier can have a dramatic effect on a and b (and the uncertainties in a and b).

In figure 6.15 the residuals increase steadily with increasing in x. This occurs when the scatter in the data (i.e. due to random errors in the y quantity) increases with increasing x. In this situation an unweighted fit of the equation to the data is not appropriate and a *weighted* fit using weighted least squares should be used instead. Weighted least squares is considered in section 6.10. If the standard deviation of the residuals is not constant as x increases (as in figure 6.15), those residuals are sometimes referred to a being *heteroscedastic*.

Though residuals can be very revealing with regard to identifying outliers, inappropriate equations and incorrect weighting, they can be misleading if there are only a few data (say, <10) as a pattern can emerge within the residuals 'by chance'.

Exercise N

The period, T, of oscillation of a body on the end of a spring is measured as the mass, m, of the body increases. Data obtained are shown in table 6.22.

(i)   Plot a graph of period versus mass of body.
(ii) Assuming that the period is linearly related to the mass, use least squares to find the equation of the best line through the points.
(iii) Calculate the residuals and plot a graph of residuals versus mass.
(iv) Is there any 'pattern' discernible in the residuals? If so, suggest a possible cause of the pattern.

Table 6.22. *Variation of period, T, of a body of mass, m, on a spring.*

m (kg)	T (s)
0.2	0.39
0.4	0.62
0.6	0.72
0.8	0.87
1.0	0.92
1.2	1.07
1.4	1.13
1.6	1.16
1.8	1.23
2.0	1.32

6.7.1 Standardised residuals

If each residual, Δy_i, is divided by the standard deviation in each y value, s_i, we refer to the quantity

$$\Delta y_{is} = \frac{\Delta y_i}{s_i} \qquad\qquad (6.42)$$

as the *standardised* residual.[26] In situations in which the standard deviation in each y value is the same, we replace s_i by s, so that

$$\Delta y_{is} = \frac{\Delta y_i}{s}, \qquad\qquad (6.43)$$

where s is given by equation 6.10.

Scaling residuals in this manner is very useful as the standardised residuals should be normally distributed with a standard deviation of 1 if the errors causing scatter in the data are normally distributed. If we were to plot Δy_s against x, we should find that, not only should the standardised residuals be scattered randomly about the $\Delta y_s = 0$ axis, but that (based on properties of the normal distribution[27]) about 70% of the standardised residuals should lie between $\Delta y_s = \pm 1$ and about 95% should lie between $\Delta y_s = \pm 2$. If this is not the case, it suggests that the scatter of y values does not follow a normal distribution.

Example 8

Consider the x–y data in table 6.23.

(i)   Assuming a linear relationship between x and y, determine the equation of the best line through the data in table 6.23.
(ii) Calculate the standard deviation, s, of the data about the fitted line.
(iii) Determine the standardised residuals and plot a graph of standardised residuals versus x.

ANSWER

(i)   Applying equations 6.6 and 6.7 to the data in table 6.23, we find $a = 7.849$ and $b = 2.830$, so that the equation of the best line can be written as

$$\hat{y} = 7.849 + 2.830x. \qquad\qquad (6.44)$$

(ii) Applying equation 6.10 gives $s = 5.376$.
(iii) Table 6.24 includes the predicted y values found using equation 6.44, the residuals and the standardised residuals.

Figure 6.16 shows a plot of the standardised residuals, Δy_s, versus x.
As anticipated, the standardised residuals lie between $\Delta y_s = \pm 2$.

---

[26] See Devore (2007) for a fuller discussion of standardised residuals.

[27] See section 3.5.

Table 6.23. *The x–y data for example 8.*

x	2	4	6	8	10	12	14	16	18	20	22
y	8.2	23.0	26.3	31.4	39.5	36.7	56.7	46.8	53.4	63.2	74.7

Table 6.24. *Predicted values, residuals and standardised residuals.*

x	2	4	6	8	10	12	14	16	18	20	22
y	8.2	23.0	26.3	31.4	39.5	36.7	56.7	46.8	53.4	63.2	74.7
$\hat{y}$	13.509	19.169	24.829	30.489	36.149	41.809	47.469	53.129	58.789	64.449	70.109
Δy	−5.309	3.831	1.471	0.911	3.351	−5.109	9.231	−6.329	−5.389	−1.249	4.591
Δy_s	−0.988	0.713	0.274	0.169	0.623	−0.950	1.717	−1.177	−1.002	−0.232	0.854

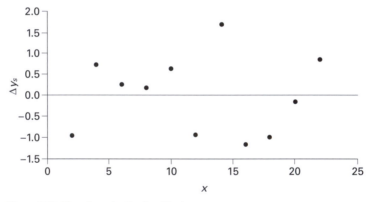

Figure 6.16. Plot of standardised residuals.

6.8 Data rejection

A topic of some importance and controversy is that of data rejection. If we gather *x–y* data and most of the points lie along a straight line, but one point lies well away from the line, should we reject that point? This is a difficult question to which to give a simple yes or no answer. It may be that a 'slip-up' occurred during the recording of the data and therefore we are justified in ignoring the point. However, it could be that there is a true deviation from straight line behaviour and that the apparent outlier is not an outlier at all but is revealing something really important – perhaps we have observed a new effect! To reject that point may be to eliminate the most important value you have gathered. If possible, repeating the experiment is always preferable to rejecting data with little justification other than it 'doesn't fit'.

The decision to reject data largely depends on what relationship we believe underlies the data that has been collected. For example, figure 6.17 shows data obtained in an experiment to study the specific heat of a metal at low temperature. Though the underlying trend between x and y may or may not be linear, it *does* appear that there is a spurious value for the specific heat at about 3 K. Performing a statistically based test, such as that described next, would indicate that the point should be removed. However, the 'jump' in specific heat at about 3 K is a real effect and to eliminate the point would mean that a vital piece of evidence giving insight into the behaviour of the material would have been discarded.

There may be situations in which large deviations can be attributed to spurious effects. In these cases it is possible to use the trend exhibited by the majority of the data as a basis for rejecting data that appear to be outliers.

In section 5.11.2 we introduced Chauvenet's criterion as a way of deciding whether a value within a sample is so far from the mean of the sample that it should be considered for rejection. The basic assumption when applying the criterion is that data are normally distributed. If the assumption is valid, the probability can be determined that a value will be obtained that is at least as far from the mean as a 'suspect' value. Multiplying that probability by the number

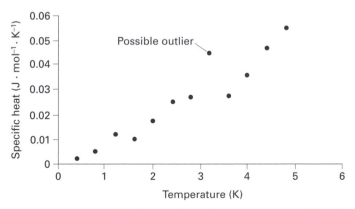

Figure 6.17. Variation of specific heat of tin with temperature. A possible outlier is indicated.

of data gives the number of values *expected* to lie at least as far from the mean as a 'suspect' value. If the expected number is less than or equal to 0.5, the suspect point should be rejected and the sample mean and standard deviation recalculated.

We can apply Chauvenet's criterion to x-y data by considering how far an observed y value is from the line of best fit. The difference between the observed value of y and the predicted value of y based on the line of best fit is expressed in terms of the number of standard deviations between observed and predicted value. Specifically, we write

$$z_{OUT} = \frac{y_{OUT} - \hat{y}}{s}, \tag{6.45}$$

where y_{OUT} is the outlier (i.e. the point furthest from the line of best fit), $\hat{y}$ is the predicted value, found using $\hat{y} = a + bx$. We calculate s using equation 6.10.

Once z_{OUT} has been calculated, we determine the expected number of values, N, at least as far away from $\hat{y}$ as y_{OUT}. To do this:

(i)   determine the probability, P, of obtaining a value of $|z| > z_{OUT}$;
(ii) calculate the expected number of values, N, greater than or equal to $|z_{OUT}|$. $N = nP$, where n is the number of data.

If N is less than 0.5, consider rejecting the point. If a point *is* rejected, then a and b should be recalculated (as well as other related quantities, such as s_a and s_b).

Example 9

Consider x-y values in table 6.25.

(i)   Plot an x-y graph and use unweighted least squares to fit a line of the form $y = a + bx$ to the data.
(ii) Identify any suspect point(s).
(iii) Calculate the standard deviation of the y values.
(iv) Apply Chauvenet's criterion to the suspect point – should it be rejected?

ANSWER

(i)   A plot of data is shown in figure 6.18 with line of best fit attached ($a = -1.610$ and $b = 2.145$).
(ii) A suspect point would appear to be $x = 6$, $y = 8.5$ as this point is furthest from the line of best fit.
(iii) Using the data in table 6.25 and equation 6.10, $s = 1.895$.
(iv) Using equation 6.45, we have

Table 6.25. *The x-y data with a 'suspect' value.*

x	y
2.0	3.5
4.0	7.2
6.0	8.5
8.0	17.1
10.0	20.0

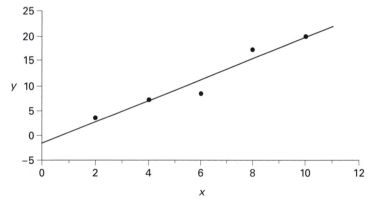

Figure 6.18. The x-y graph for data in table 6.25.

$$z_{OUT} = \frac{8.5 - (-1.610 + 2.145 \times 6)}{1.895} = -1.46.$$

Using table 1 in appendix 1, the probability of an absolute value of $z_{OUT} > 1.46 = 2 \times 0.0722 = 0.144$. The expected number of points at least this far from the line is expected to be $5 \times 0.144 = 0.72$. As this value is greater than 0.5, do not reject the point.

Exercise O

Consider the x-y data in table 6.26.

When a straight line is fitted to the data in table 6.26, it is found that $a = 4.460$ and $b = 0.6111$.

Assuming that the data point $x = 10$, $y = 13$ is 'suspect', apply Chauvenet's criterion to decide whether this point should be rejected.

Table 6.26. *The x–y data for exercise O.*

x	y
5	7
6	8
8	9
10	13
12	11
14	12
15	14

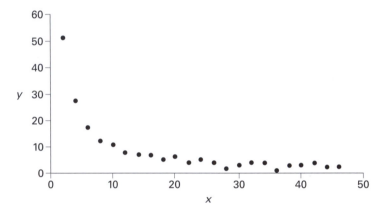

Figure 6.19. Data with non-linear relationship between x and y.

6.9 Transforming data for least squares analysis

Using the technique of least squares we can establish best estimates for the intercept, a, and slope, b, of a line through x–y data through equations 6.6 and 6.7. These equations should only be applied when we are confident that there is a linear relationship between the quantities plotted on the x and y axes. What do we do if x–y data are clearly non-linearly related, such as those in figure 6.19?

It may be possible to apply a mathematical operation to the x or y data (or to both x and y) so that the *transformed* data appear linearly related. The next stage is to fit the equation $y = a + bx$ to the transformed data. How do we choose which mathematical operation to apply? If we have little or no idea what the relationship between the dependent and independent variable is likely to be, we may be forced into a 'trial and error' approach to transforming data. As an example, the rapid decrease in y with x indicated for $x < 10$ in figure 6.19 suggests that there might be an exponential relationship between x and y, such as

$$y = Ae^{Bx},\tag{6.46}$$

where A and B are constants.

Assuming this to be the case, taking natural logs of each side of the equation gives

$$\ln y = \ln A + Bx.\tag{6.47}$$

Comparing equation 6.47 with the equation of a straight line predicts that plotting $\ln y$ versus x will produce a straight line with intercept, $\ln A$, and slope, B. Figure 6.20 shows the effect of applying the transformation suggested by equation 6.47 to the data in figure 6.19.

The transformation has not been successful in producing a linear x–y graph, indicating that equation 6.46 is not appropriate to these data and that other transformation options should be considered (for example $\ln y$ versus $\ln x$). Happily, in many situations in the physical sciences, the work of others (either experimental or theoretical) provides clues as to how the data should be treated in order to produce a linear relationship between data plotted as an x–y graph. Without such clues we must use 'intelligent guesswork'.

As an example, consider a ball falling freely under gravity. The distance, s, through which a ball falls in a time, t, is measured and the values obtained are shown in table 6.27. The data are shown in graphical form in figure 6.21.

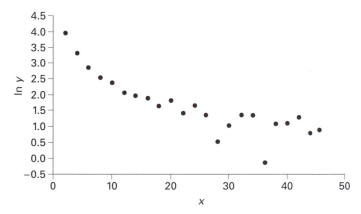

Figure 6.20. Effect of transforming y values by taking natural logarithms.

Table 6.27. *Distance–time data for a freely falling ball.*

t (s)	s (m)
1	7
2	22
3	52
4	84
5	128
6	200

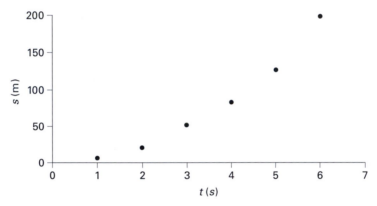

Figure 6.21. Distance–time data for a falling ball.

The relationship between s and t is not linear. Can the data be transformed so that a linear graph is produced? The starting point is to look for an equation which might describe the motion of the ball. When the acceleration, g, is constant, we can write the relationship between s and t, as[28]

$$s = ut + \tfrac{1}{2}gt^2, \tag{6.48}$$

where u is the initial velocity of the body.

To linearise equation 6.48, divide throughout by t to give,

$$\frac{s}{t} = u + \tfrac{1}{2}gt. \tag{6.49}$$

Compare the quantities in equation 6.49 with $y = a + bx$,

$$\frac{s}{t} = u + \tfrac{1}{2}g\ t$$
$$\underbrace{\phantom{\frac{s}{t}}}\ \underbrace{}\ \underbrace{\phantom{\tfrac{1}{2}g}}\ \underbrace{}.$$
$$y = a + b\ \ x$$

---

[28] See Young, Freedman and Ford (2007).

Table 6.28. *Transformation of data given in table 6.27.*

t (s)	s/t (m/s)
1	7
2	11
3	17.3
4	21
5	25.6
6	33.3

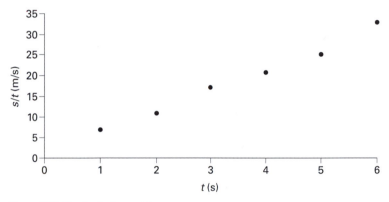

Figure 6.22. Graph of s/t versus t.

If the relationship between the displacement and time can be represented by equation 6.48, then plotting $\frac{s}{t}$ versus t should transform the experimental data so that they lie on (or close to) a straight line which has intercept, u, and slope, ½g.

Table 6.28 contains the transformed data and figure 6.22 the corresponding graph.

It certainly does appear that transforming the data has produced a graph in which the quantities plotted on the x and y axes are linearly related. Fitting a line to the transformed data using least squares gives

$$a = 1.298 \text{ m/s and } b = 5.118 \text{ m/s}^2.$$

When transforming data, the dependent variable should remain on the left hand side of the equation (so that the assumption of errors being restricted to the quantity plotted on the y axis is valid). Usually the independent variable only appears on the right hand side of the equation. However, there are situations, such as in the linearisation of equation 6.48, where this condition must be relaxed.

Exercise P

(1) Transform the equations shown in table 6.29 into the form $y = a + bx$, and indicate how the constants in each equation are related to a and b.

(2) The capacitance of a semiconductor diode decreases as the reverse bias voltage applied to the junction increases. An important diode parameter, namely the contact potential, can be found if the capacitance of the junction is measured as a function of reverse bias voltage. Table 6.30 shows experimental capacitance/voltage data for a particular diode.

Assume that the relationship between C_j and V can be written,

$$\frac{1}{C_j^2} = k(V + \phi), \tag{6.50}$$

where ϕ is the contact potential and k is a constant. Use least squares to find best estimates of k and ϕ.

Table 6.29. *Equations to be transformed into the form $y = a + bx$.*

	Equation	Dependent variable	Independent variable	Constant(s)	Hint
(i)	$R = Ae^{-BT}$	R	T	A, B	Take the natural logs of both sides of the equation.
(ii)	$PV^\gamma = C$	P	V	γ, C	Take the logs of both sides of the equation.
(iii)	$H = C(T - T_0)$	H	T	C, T_0	Multiply out the brackets.
(iv)	$T_w = T_c - kR^2$	T_w	R	T_c, k	
(v)	$T = 2\pi\sqrt{\frac{m}{k}}$	T	m	k	Square both sides of the equation.
(vi)	$E = I(R+r)$	I	R	E, r	Move I to the LHS of equation, and E to the RHS.
(vii)	$\frac{1}{u} + \frac{1}{v} = \frac{1}{f}$	v	u	f	Move $\frac{1}{u}$ to the RHS of the equation.
(viii)	$N = kC^{1/n}$	N	C	k, n	Take logs of both sides of equation.
(ix)	$n^2 = 1 + \frac{A\lambda^2}{\lambda^2 - B}$	n	λ	A, B	Subtract 1 from both sides of equation then take the reciprocals of both sides of the equation.

Table 6.29. (*cont.*)

	Equation	Dependent variable	Independent variable	Constant(s)	Hint
(x)	$t = \frac{A}{D^2}(1 + BD)$	t	D	A, B	Multiply both sides of equation by D^2 then multiply out the brackets.
(xi)	$D = kE^n$	D	E	k, n	Take logs of both sides of the equation.
(xii)	$v = \sqrt{k(A^2 - l^2)}$	v	l	k, A	Square both sides of the equation.
(xiii)	$g = g_0\left(1 - \frac{h}{2R}\right)$	g	h	g_0, R	Multiply out the brackets.
(xiv)	$v = \sqrt{\frac{2(P - P_A)}{\rho}}$	v	P	P_A, ρ	Square both sides of the equation then multiply out the brackets.

Table 6.30. *Junction capacitance, C_j, of a diode as a function of bias voltage, V.*

V (V)	C_j (pF)
6.0	248
8.1	217
10.1	196
14.1	169
18.5	149
24.6	130
31.7	115
38.1	105
45.6	96.1
50.1	92.1

6.9.1 Consequences of data transformation

Our least squares analysis to this point has assumed that the uncertainties in all the y values are the same. Is this assumption still valid if the data are transformed? Very often the answer is no and to illustrate this let us consider a situation in which data transformation requires that the natural logarithms of the y quantity be calculated.

The intensity of light, I, after it travels a distance x through a transparent medium is given by

$$I = I_0 e^{-kx},$$ (6.51)

where I_0 is the intensity of light incident on the medium (i.e. at $x = 0$) and k is the absorption coefficient.

Measurements of I are made at different values of x. We linearise equation 6.51 by taking natural logarithms of both sides of the equation, giving

$$\ln I = \ln I_0 - kx.$$ (6.52)

Assuming that equation 6.51 is valid for the experimental data, plotting $\ln I$ versus x should produce a plot in which the transformed data lie close to a straight line. A straight line fitted to the transformed data would have intercept, $\ln I_0$, and slope $-k$. As $\ln I$ is taken as the 'y quantity' when fitting a line to data using least squares, we must determine the uncertainty in $\ln I$. We write

$$y = \ln I.$$ (6.53)

If the standard uncertainty in I, $u(I)$, is small, then the standard uncertainty in y, $u(y)$, is given by[29]

$$u(y) \approx \left| \frac{\partial y}{\partial I} \right| u(I)$$ (6.54)

now, $\frac{\partial y}{\partial I} = \frac{1}{I}$, so that

$$u(y) \approx \left| \frac{u(I)}{I} \right|.$$ (6.55)

Equation 6.55 indicates that, if $u(I)$ is constant, the uncertainty in $\ln I$ decreases as I increases. The consequence of this is that the assumption of constant uncertainty in the y values used in least squares analysis is no longer valid and unweighted least squares must be abandoned in favour of an approach that takes into account changes in the uncertainties in the y values. There are many situations in which data transformation leads to a similar outcome, i.e. that the uncertainty in y values is not constant and so require a straight line to be fitted using *weighted* least squares. This is dealt with in the next section.

[29] Section 5.8.1 considers propagation of uncertainties.

Exercise Q

For the following equations, determine y and the standard uncertainty in y, $u(y)$, given that $V = 56$ and $u(V) = 2$ in each case. Express $u(y)$ to two significant figures.

(i) $y = V^{1/2}$; (ii) $y = V^2$; (iii) $y = \frac{1}{V}$; (iv) $y = \frac{1}{V^2}$; (v) $y = \log_{10} V$.

6.10 Weighted least squares

When the uncertainties in y values are constant, unweighted least squares analysis is appropriate as discussed in section 6.2.2. However, there are many situations in which the uncertainty in the y values does not remain constant. These include when:

(i) measurements are repeated at a particular value of x, thus reducing the uncertainty in the corresponding value of y;

(ii) data are transformed, for example by 'squaring' or 'taking logs';

(iii) the size of the uncertainty in y is some function of y (such as in counting experiments).

How do you know if you should use a weighted fit? A good starting point is to perform unweighted least squares to find the line of best fit through the data (data should be transformed if necessary, as discussed in section 6.9). A plot of the residuals should reveal whether a weighted fit is required. Figure 6.23 shows a plot of residuals in which the residuals decrease with increasing x (figure 6.23(a)) and increase with increasing x (figure 6.23(b)). Such patterns in residuals are 'tell tale' signs that weighted least squares fitting should be used.

In order to find the best line through the points when weighted fitting is required, we must include explicitly the standard deviation in each y value in the equations for a and b. This is due to the fact that, for weighted fitting, we must minimise the weighted sum of squares of residuals, χ^2, where[30]

$$\chi^2 = \sum \left[\frac{y_i - \hat{y}_i}{s_i} \right]^2 . \tag{6.56}$$

Here s_i is the standard deviation in y_i, and a and b are given by

$$a = \frac{\sum \frac{x_i^2}{s_i^2} \sum \frac{y_i}{s_i^2} - \sum \frac{x_i}{s_i^2} \sum \frac{x_i y_i}{s_i^2}}{\Delta} \tag{6.57}$$

---

[30] Refer to appendix 3 for more details.

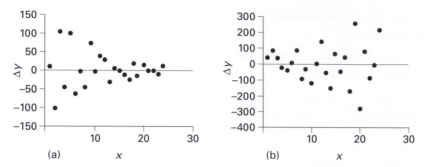

Figure 6.23. Residuals indicating weighted fit is required: (a) indicates the size of the residual decreasing with increasing x; (b) indicates the size of the residual increasing with increasing x.

$$b = \frac{\sum \frac{1}{s_i^2} \sum \frac{x_i y_i}{s_i^2} - \sum \frac{x_i}{s_i^2} \sum \frac{y_i}{s_i^2}}{\Delta},$$
(6.58)

where

$$\Delta = \sum \frac{1}{s_i^2} \sum \frac{x_i^2}{s_i^2} - \left(\sum \frac{x_i}{s_i^2}\right)^2.$$
(6.59)

Equations 6.57 and 6.58 give more weight to the points that have smaller uncertainty, thereby ensuring that the fitted line will pass closer to these points than those with large uncertainty.

A word of caution regarding weighted fitting; if the estimates of the standard deviations, s_i, are poor then points will be weighted inappropriately. In such a situation an unweighted fit may be the best option.

With weighted fitting, the best line no longer passes through $(\bar{x}, \bar{y})$ but through the *weighted* centre of gravity of the points, $(\bar{x}_w, \bar{y}_w)$, and $\bar{x}_w$ and $\bar{y}_w$ are given by[31]

$$\bar{x}_w = \frac{\sum \frac{x_i}{s_i^2}}{\sum \frac{1}{s_i^2}}$$
(6.60)

and

$$\bar{y}_w = \frac{\sum \frac{y_i}{s_i^2}}{\sum \frac{1}{s_i^2}}.$$
(6.61)

[31] For a derivation of equation 6.60 see section A3.1 in appendix 3.

One difficulty with finding a and b is that, in many situations, we do not know s_i. When calculating a and b, a knowledge of the absolute values of s_i is not crucial. As long as we know the relative magnitudes of s_i for all data, the values of a and b are unaffected.

Example 10

Some x-y data along with the standard deviation in the y values, s_i, are shown in table 6.31. Using weighted least squares, find the intercept and slope of the best line through the points.

ANSWER

Table 6.32 contains all the quantities necessary to calculate a and b.

Summing the appropriate columns gives

$$\sum 1/s_i^2 = 0.178\,403, \sum x_i/s_i^2 = 15.009\,86, \sum y_i/s_i^2 = 15.465\,28,$$
$$\sum x_i y_i/s_i^2 = 1242.986, \sum x_i^2/s_i^2 = 1379.559.$$

Substituting the summations into equations 6.57 to 6.59 gives

$$a = 128.6, b = -0.4985.$$

Table 6.31. *x-y data for example 10.*

x_i	y_i	s_i
18	125	10
42	108	8
67	91	6
89	84	4
108	76	4

Table 6.32. *Quantities for weighted least squares calculation.*

x_i	y_i	s_i	$1/s_i^2$	x_i/s_i^2	y_i/s_i^2	$x_i y_i/s_i^2$	x_i^2/s_i^2
18	125	10	0.01	0.18	1.25	22.5	3.24
42	108	8	0.015\,625	0.656\,25	1.6875	70.875	27.5625
67	91	6	0.027\,778	1.861\,111	2.527\,778	169.3611	124.6944
89	84	4	0.0625	5.5625	5.25	467.25	495.0625
108	76	4	0.0625	6.75	4.75	513	729

Exercise R

Repeat example 10 with every value of s_i multiplied by 5 (so, for example, when $y_i = 125$, $s_i = 50$). Show that the values of a and b remain unchanged (suggestion: use a spreadsheet!).

6.10.1 Weighted uncertainty in *a* and *b*

When a weighted fit is required,[32] equations 6.62 and 6.63 can be used to calculate the standard uncertainties in a and b:

$$s_a = \left(\frac{\sum \frac{x_i^2}{s_i^2}}{\Delta} \right)^{1/2} \tag{6.62}$$

$$s_b = \left(\frac{\sum \frac{1}{s_i^2}}{\Delta} \right)^{1/2}, \tag{6.63}$$

where Δ is given by equation 6.59.

Equations 6.62 and 6.63 are applicable as long as actual values of s_i are known, as relative magnitudes will not do in this case. Although this might seem unduly restrictive, there is one case in which s_i may be estimated fairly accurately and that is in counting experiments (such as those involving radioactivity or X-rays, where the Poisson distribution is valid). If the number of counts recorded is C_i, then the standard deviation in C_i, s_i, is given by

$$s_i = \sqrt{C_i}. \tag{6.64}$$

Table 6.33. *Variation of counts with depth in a solid.*

C (counts)	1.86×10^4	1.18×10^4	4.33×10^3	1.00×10^3	1.36×10^2
d (mm)	10	30	50	70	90

---

[32] See appendix 4 for more details.

Exercise S

In a diffusion experiment, a radiotracer diffuses into a solid when the solid is heated to a high temperature for a fixed period of time. The solid is sectioned and the number of gamma counts is recorded by a particle counter over a period of 1 minute for each section. Table 6.33 shows the number of counts, C, as a function of depth, d, cut in the material.

Assume that the equation that relates C to d is

$$C = A \exp(\lambda d^2),$$

where A and λ are constants.

  (i) Transform this equation into the form $y = a + bx$.

 (ii) Assuming the standard deviation in C to be $\sqrt{C}$, perform a weighted fit to find values of A, λ, s_A and s_λ.

6.10.2 Weighted standard deviation, s_w

When only the relative magnitudes of s_i are known, it is still possible to determine s_a and s_b. In this case it is necessary to calculate the weighted standard deviation of y values (i.e. the weighted equivalent to equation 6.10). Writing the weighted standard deviation as s_w, we have[33]

$$s_w = \frac{\left(\frac{n}{n-2}\right)^{1/2}}{\sum \frac{1}{s_i^2}} \left[\sum \frac{1}{s_i^2} \sum \frac{y_i^2}{s_i^2} - \left(\sum \frac{y_i}{s_i^2}\right)^2 - \frac{\left(\sum \frac{1}{s_i^2} \sum \frac{x_i y_i}{s_i^2} - \sum \frac{x_i}{s_i^2} \sum \frac{y_i}{s_i^2}\right)^2}{\Delta} \right]^{1/2}$$

(6.65)

where Δ is given by equation 6.59.

We write s_a and s_b as

$$s_a = s_w \left(\frac{\sum \frac{1}{s_i^2} \sum \frac{x_i^2}{s_i^2}}{n\Delta} \right)^{1/2}$$

(6.66)

and

$$s_b = \frac{s_w \sum \frac{1}{s_i^2}}{(n\Delta)^{1/2}}.$$

(6.67)

For completeness, we include the expression for the weighted linear correlation coefficient, r_w, which should be used whenever weighted least squares fitting occurs; r_w is given as,

---

[33] For a derivation of equation 6.65 see chapter 10 of Dietrich (1991).

$$r_w = \frac{\sum \frac{1}{s_i^2} \sum \frac{x_i y_i}{s_i^2} - \sum \frac{x_i}{s_i^2} \sum \frac{y_i}{s_i^2}}{\left[\sum \frac{1}{s_i^2} \sum \frac{x_i^2}{s_i^2} - \left(\sum \frac{x_i}{s_i^2}\right)^2\right]^{1/2} \left[\sum \frac{1}{s_i^2} \sum \frac{y_i^2}{s_i^2} - \left(\sum \frac{y_i}{s_i^2}\right)^2\right]^{1/2}}. \tag{6.68}$$

Example 11

The data shown in table 6.34 were obtained from a study of the relationship between current and voltage for a tunnel diode.[34]

For the range of voltage in table 6.34, the relationship between current, I, and voltage, V, for a tunnel diode can be written as

$$I = CV \exp\left(\frac{-V}{B}\right). \tag{6.69}$$

(i) Transform equation 6.69 into the form $y = a + bx$.
(ii) Determine the summations, $\sum \frac{1}{s_i^2}$, $\sum \frac{x_i}{s_i^2}$ etc. required in the calculation of the weighted standard deviation, s_w, given by equation 6.65. Assume that the uncertainties in the values of current are constant.
(iii) Calculate the weighted standard deviation.

ANSWER

(i) Dividing both sides of equation 6.69 by V gives

$$\frac{I}{V} = C \exp\left(\frac{-V}{B}\right). \tag{6.70}$$

Taking the natural logarithms of both sides of equation 6.70,

$$\ln \frac{I}{V} = \ln C - \frac{V}{B}. \tag{6.71}$$

Comparing equation 6.71 with $y = a + bx$, we find that $y = \ln \frac{I}{V}$, $x = V$, $a = \ln C$ and $b = -\frac{1}{B}$. Plotting $\ln \frac{I}{V}$ versus V would be expected to give a straight line with intercept equal to $\ln C$ and slope equal to $-\frac{1}{B}$.

(ii) In order to determine the standard deviation in the ith y value, s_i, we must consider the left hand side of equation 6.71:

$$y = \ln \frac{I}{V}. \tag{6.72}$$

---

[34] A tunnel diode is a semiconductor device with unusual electrical characteristics. It is sometimes used in high frequency oscillator circuits.

The standard deviation in the ith value y, s_i, is given by

$$s_i \approx \left|\frac{\partial y}{\partial I}\right| s_I, \tag{6.73}$$

now, $\frac{\partial y}{\partial I} = \frac{1}{I}$, so that

$$s_i \approx \frac{s_I}{I}. \tag{6.74}$$

Table 6.35 shows the raw data, the transformed data and the sums of the columns necessary to calculate the weighted standard deviation, s_w. For convenience we take the standard deviation in I, s_I, to be equal to 1, so that using equation 6.74, $s_i = \frac{1}{I}$.

(iii) The weighted standard deviation, calculated using equation 6.65 and the sums of numbers in appearing in the bottom row of table 6.35 is $s_w = 0.1064$.

Table 6.34. *Current–voltage data for a tunnel diode.*

V (V)	0.01	0.02	0.03	0.04	0.05	0.06	0.07	0.08
I (A)	0.082	0.130	0.136	0.210	0.181	0.180	0.190	0.172

V (V)	0.09	0.10	0.11	0.12	0.13	0.14	0.15
I (A)	0.136	0.150	0.108	0.119	0.110	0.070	0.080

Table 6.35. *Weighted fitting of data in table 6.33.*

V (V) = x	I (A)	$\ln(I/V) = y$	$s_i = \frac{1}{I}$	$\frac{1}{s_i^2}$	$\frac{x_i}{s_i^2}$	$\frac{y_i}{s_i^2}$	$\frac{x_i y_i}{s_i^2}$	$\frac{x_i^2}{s_i^2}$	$\frac{y_i^2}{s_i^2}$
0.01	0.082	2.104134	12.19512	0.006724	6.72E-05	0.014148	0.000141	6.72E-07	0.02977
0.02	0.13	1.871802	7.692308	0.0169	0.000338	0.031633	0.000633	6.76E-06	0.059212
0.03	0.136	1.511458	7.352941	0.018496	0.000555	0.027956	0.000839	1.66E-05	0.042254
0.04	0.21	1.658228	4.761905	0.0441	0.001764	0.073128	0.002925	7.06E-05	0.121263
0.05	0.181	1.286474	5.524862	0.032761	0.001638	0.042146	0.002107	8.19E-05	0.05422
0.06	0.18	1.098612	5.555556	0.0324	0.001944	0.035595	0.002136	0.000117	0.039105
0.07	0.19	0.998529	5.263158	0.0361	0.002527	0.036047	0.002523	0.000177	0.035994
0.08	0.172	0.765468	5.813953	0.029584	0.002367	0.022646	0.001812	0.000189	0.017334
0.09	0.136	0.412845	7.352941	0.018496	0.001665	0.007636	0.000687	0.00015	0.003152
0.1	0.15	0.405465	6.666667	0.0225	0.00225	0.009123	0.000912	0.000225	0.003699
0.11	0.108	−0.01835	9.259259	0.011664	0.001283	−0.00021	−2.4E-05	0.000141	3.93E-06
0.12	0.119	−0.00837	8.403361	0.014161	0.001699	−0.00012	−1.4E-05	0.000204	9.92E-07
0.13	0.11	−0.16705	9.090909	0.0121	0.001573	−0.00202	−0.00026	0.000204	0.000338
0.14	0.07	−0.69315	14.28571	0.0049	0.000686	−0.0034	−0.00048	9.6E-05	0.002354
0.15	0.08	−0.62861	12.5	0.0064	0.00096	−0.00402	−0.0006	0.000144	0.002529
			sums	0.307286	0.021316	0.290285	0.013336	0.001824	0.411229

Exercise T

Determine the intercept and slope of the transformed data in example 11. Calculate also the standard uncertainties in intercept and slope.

6.10.3 Weighted least squares and Excel

If the uncertainties in y values are small, then it matters little whether weighted or unweighted least squares is used to fit a straight line to data, as both will yield very similar values for intercept and slope. Owing to its comparative computational simplicity, it is wise to apply unweighted least squares first. Examining the residuals can help decide whether fitting an equation using weighted least squares is warranted. An added advantage of performing unweighted least squares first is that we have values for the 'unweighted' intercept and slope against which weighted estimates of intercept and slope can be compared. If there is large difference between unweighted and weighted estimates this might indicate a mistake in the weighted analysis.

If weighted fitting is required, it is quite daunting to attempt such an analysis equipped with only a pocket calculator. The evaluation of the many summations required for the determination of a and b (see equations 6.57, 6.58 and 6.59) is enough to deter all but the most tenacious person.

Most 'built in' least squares facilities available with calculators and spreadsheets offer only unweighted least squares. This is also true of the LINEST() feature in Excel which does not provide for the weighting of the fit. Nevertheless, a table, such as that appearing in tables 6.32 and 6.35, can be quite quickly drawn up in Excel. The table should include the 'raw x–y data' the standard deviation in each y value, s_i, and the other quantities, such as $x_i y_i / s_i^2$. Using AutoSum, Σ **AutoSum** ▾, which can be found in the Editing group on the Home Ribbon, permits the sums to be calculated with the minimum of effort. The calculations required for the examples in this chapter involving weighted least squares were carried out in this manner.

6.11 Review

An important goal in science is to progress from a qualitative understanding of phenomena to a quantitative description of the relationship between physical variables. Least squares is a powerful and widely used technique in the physical sciences (and in many other disciplines) for establishing a quantitative relationship between two or more variables. In this chapter we have focussed upon fitting an equation representing a straight line to data, as linearly related data frequently emerge from experiments.

Due to the ease with which modern analysis tools such as spreadsheets can fit an equation to data, it is easy to overlook the question 'should we really fit a *straight* line to data?'. To assist in answering this question we introduced the correlation coefficient and residual plots as quantitative and qualitative indicators of 'goodness of fit' and have indicated situations in which each can be misleading. We have also considered situations in which data transformation is required before least squares is applied. Often, after completing the data transformation, we must forsake unweighted least squares in favour of weighted least squares.

The technique of fitting an equation to data using least squares can be extended to situations in which there are more than two parameters to be estimated, for example $y = \alpha + \frac{\beta}{x} + \gamma x$ (where α, β, and γ are the parameters) or where there are more than two independent variables. This will be considered in the next chapter.

End of chapter problems

(1) Table 6.36 contains values of acceleration due to gravity, g, measured at various heights, h, above sea level.

Taking the height as the independent variable (x) and the acceleration due to gravity as the dependent variable (y), do the following.

Table 6.36. *Variation of acceleration due to gravity with height.*

h (km)	g (m/s^2)
10	9.76
20	9.74
30	9.70
40	9.69
50	9.73
60	9.62
70	9.59
80	9.55
90	9.54
100	9.51

(i) Plot a graph of g versus h.

(ii) Assuming g to be linearly related to h, calculate the intercept, a, and slope, b, of the line of best fit.

(iii) Calculate the sum of squares of the residuals and the standard deviation of the y values.

(iv) Calculate the standard uncertainties in a and b.

(v) Determine the correlation coefficient.

(vi) The data pair $h = 50$ km, $g = 9.73$ m/s^2 was incorrectly recorded and should be replaced by $h = 50$ km, $g = 9.65$ m/s^2. Carry out the replacement and repeat parts (i) to (v).

(2) This question uses data from question (1), but with the data pair $h = 50$ km, $g = 9.73$ m/s^2 in table 6.36 replaced by $h = 50$ km, $g = 9.65$ m/s^2.

Assume that the relationship between g and h which is applicable to the data in table 6.36 can be written

$$g = g_0 \left(1 - \frac{2h}{R_E} \right), \tag{6.75}$$

where g_0 is the acceleration due to gravity at the Earth's surface, and R_E is the radius of the Earth.

Use least squares to determine best estimates of g_0 and R_E and the standard uncertainty in each.

(3) When a rigid tungsten sphere presses on a flat glass surface, the surface of the glass deforms. The relationship between the mean contact pressure, p_m, and the indentation strain, s, is

$$p_m = ks. \tag{6.76}$$

Table 6.37 shows s–p_m data for indentations made into glass. k is a constant dependent upon the mechanical stiffness of the material.

Using least squares, determine the best estimate of k and the standard uncertainty in the best estimate.

(4) A thermal convection flowmeter measures the local speed of a fluid by measuring the heat loss from a heated element in the path of the flowing fluid. Assume the relationship between the speed of the fluid, u, and the electrical current, I, supplied to the element is

$$u = K_1 (I^2 - K_2)^2, \tag{6.77}$$

where K_1 and K_2 are constants. Thirty values for u and I obtained in a water flow experiment are shown in table 6.38.

Assume I to be the dependent variable, and u the independent variable.

Table 6.37. *Variation of contact pressure with indentation strain.*

s	p_m (MPa)
0.0280	1.0334
0.0354	1.2295
0.0383	1.1851
0.0415	1.3574
0.0521	1.6714
0.0568	1.7977
0.0570	1.9303
0.0754	2.3957
0.0764	2.4258
0.0975	3.0208
0.1028	2.8824
0.1149	3.5145
0.1210	3.5072
0.1430	3.8888
0.1703	5.0148

Table 6.38. *Speed/current values in a water flow experiment.*

u (m/s)	0.1	0.2	0.3	0.4	0.5	0.6	0.7	0.8	0.9	1.0	1.1
I (μA) $\pm$ 10 μA	270	300	300	330	300	330	350	330	330	360	340

u (m/s)	1.2	1.3	1.4	1.5	1.6	1.7	1.8	1.9	2.0	2.1	2.2
I (μA) $\pm$ 10 μA	340	360	350	370	340	370	360	360	380	380	390

u (m/s)	2.3	2.4	2.5	2.6	2.7	2.8	2.9	3.0
I (μA) $\pm$ 10 μA	390	400	410	410	400	400	410	410

(i) Show that equation 6.77 can be rearranged into the form

$$I^2 = K_2 + \frac{1}{K_1^{1/2}} u^{1/2}. \tag{6.78}$$

(ii) Compare equation 6.78 with $y = a + bx$, and use a spreadsheet to perform a weighted least squares fit to data to find best estimates of K_1 and K_2.

(iii) Plot the standardised residuals. Can anything be concluded from the pattern of the residuals (for example, is the weighting you have used appropriate?).

(5) Acetic acid is adsorbed from solution by activated charcoal. The amount of acetic acid adsorbed, Y, is given by

$$Y = kC^{1/n},\tag{6.79}$$

where C is the concentration of the acetic acid. k and n are constants.

C–Y data are presented in table 6.39.

(i) Transform equation 6.79 into the form $y = a + bx$.
(ii) Use unweighted least squares to find best estimates for k and n and the standard uncertainties in k and n.
(iii) Plot the standardised residuals. Do you think fitting by unweighted least squares is appropriate?
(iv) Determine the 95% coverage interval for Y when $C = 0.085$ mol/L.

Table 6.39. *Variation of adsorbed acetic acid with concentration.*

C (mol/L)	0.017	0.032	0.060	0.125	0.265	0.879
Y (mol)	0.46	0.61	0.79	1.10	1.53	2.45

(6) The variation of intensity, I, of light passing through a lithium fluoride crystal is measured as a function of the angular position of a polariser placed in the path of the light emerging from the crystal. Table 6.40 gives the intensity for various polariser angles, θ.

Assuming that the relationship between I and θ is

$$I = [I_{max} - I_{min}] \cos(2\theta) + I_{min},\tag{6.80}$$

where I_{max} and I_{min} are constants:

(i) perform an unweighted fit to find the intercept and slope of a graph of I versus $\cos(2\theta)$;
(ii) use the values for the slope and intercept to find best estimates of I_{max} and I_{min};
(iii) Determine the standard uncertainty in the best estimate of I_{max}.

Table 6.40. *Variation of intensity with polariser angle.*

θ (degrees)	0	20	40	60	80	100	120	140
Intensity, I (arbitrary units)	1.86	1.63	1.13	0.52	0.16	0.00	0.57	1.11

(7) The Langmuir adsorption isotherm can be written as

$$\frac{P}{X} = \frac{1}{A} - \frac{B}{A}P,$$ (6.81)

where X is the mass of gas adsorbed per unit area of surface, P, is the pressure of the gas, and A and B are constants.

Table 6.41 shows data gathered in an adsorption experiment.

(i)   Write equation 6.81 in the form $y = a + bx$. How do the constants in equation 6.81 relate to a and b?
(ii) Using least squares, find values for a and b and standard uncertainties in a and b.
(iii) Determine the standard uncertainties in A and B.

Table 6.41. *Gas adsorbed as a function of pressure.*

P (N/m^2)	X (kg/m^2)
0.27	13.9×10^{-5}
0.39	17.8×10^{-5}
0.62	22.5×10^{-5}
0.93	27.5×10^{-5}
1.72	32.9×10^{-5}
3.43	38.6×10^{-5}

(8) The relationship between mean free path, λ, of electrons moving through a gas and the pressure of the gas, P, can be written

$$\frac{1}{\lambda} = \left(\frac{\pi d^2}{4kT}\right)P,$$ (6.82)

where d is the diameter of the gas molecules, k is Boltzmann's constant ($= 1.38 \times 10^{-23}$ J/K) and T is the temperature of the gas in kelvins. Table 6.42 shows data gathered in an experiment in which λ was measured as a function of P.

(i) Fit the equation $y = a + bx$ to equation 6.82 to find b and the standard uncertainty in b.
(ii) If the temperature of the gas is 298 K, use equation 6.82 to estimate the diameter of the gas molecules and the standard uncertainty in the diameter.

(9) The intercept on the x axis x_{INT} of the best straight line through data is found by setting $\hat{y} = 0$ in equation 6.3. x_{INT} is given by

Table 6.42. *Variation of free path, λ, with pressure.*

λ (mm)	P (Pa)
35	1.5
20	5.8
30	5.8
25	6.5
16	8.0
9	13.1
10	14.5
7	18.9
6	24.7
6.5	27.6

$$x_{INT} = -\frac{a}{b}.$$

Given that standard uncertainties in $\bar{y}$ and b are

$$s_{\bar{y}} = \frac{s}{\sqrt{n}} \text{ and } s_b = \frac{sn^{1/2}}{\left[n\sum x_i^2 - \left(\sum x_i\right)^2\right]^{1/2}},$$

show that the standard uncertainty in the intercept, $s_{x_{INT}}$, can be expressed as

$$s_{x_{INT}} = \frac{s}{b}\left[\frac{1}{n} + \frac{n\bar{y}^2}{b^2\left(n\sum x_i^2 - \left(\sum x_i\right)^2\right)}\right]^{1/2}. \tag{6.83}$$

(10) In section 6.4.2, we found that

$$\hat{y}_0 = \bar{y} + b(x_0 - \bar{x}).$$

Show that if the errors in $\bar{y}$ and b are not correlated, the standard uncertainty in $\hat{y}_0$, $s_{\hat{y}_0}$, can be written

$$s_{\hat{y}_0} = s\left[\frac{1}{n} + \frac{n(x_0 - \bar{x})^2}{n\sum x_i^2 - \left(\sum x_i\right)^2}\right]^{1/2}, \tag{6.84}$$

where expressions for standard uncertainties in $\bar{y}$ and b are given in question 9.
(11) The equation of a straight line passing through the origin can be written, $y = bx$.

Use the method outlined in appendix 3 to show that the slope, b, of the best line to pass through the points and the origin is given by

$$b = \frac{\sum x_i y_i}{\sum x_i^2}.$$

Use the method outlined in appendix 4 to show that the standard uncertainty in b, s_b, is given by

$$s_b = \frac{s}{\left(\sum x_i^2\right)^{1/2}}, \tag{6.85}$$

where s is given by equation 6.10.

A word of caution: in situations involving 'real' data it is advisable not to treat the origin as a special point by forcing a line through it, even if 'theoretical' considerations predict that the line *should* pass through the origin. In most situations, random and systematic errors conspire to move the line so that it does not pass through the origin.

(12) Data obtained in an experiment designed to study whether the amount of CaO in rocks drawn from a geothermal field is correlated to the amount of MgO in those rocks are shown in table 6.43.

   (i) Plot a graph of concentration of CaO versus concentration of MgO.
  (ii) Calculate the linear correlation coefficient, r.
 (iii) Is the value of r obtained in (ii) significant?

Table 6.43. *Concentrations of CaO and MgO in rock specimens.*

CaO (wt%)	MgO (wt%)
2.43	0.72
4.90	2.42
10.43	4.75
15.64	3.99
16.62	5.39
21.12	8.65

(13) A thermoelectric generator (TEG) is a device capable of converting thermal energy to electrical energy. Table 6.44 shows how the voltage, V, at the terminals of the TEG varies with the electrical load, R, to which the TEG is attached.

The relationship between the current, I, through the load and R can be written as

$$\frac{1}{I} = \frac{r}{E} + \frac{R}{E}. \qquad\qquad (6.86)$$

(i) Use Excel to complete the third column in table 6.44.
(ii) Plot a graph of $1/I$ versus R.
(iii) By comparing $y = a + bx$ with equation 6.86, show that $a = r/E$ and $b = 1/E$.
(iv) Use unweighted least squares to determine the best estimates of r and E and the standard uncertainty in each.
(v) Is fitting by unweighted least squares justified in this situation?

Table 6.44. *Relationship between load attached to a TEG and the voltage across the terminals of the TEG.*

R (Ω)	V (mV)	$1/I = R/V$ (mA^{-1})
1	11.2	0.089 286
2	16.9	0.118 343
3	21.1	
4	24.1	
5	26.1	
6	28.8	
7	30.1	
8	30.6	
9	31.5	
10	32.4	

(14) A *Spathiphyllum wallisii* 'Petite' indoor potted plant is sealed in a perspex chamber with field capacity water[35] at 23 °C. The amount of CO_2 in the chamber is measured over a period of 6 hours using an infrared gas analyser. Data gathered are shown in table 6.45.

Assuming that the relationship between CO_2 concentration and time is linear:

(i) fit a straight line to the data using unweighted least squares and determine the slope, intercept and the standard uncertainties in slope and intercept;
(ii) calculate the correlation coefficient, r;
(iii) use the line of best fit to estimate the concentration of CO_2 at $t = 150$ min and $t = 400$ min;
(iv) calculate the 95% coverage interval for the true value of the concentration of CO_2 at $t = 150$ min and $t = 400$ min.

---

[35] Field capacity water is the amount of water held in the soil after excess water has drained away.

Table 6.45. *Variation of CO_2 concentration with time.*

t (min)	CO_2 concentration (ppm)
0	404
60	478
120	556
180	633
240	716
300	796
360	880

(15) In a study on skin cancer, the concentration of antimony in lymph nodes was established using the technique of Inductivity Coupled Plasma Mass Spectrometry (ICP-MS). An ICP-MS instrument was calibrated using samples containing known amounts of antimony. Table 6.46 shows the output of the instrument (in counts per second, cps) for various antimony concentrations (antimony concentration is expressed in parts per billion, ppb).

   (i) Plot a calibration graph of instrument output versus antimony concentration.
  (ii) Fit a straight line to the data in table 6.46 using unweighted least squares and determine the slope and intercept, as well as the standard uncertainties in the slope an intercept.
 (iii) Three repeat measurements were made on a sample containing unknown antimony concentration. The mean of the three values is 6255 cps. Use this information to find the best estimate of the antimony concentration in the sample and the 95% coverage interval for the true concentration.

Table 6.46. *Calibration data for an ICP-MS.*

Concentration of antimony (ppb)	Output of ICP-MS (cps)
0	71
0.5	1426
1.0	2707
2.0	4974
5.0	12982
10.0	26062
20.0	53135

(16) When a photon is emitted from a nucleus, the photon's energy is shifted due to the recoil of the nucleus. The relationship between the energy of the photon, E, and the angle, θ, between the direction of the recoil of the nucleus and the focal plane containing the photon detectors may be written

$$E = E_o \left(1 + \frac{v}{c} \cos(\theta) \right), \tag{6.87}$$

where E_o is the energy of the photon in the absence of recoil, v is the recoil velocity of the nucleus, and c is the speed of light. In this question take c to be equal to 2.998×10^8 m/s. See also the data in Table 6.47.

 (i) Plot a graph of E versus $\cos(\theta)$.
 (ii) Use unweighted least squares to fit equation 6.87 to the E versus $\cos(\theta)$ data.
 (iii) Using the fit to data, estimate E_o and v and the standard uncertainties in these.
 (iv) Use a plot of residuals to provide evidence as to the goodness of fit of the equation to data.

Table 6.47. *Variation of photon energy, E, with angle, θ.*

angle, θ (degrees)	energy, E (keV)
31.72	468.10
37.38	467.61
50.07	464.95
58.28	463.04
69.82	460.19
79.19	457.60
80.71	457.20
90.00	454.87
99.29	452.40
100.81	451.90
110.18	449.55
121.72	446.90
129.93	445.31
142.62	443.13
148.28	442.17
162.73	440.50

(17) As part of a study on the effect of climate change on the survival of Australian plant species, leaf temperature was recorded for leaves of varying size. Table 6.48 shows how the temperature ΔT, where $\Delta T = $ (leaf temperature – ambient temperature) varies with the surface area, A, of leaves as recorded during on a hot summer's day.

Assuming that the relationship between ΔT and A can be written

$$\Delta T = k \log_{10} A + D. \tag{6.88}$$

- (i) Plot a graph of ΔT versus $\log_{10} A$.
- (ii) Fit a straight line to ΔT versus $\log_{10} A$ data using unweighted least squares and find best estimates for k and D.
- (iii) Determine standard uncertainties in best estimates of k and D.
- (iv) Plot the residuals. Does it appear as though a weighted or unweighted fit is appropriate?

Table 6.48. *Variation of ΔT with surface area of leaves.*

Leaf area, A, (cm^2)	ΔT (°C)	Leaf area, A (cm^2)	ΔT (°C)
1.19	3.5	54.89	6.3
1.58	2.7	56.09	5.0
1.69	3.3	70.82	4.0
3.50	3.1	72.24	4.5
8.92	4.2	83.80	7.1
11.23	3.8	115.94	4.7
14.48	5.6	133.57	7.0
17.32	5.3	151.90	9.7
18.68	5.4	229.78	5.7
20.56	5.1	238.66	4.8
21.63	3.4	300.01	7.9
21.79	3.7	314.41	8.8
38.93	6.3	360.05	6.4

(18) As the random error of individual y values increases, so do the standard uncertainties in slope and intercept. Figure 6.24 shows an Excel spreadsheet in which 'error-free' y values have been generated in the B column using the relationship $y = 1.54 x + 4$. Errors, which have a normal distribution with mean of zero and standard deviation of 1, were generated in the C column using the Random Number Generator in the Analysis Toolpak. The D column contains the sum of the errors and the error-free y values. The LINEST() function was used to determine the best estimates of slope and intercept and standard uncertainties in these.

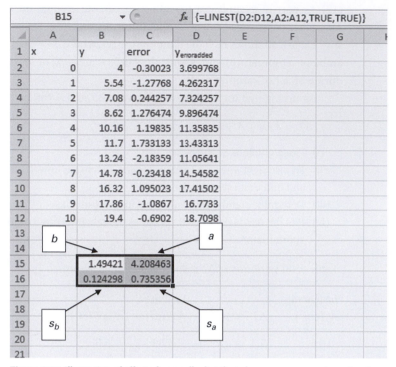

Figure 6.24. Illustration of effect of normally distributed errors on parameter estimates.

(i) Keeping the range of x values the same as in figure 6.24, change the interval between successive x values from 1 to 0.2, so that the total number of x values in the range 0 to 10 increases from 11 to 51. Repeat the procedure described above of adding errors to the y values of mean = 0 and standard deviation = 1.

(ii) What is the effect on the standard uncertainties in the slope and intercept of increasing the number of points?

(iii) By examining the relationship between the standard deviation of sample means and the number of values used to calculate the mean of a sample (see equation 3.21) how might you expect the standard uncertainties in slope and intercept to depend on the number of data points?

(iv) If you increase the number of points from 51 to 501 (while keeping the range and the error distribution as before), how would you expect the standard uncertainties in slope and intercept to change? Try this.

Chapter 7

Least squares II

7.1 Introduction

In chapter 6 we considered fitting a straight line to x–y data in situations in which the equation representing the line is written

$$y = a + bx,$$

where a is the intercept of the line and b is its slope. Other situations that we need to consider, as they occur regularly in the physical sciences, require fitting equations to x–y data where the equations:

- possess more than two parameters that must be estimated, for example, $y = a + bx + cx^2$ or $y = a + \dfrac{b}{x} + cx$;
- contain more than one independent variable, for example $y = a + bx + cz$ where x and z are the independent variables;
- cannot be written in a form suitable for analysis by linear least squares, for example $y = a + be^{cx}$.

As in chapter 6, we apply the technique of least squares to obtain best estimates of the parameters in equations to be fitted to data. As the number of parameters increases, so does the number of the calculations required to estimate those parameters. Matrices are introduced as a means of solving for parameter estimates efficiently. Matrix manipulation is tedious to carry out 'by hand', but the built in matrix functions in Excel make it well suited to assist in extending the least squares technique. In addition, Excel's dedicated least squares function, LINEST(), is able to fit equations to data where those equations contain more than two parameters.

We also consider in this chapter the important issue of choosing the best equation to fit to data, and what steps to take if two (or more) equations fitted to the same data must be compared.

7.2 Extending linear least squares

In chapter 6 we introduced α and β as (population) parameters which, in the absence of systematic errors, are regarded as the 'true' intercept and slope respectively of a straight line drawn through x–y data. Estimates of these parameters, obtained using least squares, were written as a and b. As we extend the least squares technique, we must deal with an increased number of parameters. Consistent with our earlier choice of symbols, we write estimates of parameters obtained by least squares as a, b, c, d and so on.

Fitting a straight line to data has a central position in the analysis of experimental data in the physical sciences. This is due to the fact that many physical quantities (at least over a limited range) are linearly related to each other. For example, consider the relationship between the output voltage of a thermocouple and the temperature of one junction of a thermocouple[1] as shown in figure 7.1.

There is little departure from linearity exhibited by the data in figure 7.1. However, the linearity of output voltage with temperature for the thermocouple is no longer maintained when a much larger range of temperature is considered, as shown in figure 7.2.

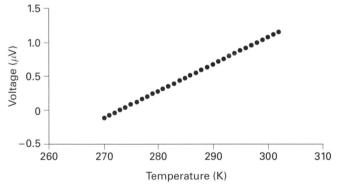

Figure 7.1. Output voltage of a thermocouple between 270 K and 303 K.

---

[1] The graph shown is for a Type K thermocouple. The reference junction of the thermocouple was held at 273 K.

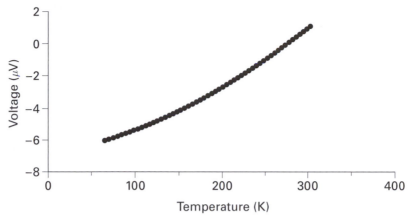

Figure 7.2. Output voltage of a thermocouple between 70 K and 303 K.

If x–y data recorded during an experiment are not linearly related, then they can often be 'linearised' by some straightforward transformation (such as taking the logarithms of the dependent variable, as described in section 6.9). In other situations, such as for the thermocouple voltage–temperature-values shown in figure 7.2, no amount of legitimate mathematical manipulation will linearise these data, and we must consider fitting an equation other than $y = a + bx$ to the data.

The technique of linear least squares is not confined to finding the best straight line through data, but can be extended to any function linear in the parameters appearing in the equation. As examples, a, b, and c in the following equations can be determined using linear least squares:

$$y = a + bx + cx^2;$$
$$y = a + bx + cx \ln x;$$
$$y = a + \frac{b}{x} + cx;$$
$$y = \frac{a}{x} + bx + cx^2.$$

In fitting functions with more than two parameters to data, we continue to assume that:

- random errors affecting y values are normally distributed;
- there are no random errors in the independent (x) values;
- systematic errors in both x and y values are negligible.

The approach to finding an equation which best fits the data, where the equation incorporates more than two parameters follows that described in

appendix 3. Though the complexity of the algebraic manipulations increases as
the number of parameters increases, this can be overcome by using matrices to
assist in parameter estimation.

7.3 Formulating equations to solve for parameter estimates

To determine best estimates of the parameters in an equation using the
technique of least squares, we begin with the weighted sum of squares of
residuals, χ^2, given by[2]

$$\chi^2 = \sum \left[\frac{y_i - \hat{y}_i}{s_i} \right]^2,$$

(7.1)

where y_i is the ith value of y (obtained through measurement), $\hat{y}_i$ is the
corresponding predicted value of y found using an equation relating y to x,
and s_i is the standard deviation in the ith value of y.

   As an example of fitting an equation with more than two parameters to
data, consider the equation

$$y = a + bx + cx^2.$$

(7.2)

To determine a, b and c, replace $\hat{y}_i$ in equation 7.1 by $a + bx_i + cx_i^2$, so that

$$\chi^2 = \sum \left[\frac{y_i - (a + bx_i + cx_i^2)}{s_i} \right]^2.$$

(7.3)

If the standard deviation in y_i is the same for all x_i then s_i is replaced by s and
brought outside the summation, so that

$$\chi^2 = \frac{1}{s^2} \sum (y_i - a - bx_i - cx_i^2)^2.$$

(7.4)

To find a, b and c that minimise χ^2, differentiate χ^2 with respect to a, b and c in
turn then set the resulting equations equal to zero.

   Differentiating equation 7.4 gives

$$\frac{\partial \chi^2}{\partial a} = \frac{-2}{s^2} \sum (y_i - a - bx_i - cx_i^2) = 0$$

(7.5)

$$\frac{\partial \chi^2}{\partial b} = \frac{-2}{s^2} \sum x_i(y_i - a - bx_i - cx_i^2) = 0$$

(7.6)

$$\frac{\partial \chi^2}{\partial c} = \frac{-2}{s^2} \sum x_i^2(y_i - a - bx_i - cx_i^2) = 0.$$

(7.7)

---

[2] Appendix 3 explains the origin of χ^2.

Rearranging equations 7.5 to 7.7 gives[3]

$$na + b\sum x_i + c\sum x_i^2 = \sum y_i \tag{7.8}$$

$$a\sum x_i + b\sum x_i^2 + c\sum x_i^3 = \sum x_i y_i \tag{7.9}$$

$$a\sum x_i^2 + b\sum x_i^3 + c\sum x_i^4 = \sum x_i^2 y_i. \tag{7.10}$$

We can rearrange equations 7.8 to 7.10 and substitute from one equation into another to solve for a, b and c. However, this approach is labour intensive, time consuming, and the likelihood of a numerical or algebraic mistake is quite high. If the number of parameters to be estimated increases to four or more, then solving for these estimates by 'elimination and substitution' becomes even more formidable. An effective way to proceed is to write the equations in matrix form, as solving linear equations using matrices is efficient, especially if software is available that can manipulate matrices.

Writing equations 7.8 to 7.10 in matrix form gives

$$\begin{bmatrix} n & \sum x_i & \sum x_i^2 \\ \sum x_i & \sum x_i^2 & \sum x_i^3 \\ \sum x_i^2 & \sum x_i^3 & \sum x_i^4 \end{bmatrix} \begin{bmatrix} a \\ b \\ c \end{bmatrix} = \begin{bmatrix} \sum y_i \\ \sum x_i y_i \\ \sum x_i^2 y_i \end{bmatrix}. \tag{7.11}$$

Equation 7.11 can be written concisely as[4]

$$\mathbf{AB} = \mathbf{P}, \tag{7.12}$$

where

$$\mathbf{A} = \begin{bmatrix} n & \sum x_i & \sum x_i^2 \\ \sum x_i & \sum x_i^2 & \sum x_i^3 \\ \sum x_i^2 & \sum x_i^3 & \sum x_i^4 \end{bmatrix} \quad \mathbf{B} = \begin{bmatrix} a \\ b \\ c \end{bmatrix} \quad \mathbf{P} = \begin{bmatrix} \sum y_i \\ \sum x_i y_i \\ \sum x_i^2 y_i \end{bmatrix}.$$

To determine elements, a, b, and c of $\mathbf{B}$ (which are the parameter estimates appearing in equation 7.2), equation 7.12 is manipulated to give[5]

$$\mathbf{B} = \mathbf{A}^{-1}\mathbf{P}, \tag{7.13}$$

where $\mathbf{A}^{-1}$ is the inverse matrix of the matrix, $\mathbf{A}$. Matrix inversion and matrix multiplication are tedious to perform without the aid of a computer, especially if matrices are large. The built in matrix functions in Excel are well suited to estimating parameters in linear least squares problems.

---

[3] Note that $\sum_{i=1}^{i=n} a = a + a + a + \cdots + a = na.$

[4] By convention, matrices are indicated in bold type.

[5] See appendix 5 for an introduction to matrices.

Exercise A

The relationship between voltage across a semiconductor diode, V, and temperature (in kelvin) of the diode, T, is given by

$$V = \alpha + \beta T + \gamma T \ln T, \tag{7.14}$$

where α, β and γ are constants.

An experiment is performed in which the voltage across the diode is measured as the temperature of the diode changes. Express, in matrix form, the equations that must be solved to obtain best estimates of α, β and γ (assume an unweighted fit is appropriate).

7.4 Matrices and Excel

Among the many built in functions provided by Excel are those related to matrix inversion and matrix multiplication. These functions can be used to solve for the parameter estimates (and the standard uncertainties in the estimates) appearing in an equation fitted to data.

7.4.1 The MINVERSE() function

An important step in parameter estimation using matrices is that of matrix inversion. The Excel function MINVERSE() may be used to invert a matrix such as that appearing in sheet 7.1.

The syntax of the function is

$= \text{MINVERSE(array)}$

where array contains the elements of the matrix to be inverted.

Sheet 7.1. *Cells containing a square matrix.*

	A	B	C
1	2.6	7.3	3.4
2	9.5	4.5	5.5
3	6.7	2.3	7.8
4			

Example 1

Use the MINVERSE() function to invert the matrix shown in sheet 7.1.

ANSWER

Inverting a 3×3 matrix creates another 3×3 matrix. As the MINVERSE() function returns an array with nine elements, we highlight nine cells (cells E1:G3) into which Excel can return those elements. Sheet 7.2 shows the highlighted cells along with the function =MINVERSE(A1:C3) typed into cell E1.

To complete the inversion of the matrix shown in columns A to C in sheet 7.2, it is necessary to hold down the CTRL and SHIFT keys together then press the ENTER key. Sheet 7.3 shows the elements of the inverted matrix in columns E to G.

Sheet 7.2. *Example of matrix inversion using the MINVERSE() function in Excel.*

	A	B	C	D	E	F	G
1	2.6	7.3	3.4		=MINVERSE(A1:C3)		
2	9.5	4.5	5.5				
3	6.7	2.3	7.8				
4							

Sheet 7.3. *Outcome of matrix inversion.*

	A	B	C	D	E	F	G
1	2.6	7.3	3.4		−0.09285	0.203164	−0.10278
2	9.5	4.5	5.5		0.154069	0.01034	−0.07445
3	6.7	2.3	7.8		0.034329	−0.17756	0.238445
4							

Exercise B

Use Excel to invert the following matrices:

$$\text{(i)} \begin{bmatrix} 2.5 & 9.9 & 6.7 \\ 7.8 & 3.0 & 6.9 \\ 8.8 & 4.5 & 3.2 \end{bmatrix}; \quad \text{(ii)} \begin{bmatrix} 62 & 41 & 94 & 38 \\ 42 & 65 & 41 & 87 \\ 102 & 45 & 76 & 32 \\ 91 & 83 & 14 & 21 \end{bmatrix}; \quad \text{(iii)} \begin{bmatrix} 2.3 & 1.8 & 4.4 & 9.4 & 6.9 \\ 6.7 & 9.9 & 3.4 & 3.3 & 7.0 \\ 8.4 & 3.4 & 8.2 & 3.9 & 2.9 \\ 9.4 & 6.6 & 9.0 & 6.6 & 5.6 \\ 10.4 & 1.5 & 1.7 & 5.7 & 4.6 \end{bmatrix}.$$

7.4.2 The MMULT() function

Matrix multiplication is required when performing least squares using matrices. The MMULT() function in Excel is an array function which returns the product of two matrices.

Suppose $\mathbf{P} = \mathbf{A}\,\mathbf{B}$, where

$$\mathbf{A} = \begin{bmatrix} 2.6 & 7.3 & 3.4 \\ 9.5 & 4.5 & 5.5 \\ 6.7 & 2.3 & 7.8 \end{bmatrix} \quad \text{and} \quad \mathbf{B} = \begin{bmatrix} 34.4 \\ 43.7 \\ 12.3 \end{bmatrix}.$$

The elements of the **P** matrix can be found using the MMULT() function. The syntax of the function is

$$= \text{MMULT}(\text{array1}, \text{array2})$$

where array1 and array2 contain the elements of the matrices to be multiplied together.

Example 2

Use the MMULT() function to determine the product, **P**, of the matrices **A** and **B** shown in sheet 7.4.

ANSWER

Multiplying the 3×3 matrix by the 3×1 matrix appearing in sheet 7.4, produces another matrix of dimension 3×1. As the MMULT() function returns an array containing the elements of the matrix **P**, we highlight cells (shown in column G of sheet 7.5) into which those elements can be returned.

To determine **P**, type **=MMULT(A2:C4,E2:E4)** into cell G2. Holding down the CTRL and SHIFT keys, then pressing the Enter key returns the elements of the **P** matrix into cells G2 to G4 as shown in sheet 7.6.

Sheet 7.4. *Two matrices, **A** and **B** to be multiplied.*

	A	B	C	D	E	F	G
1		A			B		
2	2.6	7.3	3.4		34.4		
3	9.5	4.5	5.5		43.7		
4	6.7	2.3	7.8		12.4		

Sheet 7.5. *Matrix multiplication using MMULT().*

	A	B	C	D	E	F	G	H
1		A			B		P	
2	2.6	7.3	3.4		34.4		=MMULT(A2:C4,E2:E4)	
3	9.5	4.5	5.5		43.7			
4	6.7	2.3	7.8		12.4			

Sheet 7.6. *Outcome of multiplying* **A**
and **B** *appearing in sheet 7.4.*

	G
1	**P**
2	450.61
3	591.65
4	427.71

Exercise C

Use Excel to carry out the following matrix multiplications:

(i) $\begin{bmatrix} 56.8 & 123.5 & 67.8 \\ 87.9 & 12.5 & 54.3 \\ 23.6 & 98.5 & 56.7 \end{bmatrix} \begin{bmatrix} 23.1 \\ 34.6 \\ 56.8 \end{bmatrix}$; (ii) $\begin{bmatrix} 12 & 45 & 67 & 56 \\ 34 & 54 & 65 & 43 \\ 12 & 54 & 49 & 31 \\ 84 & 97 & 23 & 99 \end{bmatrix} \begin{bmatrix} 32 \\ 19 \\ 54 \\ 12 \end{bmatrix}$.

7.4.3 Fitting the polynomial $y = a + bx + cx^2$ to data

We bring together the matrix approach to solving for estimated parameters discussed in section 7.3 with the built in matrix functions in Excel to determine a, b, and c in the equation $y = a + bx + cx^2$. Consider the data in table 7.1.

Assuming it is appropriate to fit the equation $y = a + bx + cx^2$ to the data in table 7.1, we can find a, b and c by constructing the matrices

$$\mathbf{A} = \begin{bmatrix} n & \sum x_i & \sum x_i^2 \\ \sum x_i & \sum x_i^2 & \sum x_i^3 \\ \sum x_i^2 & \sum x_i^3 & \sum x_i^4 \end{bmatrix} \qquad (7.15)$$

$$\mathbf{B} = \begin{bmatrix} a \\ b \\ c \end{bmatrix} \qquad (7.16)$$

and

$$\mathbf{P} = \begin{bmatrix} \sum y_i \\ \sum x_i y_i \\ \sum x_i^2 y_i \end{bmatrix} . \qquad (7.17)$$

Using the data in table 7.1 (and with the assistance of Excel), we find

$$\sum x_i = 108; \quad \sum y_i = 2845.6; \quad \sum x_i^2 = 1536; \quad \sum x_i^3 = 24192;$$
$$\sum x_i^4 = 405312; \quad \sum x_i y_i = 45206.6; \quad \sum x_i^2 y_i = 761100.4; \quad n = 9.$$

Table 7.1. *x–y data.*

x	y
4	30.4
6	51.2
8	101.6
10	184.4
12	262.6
14	369.6
16	479.4
18	601.5
20	764.9

Matrices **A** and **P** become

$$\mathbf{A} = \begin{bmatrix} 9 & 108 & 1536 \\ 108 & 1536 & 24192 \\ 1536 & 24192 & 405312 \end{bmatrix} \quad \text{and} \quad \mathbf{P} = \begin{bmatrix} 2845.6 \\ 45206.6 \\ 761100.4 \end{bmatrix}.$$

As $\mathbf{B} = \mathbf{A}^{-1}\mathbf{P}$, we find (using Excel for matrix inversion and multiplication)[6]

$$\mathbf{B} = \begin{bmatrix} a \\ b \\ c \end{bmatrix}$$

$$= \begin{bmatrix} 3.505 & -6.214 \times 10^{-1} & 2.381 \times 10^{-2} \\ -6.214 \times 10^{-1} & 1.210 \times 10^{-1} & -4.870 \times 10^{-3} \\ 2.381 \times 10^{-2} & -4.870 \times 10^{-3} & 2.029 \times 10^{-4} \end{bmatrix} \begin{bmatrix} 2845.6 \\ 45206.6 \\ 761100.4 \end{bmatrix}$$

$$= \begin{bmatrix} 1.9157 \\ -2.7458 \\ 2.0344 \end{bmatrix},$$

so that the relationship between x and y can be written,

$$y = 1.916 - 2.746x + 2.034x^2.$$

7.4.4 The X and X^T matrices

The elements of the **A** matrix given by equation 7.15 can be determined by constructing an Excel worksheet to calculate Σx, Σx^2 and so on, column by

---

[6] For convenience, the elements of $\mathbf{A}^{-1}$ are shown to four figures, but full precision (to 15 figures) is used 'internally' by Excel when calculations are carried out.

column, However there is a more efficient way to determine the elements in equation 7.15. We introduce the matrix, $\mathbf{X}$, given by

$$\mathbf{X} = \begin{pmatrix} 1 & x_1 & x_1^2 \\ 1 & x_2 & x_2^2 \\ 1 & x_3 & x_3^2 \\ 1 & x_i & x_i^2 \\ \vdots & \vdots & \vdots \\ 1 & x_n & x_n^2 \end{pmatrix}. \tag{7.18}$$

Transposing the matrix, $\mathbf{X}$, gives[7]

$$\mathbf{X}^T = \begin{pmatrix} 1 & 1 & 1 & 1 & \cdots & 1 \\ x_1 & x_2 & x_3 & x_i & \cdots & x_n \\ x_1^2 & x_2^2 & x_3^2 & x_i^2 & \cdots & x_n^2 \end{pmatrix}. \tag{7.19}$$

The elements of $\mathbf{A}$ matrix emerge by multiplying $\mathbf{X}^T$ and $\mathbf{X}$ together, i.e.

$$\mathbf{A} = \mathbf{X}^T\mathbf{X}. \tag{7.20}$$

The matrix $\mathbf{X}$ may be formed in Excel and Excel's TRANSPOSE() function may be used to form $\mathbf{X}^T$.

Similarly, if the $\mathbf{Y}$ matrix is given by

$$\mathbf{Y} = \begin{pmatrix} y_1 \\ y_2 \\ y_3 \\ y_i \\ \vdots \\ y_n \end{pmatrix} \tag{7.21}$$

then the $\mathbf{P}$ matrix given by equation 7.17 may be calculated using

$$\mathbf{P} = \mathbf{X}^T\mathbf{Y}. \tag{7.22}$$

As $\mathbf{B} = \mathbf{A}^{-1}\mathbf{P}$ (see section 7.3), then

$$\mathbf{B} = \left(\mathbf{X}^T\mathbf{X}\right)^{-1}\mathbf{X}^T\mathbf{Y}. \tag{7.23}$$

Example 3

Given that

$$\mathbf{X} = \begin{pmatrix} 1 & 2 & 4 \\ 1 & 3 & 9 \\ 1 & 4 & 16 \\ 1 & 5 & 25 \\ 1 & 6 & 36 \\ 1 & 7 & 49 \end{pmatrix} \quad \text{and} \quad \mathbf{Y} = \begin{pmatrix} 3.5 \\ 4.7 \\ 5.8 \\ 6.6 \\ 7.9 \\ 9.0 \end{pmatrix},$$

(1) determine the matrices, $\mathbf{X}^T\mathbf{X}$ and $\mathbf{X}^T\mathbf{Y}$ and $\left(\mathbf{X}^T\mathbf{X}\right)^{-1}$;

(2) use equation 7.23 to find the elements of $\mathbf{B}$.

---

[7] $\mathbf{X}^T$ is used to represent the transpose of $\mathbf{X}$.

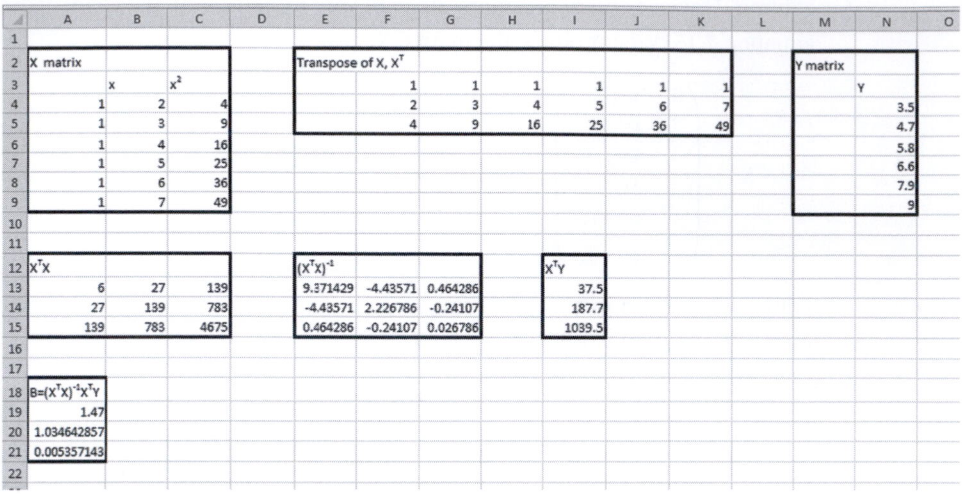

Figure 7.3. Use of Excel to transpose, invert and multiply matrices.

ANSWER

Figure 7.3 shows an Excel worksheet used to calculate $\mathbf{X}^T\mathbf{X}$, $\mathbf{X}^T\mathbf{Y}$, $(\mathbf{X}^T\mathbf{X})^{-1}$ and $(\mathbf{X}^T\mathbf{X})^{-1}\mathbf{X}^T\mathbf{Y}$.

Cells A19 to A21 shown in figure 7.3 contain the elements of the **B** matrix.

Exercise D

The variation of the electrical resistance, R, with temperature, T, of a wire made from high purity platinum is shown in table 7.2.

Assuming the relationship between R and T can be written

$$R = A + BT + CT^2, \tag{7.24}$$

use least squares to determine best estimates for A, B and C.

7.5 Fitting equations with more than one independent variable[8]

To this point we have considered situations in which the dependent variable, y, is a function of only one independent variable, x. However, there are situations

---

[8] Such fitting is often referred to as 'multiple regression'.

Table 7.2. *Variation of resistance with temperature of a platinum wire.*

T (K)	R (Ω)
70	17.1
100	30.0
150	50.8
200	71.0
300	110.5
400	148.6
500	185.0
600	221.5
700	256.2
800	289.8
900	322.2
1000	353.4

in which the effect of two or more independent variables must be accounted for. For example, when vacuum depositing a thin film, the thickness, T, of the film may depend of several variables including:

(1) deposition time, t;
(2) gas pressure in vacuum chamber, P;
(3) distance, d, between deposition source and surface to be coated with the thin film.

To find the relationship between T and the independent variables, we could consider one independent variable at a time and experimentally determine the relationship between that variable and T while holding the other variables fixed. Another approach is to allow all the variables t, P and d to vary simultaneously from one experiment to another and use least squares to establish the relationship between T and all the independent variables.

An example of an equation possessing two independent variables is

$$y = a + bx + cz. \tag{7.25}$$

As usual, y is the dependent variable. Here x and z are independent variables. Can we determine a, b and c by fitting equation 7.25 to experimental data using linear least squares? The answer is yes, and the problem is no more complicated than fitting a polynomial to data.

We begin with the weighted sum of squares of residuals, χ^2, given by

$$\chi^2 = \sum \left[\frac{y_i - \hat{y}_i}{s_i}\right]^2.\tag{7.26}$$

To fit the equation $y = ax + bz + c$ to data, replace $\hat{y}_i$ in equation 7.26 by $a + bx_i + cz_i$ to give

$$\chi^2 = \sum \left[\frac{y_i - (a + bx_i + cz_i)}{s_i}\right]^2.\tag{7.27}$$

If the uncertainty in the y values is constant, s_i is replaced by s, so that

$$\chi^2 = \frac{1}{s^2} \sum (y_i - a - bx_i - cz_i)^2.\tag{7.28}$$

To minimise χ^2, differentiate χ^2 with respect to a, b and c in turn and set the resulting equations equal to zero. Following the approach described in section 7.2, the matrix equation to be solved for a, b and c, is

$$\underbrace{\begin{bmatrix} n & \sum x_i & \sum z_i \\ \sum x_i & \sum x_i^2 & \sum x_i z_i \\ \sum z_i & \sum x_i z_i & \sum z_i \end{bmatrix}}_{\mathbf{A}} \underbrace{\begin{bmatrix} a \\ b \\ c \end{bmatrix}}_{\mathbf{B}} = \underbrace{\begin{bmatrix} \sum y_i \\ \sum x_i y_i \\ \sum z_i y_i \end{bmatrix}}_{\mathbf{P}}.\tag{7.29}$$

To solve for a, b and c, we determine $\mathbf{B} = \mathbf{A}^{-1}\mathbf{P}$.

Example 4

In an experiment to study waves on a stretched string, the frequency, f, at which the string resonates is measured as a function of the tension, T, of the string and the length, L, of the string. The data gathered in the experiment are shown in table 7.3.

Assume that the frequency, f, can be written as

$$f = KT^B L^C,\tag{7.30}$$

where K, B and C are constants to be estimated using least squares.

Taking natural logarithms of both sides of equation 7.30 gives

$$\ln f = \ln K + B \ln T + C \ln L;\tag{7.31}$$

use least squares to determine best estimates of K, B and C.

ANSWER

Equation 7.31 can be written as $y = a + bx + cz$, where

$$y = \ln f,\tag{7.32}$$
$$x = \ln T\tag{7.33}$$

Table 7.3. *Variation of resonance frequency with string tension and string length.*

f (Hz)	T (N)	L (m)
28	0.49	1.72
48	0.98	1.43
66	1.47	1.24
91	1.96	1.06
117	2.45	0.93
150	2.94	0.82
198	3.43	0.67

and

$$z = \ln L. \tag{7.34}$$

Also, $a = \ln K$, $b = B$ and $c = C$.

Using the data in table 7.3, and the transformations given by equations 7.32 to 7.34, the matrices $\mathbf{A}$ and $\mathbf{P}$ in equation 7.29 become

$$\mathbf{A} = \begin{bmatrix} 7 & 3.532 & 0.5019 \\ 3.532 & 4.596 & -1.045 \\ 0.5019 & -1.045 & 0.6767 \end{bmatrix} \quad \mathbf{P} = \begin{bmatrix} 30.97 \\ 18.38 \\ 0.8981 \end{bmatrix}.$$

The inverse of matrix $\mathbf{A}$ is

$$\mathbf{A}^{-1} = \begin{bmatrix} 2.433 & -3.512 & -7.226 \\ -3.512 & 5.406 & 10.95 \\ -7.226 & 10.95 & 23.74 \end{bmatrix}.$$

Determining $\mathbf{B} = \mathbf{A}^{-1}\mathbf{P}$, gives

$$\mathbf{B} = \begin{bmatrix} 4.281 \\ 0.4476 \\ -1.157 \end{bmatrix}, \text{ so that}$$

$a = 4.281 = \ln K$, so $K = 72.31$,

$b = 0.4476 = B$, and

$c = -1.157 = C$,

so that $\ln f = 4.281 + 0.4476 \ln T - 1.157 \ln L$.

Finally, we write,

$$f = 72.31 T^{0.4476} L^{-1.157}.$$

Table 7.4. *The y, x and z data for exercise E.*

y	x	z
20.7	1	1
23.3	2	2
28.2	3	4
35.7	4	5
48.2	5	11
56.0	6	14

Exercise E

Consider the data shown in table 7.4.

Assuming that y is a function of x and z such that $y = a + bx + cz$, use linear least squares to solve for a, b and c.

7.6 Standard uncertainties in parameter estimates

We can extend the calculation of standard uncertainties in intercept and slope given in appendix 4 to include situations in which there are more than two parameters. In section 7.3 we found, $\mathbf{B} = \mathbf{A}^{-1}\mathbf{P}$, where the elements of $\mathbf{B}$ are the parameter estimates. The standard uncertainty in each parameter estimate is obtained by forming the product of the diagonal elements of $\mathbf{A}^{-1}$ and the standard deviation, s, in each value, y_i. The diagonal elements of $\mathbf{A}^{-1}$ are written as $A_{11}^{-1}, A_{22}^{-1}, A_{33}^{-1}$. The standard uncertainties of the elements, a, b and c of $\mathbf{B}$, written as s_a, s_b, s_c respectively, are given by[9]

$$s_a = s\left(A_{11}^{-1}\right)^{1/2} \tag{7.35}$$

$$s_b = s\left(A_{22}^{-1}\right)^{1/2} \tag{7.36}$$

$$s_c = s\left(A_{33}^{-1}\right)^{1/2}, \tag{7.37}$$

and s is determined by using

$$s = \left[\frac{1}{n-M}\sum (y_i - \hat{y}_i)^2\right]^{1/2}, \tag{7.38}$$

where n is the number of data and M is the number of parameters in the equation.

---

[9] See Bevington and Robinson (2002) for a derivation of equations 7.35 to 7.37.

Example 5

Consider the data in table 7.5.

Assuming it is appropriate to fit the function $y = a + bx + cx^2$ to the data in table 7.5, determine, using linear least squares:

(i) a, b and c;

(ii) s;

(iii) standard uncertainties, s_a, s_b and s_c.

ANSWER

(i) Writing equations to solve for a, b and c in matrix form gives (see section 7.3)

$$\underbrace{\begin{bmatrix} n & \sum x_i & \sum x_i^2 \\ \sum x_i & \sum x_i^2 & \sum x_i^3 \\ \sum x_i^2 & \sum x_i^3 & \sum x_i^4 \end{bmatrix}}_{A} \underbrace{\begin{bmatrix} a \\ b \\ c \end{bmatrix}}_{B} = \underbrace{\begin{bmatrix} \sum y_i \\ \sum x_i y_i \\ \sum x_i^2 y_i \end{bmatrix}}_{P}. \tag{7.39}$$

To find the elements of **B**, write $\mathbf{B} = \mathbf{A}^{-1}\mathbf{P}$. Using MINVERSE() and MMULT() functions in Excel gives,

$$\underbrace{\begin{bmatrix} 1.206 & -0.2091 & 0.007576 \\ -0.2091 & 0.04423 & -0.001748 \\ 0.007576 & -0.001748 & 7.284 \times 10^{-5} \end{bmatrix}}_{A^{-1}} \underbrace{\begin{bmatrix} 142.6 \\ 4210 \\ 94510 \end{bmatrix}}_{P} = \underbrace{\begin{bmatrix} 7.847 \\ -8.862 \\ 0.6058 \end{bmatrix}}_{B} \tag{7.40}$$

so that $a = 7.847$, $b = $ -8.862 and $c = 0.6058$.

(ii) s is determined using equation 7.38 with, $n = 11$, $M = 3$ and $\hat{y}_i = a + bx_i + cx_i^2$ so that

Table 7.5. *The x–y data for example 5.*

x	y
2	−7.27
4	−17.44
6	−25.99
8	−23.02
10	−16.23
12	−15.29
14	1.85
16	21.26
18	46.95
20	71.87
22	105.97

$$s = \left[\frac{1}{8}\sum (y_i - (a + bx_i + cx_i^2))^2\right]^{1/2} = 2.423.$$

(iii) s_a, s_b and s_c are found by using equations 7.35 to 7.37:

$$s_a = s\left(A_{11}^{-1}\right)^{1/2} = 2.423 \times (1.2061)^{1/2} = 2.7$$

$$s_b = s\left(A_{22}^{-1}\right)^{1/2} = 2.423 \times (0.04423)^{1/2} = 0.51$$

$$s_c = s\left(A_{33}^{-1}\right)^{1/2} = 2.423 \times \left(7.284 \times 10^{-5}\right)^{1/2} = 0.021.$$

Exercise F

Consider the thermocouple data in table 7.6.

Assuming the relationship between V and T can be written as

$$V = A + BT + CT^2 + DT^3, \tag{7.41}$$

use linear least squares to determine best estimates for A, B, C, D and the standard uncertainties in these estimates.

7.6.1 Coverage intervals for parameters

If a, b and c are estimates of the parameters α, β and γ respectively obtained using least squares, then the $X\%$ coverage interval for each parameter can be written as

Table 7.6. *Output voltage of a thermocouple as a function of temperature.*

T (K)	V (μV)
75	−5.936
100	−5.414
125	−4.865
150	−4.221
175	−3.492
200	−2.687
225	−1.817
250	−0.892
275	0.079
300	1.081

$$\alpha = a \pm t_{X\%,\nu} s_a$$
$$\beta = b \pm t_{X\%,\nu} s_b$$
$$\gamma = c \pm t_{X\%,\nu} s_c,$$

where s_a, s_b and s_c are the standard uncertainties in a, b and c respectively, and $t_{X\%,\nu}$ is the critical t value for ν degrees of freedom at the $X\%$ level of confidence.[10] We can find ν by using

$$\nu = n - M, \tag{7.42}$$

where n is the number of data and M is the number of parameters in the equation.

Example 6

In fitting the equation $y = \alpha + \beta x + \gamma x^2 + \delta x^3$ to 15 data points (not shown here), it is found that the best estimates of the parameters α, β, γ and δ (written as a, b, c and d respectively) and their standard uncertainties[11] are

$$a = 14.65, \; s_a = 1.727$$
$$b = 0.4651, \; s_b = 0.02534$$
$$c = 0.1354, \; s_c = 0.02263$$
$$d = 0.04502, \; s_d = 0.01018.$$

Use this information to determine the 95% coverage interval for each parameter.

ANSWER

For the parameter α we write, $\alpha = a \pm t_{X\%,\nu} \, s_a$.

The number of degrees of freedom, $\nu = n - M = 15 - 4 = 11$. Referring to table 2 in appendix 1, we have for a level of confidence of 95%,

$$t_{95\%,11} = 2.201.$$

It follows that

$$\alpha = 14.65 \pm 2.201 \times 1.727 = 14.65 \pm 3.801 \text{ (which we normally round to } 14.7 \pm 3.8).$$

Similarly for β, γ and δ,

$$\beta = 0.4651 \pm 2.201 \times 0.02534 = 0.465 \pm 0.056$$
$$\gamma = 0.1354 \pm 2.201 \times 0.02263 = 0.135 \pm 0.050$$
$$\delta = 0.04502 \pm 2.201 \times 0.01018 = 0.0450 \pm 0.022.$$

---

[10] $t_{X\%,\nu}$ is equivalent to the coverage factor, k, at the $X\%$ level of confidence.

[11] Standard uncertainties are given to four significant figures to avoid rounding errors in subsequent calculations.

7.7 Weighting the fit

If a weighted fit is required,[12] we return to the weighted sum of squares, χ^2, given by

$$\chi^2 = \sum \left[\frac{y_i - \hat{y}_i}{s_i} \right]^2. \tag{7.44}$$

To allow for uncertainty that varies from one y value to another, we retain s_i^2 in the subsequent analysis. Equation 7.44 is differentiated with respect to each parameter in turn and the resulting equations are set equal to zero. Next we solve the equations for estimates of the parameters which minimise χ^2. As an example, if the function to be fitted to the data is the polynomial, $y = a + bx + cx^2$, then a, b and c may be determined using matrices by writing $\mathbf{B} = \mathbf{A}^{-1}\mathbf{P}$, where

$$\mathbf{B} = \begin{bmatrix} a \\ b \\ c \end{bmatrix}. \tag{7.45}$$

$\mathbf{A}^{-1}$ is the inverse of, $\mathbf{A}$, where $\mathbf{A}$ is written

$$\mathbf{A} = \begin{bmatrix} \sum \frac{1}{s_i^2} & \sum \frac{x_i}{s_i^2} & \sum \frac{x_i^2}{s_i^2} \\ \sum \frac{x_i}{s_i^2} & \sum \frac{x_i^2}{s_i^2} & \sum \frac{x_i^3}{s_i^2} \\ \sum \frac{x_i^2}{s_i^2} & \sum \frac{x_i^3}{s_i^2} & \sum \frac{x_i^4}{s_i^2} \end{bmatrix} \tag{7.46}$$

---

[12] Weighted fitting is required if the uncertainty in each y value is not constant – see section 6.10.

and **P** is given by

$$
\mathbf{P} = \begin{bmatrix} \sum \frac{y_i}{s_i^2} \\ \sum \frac{x_i y_i}{s_i^2} \\ \sum \frac{x_i^2 y_i}{s_i^2} \end{bmatrix}.
\tag{7.47}
$$

The diagonal elements of the $\mathbf{A}^{-1}$ matrix can be used to determine the standard uncertainties in a, b and c. We have[13]

$$
s_a = \left(A_{11}^{-1}\right)^{1/2}
\tag{7.48}
$$

$$
s_b = \left(A_{22}^{-1}\right)^{1/2}
\tag{7.49}
$$

$$
s_c = \left(A_{33}^{-1}\right)^{1/2}.
\tag{7.50}
$$

Exercise H

Consider the data in table 7.7.

Assuming that it is appropriate to fit the equation $y = a + bx + cx^2$ to these data and that the standard deviation, s_i in each y value is given by $s_i = 0.1y_i$:

  (i) use weighted least squares to determine a, b and c;
  (ii) determine the standard uncertainties in a, b and c.

Table 7.7. *The x–y data for exercise H.*

x	y
−10.5	143
−9.5	104
−8.3	79
−5.3	34
−2.1	12
1.1	19
2.3	29

[13] See Kutner, Nachtsheim and Neter (2003).

7.7.1 More on weighted fitting using matrices

When weighted fitting by least squares is undertaken, best estimates in parameters and standard uncertainties in those best estimates can be established by introducing an n by n 'weight' matrix, $\mathbf{W}$, given by[14]

$$\mathbf{W} = \begin{bmatrix} \frac{1}{s_1^2} & 0 & \cdots & 0 \\ 0 & \frac{1}{s_2^2} & \cdots & 0 \\ \vdots & \vdots & & \vdots \\ 0 & 0 & \cdots & \frac{1}{s_n^2} \end{bmatrix}. \tag{7.51}$$

The matrix given by equation 7.46 can be determined using

$$\mathbf{A} = \mathbf{X}^T\mathbf{W}\mathbf{X}. \tag{7.52}$$

Similarly, the $\mathbf{P}$ matrix given by equation 7.47 can be determined using

$$\mathbf{P} = \mathbf{X}^T\mathbf{W}\mathbf{Y}. \tag{7.53}$$

The $\mathbf{B}$ matrix containing the best parameter estimates is found using

$$\mathbf{B} = \mathbf{A}^{-1}\mathbf{P} = \left(\mathbf{X}^T\mathbf{W}\mathbf{X}\right)^{-1}\mathbf{X}^T\mathbf{W}\mathbf{Y}. \tag{7.54}$$

The matrix containing the standard uncertainties in the best estimates of the parameters (as diagonal elements of the matrix) can be written as $\mathbf{s}^2(\mathbf{B})$, where[15]

$$\mathbf{s}^2(\mathbf{B}) = \left(\mathbf{X}^T\mathbf{W}\mathbf{X}\right)^{-1}. \tag{7.55}$$

7.8 Coefficients of multiple correlation and multiple determination

In section 6.6 we introduced the linear correlation coefficient, r, which quantifies how well x and y are correlated. r close to 1 or -1 indicates excellent correlation, while r close to zero indicates poor correlation. We can generalise the correlation coefficient to take into account equations fitted to data which incorporate more than one independent variable or where there are more than two parameters in the equation. The coefficient of multiple correlation, R, is written

$$R = \left(1 - \frac{\sum (y_i - \hat{y}_i)^2}{\sum (y_i - \bar{y}_i)^2}\right)^{1/2}, \tag{7.56}$$

where y_i is the ith observed value of y, $\hat{y}_i$ is the ith predicted y value found using the equation representing the best line through the points and $\bar{y}$ is the mean of

---

[14] n is the number of data points.

[15] $\mathbf{s}^2(\mathbf{B})$ is often referred to as the variance–covariance matrix.

the observed y values.[16] We argue that equation 7.56 is plausible by considering the summation $\sum (y_i - \hat{y}_i)^2$ which is the sum of squares of the residuals, SSR. If the line passes near to all the points then SSR is close to zero (and is equal to zero if all the points lie on the line) and so R tends to one, as required by perfectly correlated data.

The square of the coefficient of multiple correlation is termed the *coefficient of multiple determination, R^2*, and is written

$$R^2 = 1 - \frac{\sum (y_i - \hat{y}_i)^2}{\sum (y_i - \bar{y}_i)^2}. \tag{7.57}$$

R^2 gives the fraction of the square of the deviations of the observed y values about their mean which can be explained by the equation fitted to the data. So for example, if $R^2 = 0.92$ this indicates that 92% of the scatter of the deviations can be explained by the equation fitted to the data.

Exercise I

Consider the data in table 7.8.

Fitting the equation $y = a + bx + cz$ to the data in table 7.8 using least squares gives

$\hat{y}_i = 2.253 + 1.978x_i - 0.3476z_i,$

where $\hat{y}_i$ is the predicted value of y for a given x_i.

Use this information to determine the coefficient of multiple determination, R^2, for the data in table 7.8.

Table 7.8. *Data for exercise I.*

y	x	z
0.84	0.24	5.5
1.47	0.35	4.2
2.23	0.55	3.2
2.82	0.78	2.8
3.60	0.98	1.5
4.25	1.1	0.7

---

[16] See Walpole, Myers and Myers (1998) for a discussion of equation 7.56.

7.9 Estimating more than two parameters using the LINEST() function

The LINEST() function introduced in section 6.3 can be used to determine:

- estimates of parameters;
- standard uncertainties in the estimates;
- the coefficient of multiple determination, R^2;
- the standard deviation in the y values, s, as given by equation 7.38.

When fitting a straight line to data using LINEST(), a single column of numbers representing the x values is entered into the function whose syntax is as follows.[17]

LINEST(y values, x values, constant, statistics)

By contrast, when there are two independent variables (such as x and z), an array with two columns of numbers must be entered into LINEST(). That array consists of one column of x values and an adjacent column of z values in the case where $y = a + bx + cz$ is fitted to data. Where the same independent variable appears in two terms of an equation, such as in $y = a + bx + cx^2$, one column in Excel contains x values and an adjacent column contains x^2.

The use of the LINEST() function to fit an equation containing more than two parameters is best illustrated by example. Consider the data in sheet 7.7.

Sheet 7.7. *Using the LINEST() function.*

	A	B	C
1	y	x	z
2	28.8	1.2	12.8
3	42.29	3.4	9.8
4	50.69	4.5	6.7
5	66.22	7.4	4.5
6	73.12	8.4	2.2
7	81.99	9.9	1
8			
9	=LINEST(A2:A7, B2:C7,TRUE,TRUE)		
10			
11			
12			
13			

Assuming it is appropriate to fit the equation $y = a + bx + cz$ to the data in sheet 7.7, we proceed as follows.

---

[17] More details of the syntax of this function appear in section 6.3.

- Highlight a range of cells (cells A9 to C13 in sheet 7.7) into which Excel can return the estimates a, b and c, the standard uncertainties in these estimates, the coefficient of multiple determination and the standard deviation in the y values.
- Type =**LINEST(A2:A7,B2:C7,TRUE,TRUE)** into cell A9.
- Hold down the Ctrl and Shift keys together, then press the Enter key.

The numbers returned in cells A9 to C13 are shown in sheet 7.8.

Sheet 7.8. *Numbers returned by the LINEST() function.*

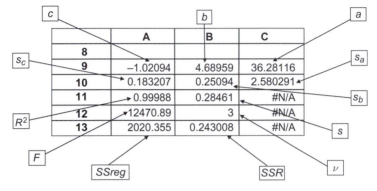

	A	B	C
8			
9	−1.02094	4.68959	36.28116
10	0.183207	0.25094	2.580291
11	0.99988	0.28461	#N/A
12	12470.89	3	#N/A
13	2020.355	0.243008	#N/A

For completeness we state what each value returned by Excel represents when using the LINEST() function though we do not explain them in detail. The F statistic and its value to fitting using least squares is discussed in chapter 9.

Note that a, b and c are the best estimates of the parameters of the equation fitted to the data and s_a, s_b and s_c are the standard uncertainties in a, b and c respectively; s is the standard deviation in the y values; R^2 is the coefficient of multiple determination, v is the number of degrees of freedom (which is equal to the number of data points, n, minus the number of parameters estimated using least squares) and SSR is the sum of squares of residuals. $SSReg$ is the regression sum of squares, given by

$$SSreg = \sum (\hat{y}_i - \bar{y}_i)^2. \tag{7.58}$$

F is the F-statistic given by[18]

$$F = \frac{SSreg/(n - v - 1)}{SSR/v}. \tag{7.59}$$

---

[18] If fitting an equation to data forces the line through the origin (i.e. $a = 0$) then equation 7.59 should be replaced by $F = \frac{SSreg/(n-v)}{SSR/v}$.

Table 7.9. *Data for exercise J.*

x	y
0.02	−0.0261
0.04	−0.0202
0.06	−0.0169
0.08	−0.0143
0.10	−0.0124
0.12	−0.0105
0.14	−0.0094
0.16	−0.0080
0.18	−0.0067

A limitation of the LINEST() function is that it does not allow for weighted fitting. If weighted fitting of linear functions containing several parameters is required, then the matrix method described in section 7.7 is recommended.

Exercise J

Consider the data in table 7.9.

Use LINEST() to fit the equation $y = a + bx + c \ln x$ to the data in table 7.9 and so determine:

(i) a, b and c;
(ii) the standard uncertainties in a, b and c;
(iii) the coefficient of multiple determination, R^2;
(iv) the standard deviation, s, in the y values.

7.10   Choosing equations to fit to data

In some situations, selecting an equation to fit to data is not difficult. If consideration of the physical principles underlying the x–y data predicts that the relationship between y and x should be linear, and if a visual inspection of the data lends support to that prediction, then fitting the equation $y = a + bx$ is appropriate.

If there is little or no past experience to act as a guide as to which equation to fit to data we may need to rely on an examination of the data presented in graphical form to suggest one or more equations which can be trialled. However, it is worth emphasising that an equation relating the dependent to independent variables which derives from a consideration of physical principles is to be preferred to one that simply gives the smallest sum of squares of residuals.

Another situation in which the choice of equation to fit to data is 'clear cut' is when the purpose of data gathering is to calibrate an instrument. Previous experience, manufacturers guidelines or other published information are determining factors in the choice of the equation. When calibration is the aim of the least squares analysis, the physical interpretation of the parameter estimates that emerge from the fitting process is often not important, as the main aim is to use the fitted function for the purpose of interpolation (and occasionally extrapolation).[19]

Whatever equation is fitted to data, we should be suspicious if standard uncertainties of the parameter estimates are comparable in size (or larger than) the estimates themselves. For example, suppose some parameter estimate, b, is estimated by least squares to be $b = 3.6$ with a standard uncertainty, $s_b = 5.5$. It is likely that the term containing b can be eliminated with little adverse effect on the quality of the fit of the equation to data. In chapter 9 we will discover that it possible to make a judgement, based upon statistical considerations, as to whether a particular parameter estimate is 'significant' and we will leave discussion of this until we have considered 'tests of significance'.

7.10.1 Comparing equations fitted to data

Occasionally a situation arises in which, based on a consideration of the physical processes believed to be underlie the relationship being studied, two or more equations may reasonably be fitted to data. For example, the electrical resistance, r, of a material at a temperature, T, may be described by

$$r = A + BT \tag{7.60}$$

or

$$r = \alpha + \beta T + \gamma T^2, \tag{7.61}$$

where A, B, α, β, and γ are constants.

If we use χ^2 as given by equation 7.1, as a means of determining which of the two equations better fits the data, we will always favour equation 7.61 over equation 7.60. This is due to the fact that as the addition of the term in T^2 in equation 7.61 gives extra flexibility to the line so that predicted values found using the equation lie closer to the data values than if that term were omitted. However, the sum of squares of residuals obtained when fitting equation 7.61 to data may only be marginally less than when equation 7.60 is fitted to the same data.

---

[19] For details on fitting for calibration purposes, see Kirkup and Mulholland (2004).

Another possibility as an indicator of the equation that better fits the data is the coefficient of multiple determination, R^2, such that the equation yielding the larger value of R^2 is regarded as the better equation. Regrettably, as with the sum of squares of residuals, R^2 favours the equation with the greater number of parameters. For example, the equation $y = a + bx + cx^2 + dx^3$ would always be favoured over the equation $y = a + bx + cx^2$ when both equations are fitted to the same data. To take into account the number of parameters such that any marginal increase in R^2 is offset by the number of parameters used, the adjusted coefficient of multiple determination, R^2_{ADJ} is sometimes used, where[20]

$$R^2_{ADJ} = \frac{(n-1)R^2 - (M-1)}{n-M}.$$

(7.62)

R^2 is given by equation 7.57, n is the number of data and M is the number of parameters.

Once R^2_{ADJ} is calculated for each equation fitted to data, the equation is preferred that has the larger value of R^2_{ADJ}.

Another way of comparing two (or more) equations fitted to data where the equations have different numbers of parameters is to use Akaike's Information Criterion[21] (AIC). This criterion takes into account the sum of squares of residuals SSR, but also includes a term proportional to the number of parameters used. AIC may be written

$$AIC = n \ln\left(\frac{SSR}{n}\right) + 2K,$$

(7.63)

where n is the number of data and $K = (M + 1)$ in the equation.[22]

AIC depends on the sum of squares of residuals which appears in the first term of the right hand side of equation 7.63. The second term on the right hand side can be considered as a 'penalty' term. If the addition of another parameter in an equation reduces SSR then the first term on the right hand side of equation 7.63 becomes smaller. However, the second term on the right hand side increases by two for every extra parameter used. It follows that a modest decrease in SSR which occurs when an extra term is introduced into an equation may be more than offset by the increase in AIC by using another parameter. We conclude that, if two or more equations are fitted to data, then the equation producing the *smallest* value for AIC is preferred.

---

[20] See Kutner, M. J., Nachtsheim, C. J. and Neter, J. (2003) for a discussion of equation 7.62.

[21] See Akaike (1974) for a discussion of model identification.

[22] See Motulsky and Christopoulos (2005).

When the number of data is small, such that n/K is less than ≈ 40, a correction should be applied to equation 7.63. The corrected criterion, written as AIC_c, is given by

$$AIC_c = AIC + \frac{2K(K+1)}{n-K-1},\tag{7.64}$$

where AIC is given by equation 7.63.

If weighted least squares is required, we replace SSR in equation 7.63 by the weighted sum of squares of residuals, χ^2, where χ^2 is given by equation 7.1.

Example 7

Table 7.10 shows the variation of the resistance of an alloy with temperature.

Using (unweighted) linear least squares, fit equations 7.60 and 7.61 to the data in table 7.10 and determine for each equation:

  (i) estimates for the parameters;
  (ii) the standard uncertainty in each estimate;
  (iii) the standard deviation, s, in each y value;
  (iv) the coefficient of multiple determination, R^2;
  (v) the adjusted coefficient of multiple determination, R^2_{ADJ};
  (vi) the sum of squares of the residuals, SSR;
  (vii) the Akaike's information criterion, AIC;
  (viii) the corrected Akaike's information criterion, AIC_c.

ANSWER

The extent of computation required in this problem is strong inducement to use the LINEST() function in Excel. As usual, we write estimates of parameters as a, b and c etc. In the case of equation 7.60 we determine a and b where the equation is of the form, $y = a + bx$, and for equation 7.61, we determine a, b and c where $y = a + bx + cx^2$. The numbers shown in table 7.11 are rounded to four significant figures, though intermediate calculations utilise the full precision of Excel.

R^2_{ADJ} for equation 7.60 fitted to data is greater than R^2_{ADJ} for equation 7.61, indicating that equation 7.60 is the better fit. The AIC for equation 7.60 fitted to data is lower than for equation 7.61, further supporting equation 7.60 as the more appropriate equation. Finally, an inspection of the standard uncertainties of the parameter estimates suggests that, for the equation with three parameters, the standard uncertainty in c is so large in comparison to c that y in equation 7.61 is likely to be 'redundant'.

Table 7.10. *Data for example 7.*

r (Ω)	19.5	18.4	20.2	20.1	20.9	20.8	21.2	21.8	21.9	23.6	23.2
T (K)	150	160	170	180	190	200	210	220	230	240	250

r (Ω)	23.9	23.2	24.1	24.2	26.3	25.5	26.1	26.3	27.1	28.0
T (K)	260	270	280	290	300	310	320	330	340	350

Table 7.11. *Parameter estimates found by fitting equations 7.60 and 7.61 to data in table 7.10*[23]

	fitting $y = a + bx$ to data	fitting $y = a + bx + cx^2$ to data
parameter estimates	$a = 12.41, b = 4.299 \times 10^{-2}$	$a = 13.97, b = 2.970 \times 10^{-2}, c = 2.658 \times 10^{-5}$
standard	$s_a = 0.49, s_b = 1.9 \times 10^{-3}$	$s_a = 2.166, s_b = 1.8 \times 10^{-2}, s_c = 3.6 \times 10^{-5}$
uncertainties		
s	0.5304	0.5368
R^2	0.9638	0.9649
R^2_{ADJ}	0.9619	0.9610
SSR	5.344	5.186
AIC	−22.74	−21.37
AIC_c	−21.33	−18.87

Exercise K

Using the data in table 7.10 verify that the parameter estimates and the standard uncertainties in the estimates appearing in table 7.11 are correct.

7.10.2   Akaike's weights

When, say, two equations are fitted to data, the equation with the lowest *AIC* is preferred. If the *AIC*s of the two equations are comparable the extent to which the equation with the smaller *AIC* is preferred may be quite marginal. We can put it this way: in the situation in which two equations are fitted to the same data, what is the probability that the equation with the lower *AIC* is the better of the two?

Suppose the *AIC* value for the first equation fitted to data is AIC_1 and for the second equation is AIC_2. Also assume that $AIC_1 < AIC_2$ (i.e. the first equation is the preferred equation).

The difference between the *AIC* values for the two equations as $\Delta(AIC)$ is

---

[23] In the interests of conciseness, units of measurement have been omitted from the table.

$$\Delta(AIC) = AIC_1 - AIC_2. \tag{7.65}$$

The probability, p, that equation 1 is better equation that equation 2 is given by[24]

$$p = \frac{\exp[-\frac{1}{2}\Delta(AIC)]}{1 + \exp[-\frac{1}{2}\Delta(AIC)]}. \tag{7.66}$$

Example 8

Two equations are fitted to data. Fitting equation 1 gives an AIC value of -142.1 and fitting equation 2 gives an AIC value of -139.7. What is the probability that equation 1 is the better of the two?

ANSWER

Using equation 7.65, $\Delta(AIC) = -142.1 - (-139.7) = -2.40$.

Now, using equation 7.66, $p = \frac{\exp[-1/2 \times -2.40]}{1+\exp[-1/2 \times -2.40]} = 0.769$, or expressed as a percentage, 76.9%.

Exercise L

(i) Two equations are fitted to the same data. The first equation gives an AIC value of -92.1; the second equation gives an AIC of -88.2. What is the probability that the first equation is a better fit to the data than the second equation?

(ii) Using the relationship between p and $\Delta(AIC)$ given by equation 7.66, plot a graph of p versus $\Delta(AIC)$ for $-5 < \Delta(AIC) < 0$.

7.11 Review

Fitting equations to experimental data is a common occupation of workers in the physical sciences. Whether the aim of the comparison is to establish the values of parameters so that they can be compared with those predicted by theory, or to use the parameter estimates for calibration purposes, finding 'best' estimates of the parameters and the standard uncertainties in the estimates is very important.

In this chapter we considered how the technique of least squares can be extended beyond fitting the equation $y = a + bx$ to data. Increasing the number of parameters that must be estimated increases the number of the calculations that must be carried out in order to determine the best estimates of the parameters

---

[24] Note that this approach can be extended to the comparison of more than two equations, see Burnham and Anderson (2002) and Wagenmakers and Farrel (2004).

and their standard uncertainties. Applying matrix methods considerably eases the process of estimating parameters. The facility of being able to fit equations with more and more terms to produce a better fit must be applied cautiously as adding more terms reduces the sum of squares of residuals, but the reduction may only be marginal. We introduced the adjusted coefficient of multiple determination and the Akaike's Information Criterion as means of establishing which equation best fits when there are several competing equations.

In the next chapter we will consider fitting equations to data where those equations cannot be fitted using linear least squares. In particular, we will investigate Excel's Solver which is a useful utility that allows for the fitting of equations to data using non-linear least squares.

End of chapter problems

(1) The equation

$$y = a + bx + c \exp x \tag{7.67}$$

is to be fit to x–y data.

   (i)   Substitute equation 7.67 into equation 7.1. By differentiating χ^2 with respect to each parameter in turn, obtain three equations that may be solved for a, b and c. Assume that an unweighted fit using least squares is appropriate.
   (ii) Write the equations obtained in part (i) in matrix form.
         Consider the data in table 7.12.
   (iii) Assuming that equation 7.67 is appropriate to the data in table 7.12, use matrices to find a, b and c.

Table 7.12. *The x–y data.*

x	y
1	8.37
2	3.45
3	1.70
4	0.92
5	0.53
6	0.62
7	−0.06
8	0.63
9	0.16
10	−0.27

Table 7.13. *Variation of plate height with flow rate of solute.*

v (mL/minute)	H (mm)
3.5	9.52
7.5	5.46
15.7	3.88
20.5	3.48
25.8	3.34
36.7	3.31
41.3	3.13
46.7	3.78
62.7	3.55
78.4	4.24
96.7	4.08
115.7	4.75
125.5	4.89

(2) The movement of a solute through a chromatography column can be described by the van Deemter equation,

$$H = A + \frac{B}{v} + Cv, \tag{7.68}$$

where H is the plate height, and v is the rate at which the mobile phase of the solute flows though the column. A, B, and C are constants. For a particular column, H varies with v as given in table 7.13.

(i) Write the equations in matrix form that must be solved for best estimates of A, B and C, assuming unweighted fitting using least squares is appropriate (hint: follow the steps given in section 7.2).

(ii) Solve for best estimates of A, B and C.

(iii) Determine the standard uncertainties for the estimates obtained in (ii).

(3) An object is thrown off a building and its vertical displacement above the ground at various times after it is released is shown in table 7.14.

The predicted relationship between s and t is

$$s = s_0 + ut + \frac{1}{2}gt^2, \tag{7.69}$$

where s_0 is the vertical displacement of the object at $t = 0$, u is its initial vertical velocity and g is the acceleration of the body falling freely under the action of gravity.

Table 7.14. *Vertical displacement of an object with time.*

t (s)	s (m)
0.0	135.2
0.5	141.7
1.0	142.3
1.5	148.2
2.0	143.7
2.5	140.3
3.0	134.1
3.5	126.7
4.0	114.5
4.5	100.8
5.0	85.7
5.5	66.7
6.0	46.5

Use least squares to determine best estimates for s_0, u and g and the standard uncertainties in the estimates.

(4) To illustrate the way in which a real gas deviates from perfect gas behaviour, $\frac{PV}{RT}$ is often plotted against $\frac{1}{V}$, where P is the pressure, V the volume and T the temperature (in kelvins) of the gas. R is the gas constant. Values for $\frac{PV}{RT}$ and V are shown in table 7.15 for argon gas at $T = 150$ K.

Assuming the relationship between $\frac{PV}{RT}$ and $\frac{1}{V}$ can be written as

$$\frac{PV}{RT} = A + \frac{B}{V} + \frac{C}{V^2} + \frac{D}{V^3}, \tag{7.70}$$

use least squares to obtain best estimates for A, B, C and D and standard uncertainties in the estimates.

(5) Consider the x–y data in table 7.16.

Fit equations $y = a + bx$ and $y = a + bx + cx^2$ to these data.

   (i) Use the adjusted multiple correlation coefficient and corrected Akaike's Information Criterion to establish which equation better fits the data.

   (ii) Calculate the probability that the equation giving the lowest value of AIC is the better equation to fit to the data in table 7.16.

(6) Table 7.17 shows data of the variation of the molar heat capacity (at constant pressure), C_p, of oxygen with temperature, T.

Table 7.15. *Variation of* $\frac{PV}{RT}$
with V for argon gas.

$V \, (\text{cm}^3)$	$\frac{PV}{RT}$
35	1.21
40	0.89
45	0.72
50	0.62
60	0.48
70	0.45
80	0.51
90	0.50
100	0.53
120	0.57
150	0.61
200	0.69
300	0.76
500	0.84
700	0.89

Table 7.16. *x–y data.*

x	y
5	361.5
10	182.8
15	768.6
20	822.5
25	1168.2
30	1368.6
35	1723.3
40	1688.7
45	1800.9
50	2124.5
55	2437.9
60	2641.2

Table 7.17. *Measured molar heat capacity of oxygen in the temperature range 300 K to 1000 K.*

$T(\mathrm{K})$	$C_p\,(\mathrm{J \cdot mol^{-1} \cdot K^{-1}})$
300	29.43
350	30.04
400	30.55
450	30.86
500	31.52
550	31.71
600	32.10
650	32.45
700	32.45
750	32.80
800	33.11
850	33.38
900	33.49
950	33.85
1000	34.00

Assuming that the relationship between C_p and T can be written as

$$C_p = A + BT + \frac{C}{T^2}, \tag{7.71}$$

determine:

 (i) best estimates for A, B and C;
 (ii) the standard uncertainties in the estimates;
 (iii) the 95% coverage intervals for A, B and C.

(7) As part of a study into the behaviour of electrical contacts made to a ceramic conductor, the data in table 7.18 were obtained for the temperature variation of the electrical resistance of the contacts.

It is suggested that there are two possible models that can be used to describe the variation of the contact resistance with temperature.

Model 1

The first model assumes that contacts show semiconducting behaviour, where the relationship between R and T can be written

Table 7.18. *Resistance versus temperature for electrical contacts on a ceramic.*

T (K)	R (Ω)	T (K)	R (Ω)
50	4.41	190	0.69
60	3.14	200	0.85
70	2.33	210	0.94
80	2.08	220	0.78
90	1.79	230	0.74
100	1.45	240	0.77
110	1.36	250	0.68
120	1.20	260	0.66
130	0.86	270	0.84
140	1.12	280	0.77
150	1.05	290	0.75
160	1.05	300	0.86
170	0.74		
180	0.88		

$$R = A \exp\left(\frac{B}{T}\right), \tag{7.72}$$

where A and B are constants.

Model 2

Another equation proposed to describe the data is

$$R = \alpha + \beta T + \gamma T^2, \tag{7.73}$$

where α, β and γ are constants.

Using plots of residuals, the adjusted coefficient of multiple determination, corrected Akaike's Information Criterion and any other indicators of 'goodness of fit', determine whether equation 7.72 or equation 7.73 better fits the data.

Assistance: To allow a straight line based on equation 7.72 to be fitted to data, the data need to be transformed. Transform back to the original units after fitting a line to data before calculating SSR, R^2_{ADJ}, and AIC_c. This will allow for direct comparison between SSR, R^2_{ADJ}, and AIC_c between equations 7.72 and 7.73.

Table 7.19. *Flow rate dependence on Δp, d, and l.*

Q (ml/s)	Δp (Pa)	d (mm)	l (mm)
0.0092	5488	0.265	43
0.080	5488	0.395	43
0.22	4753	0.82	105
0.37	4753	0.82	80
0.53	4753	0.82	56
0.60	4753	0.82	43
0.67	5488	0.82	43
0.76	2842	1	43
0.90	3626	1	43
1.00	4312	1	43
1.02	4704	1	43
1.14	5488	1	43

(8) The flow of fluid, Q, through a hollow needle (of circular cross-section) depends on:

- the pressure difference, Δp, between the ends of the needle;
- the internal diameter of the needle, d;
- the length of the tube, l.

Table 7.19 shows data obtained for Q for various combinations of d, l and Δp for a range of hollow steel needles.

Assume that the relationship between Q, d, l and Δp can be written as

$$Q = k\Delta p^r d^s l^t, \tag{7.74}$$

where k, r, s and t are constants.

Use least squares to find best estimates for k, r, s and t and standard uncertainties in the best estimates.

Chapter 8

Non-linear least squares

8.1 Introduction

In chapters 6 and 7 we considered fitting equations such as $y = a + bx$ and $y = a + b \ln x + cx^2$ to data by using linear least squares. The equations we fitted were linear in the parameters, a, b, etc.

Section 6.9 showed that, where equations are not linear in the parameters, a transformation (such as taking the logarithm of the dependent variable), may allow fitting by linear least squares to proceed.

Though transforming equations can assist in many situations, there are some equations that cannot be transformed into a form suitable for fitting by linear least squares. Examples of such equations are:

$$y = a + b \sin cx \tag{8.1}$$

$$y = a + b \ln cx \tag{8.2}$$

$$y = a + b[1 - \exp cx] \tag{8.3}$$

$$y = a \exp bx + c \exp dx. \tag{8.4}$$

For equations 8.1 to 8.4 it is not possible to obtain a set of linear equations that may be solved for best estimates of the parameters. We therefore resort to another method of finding best estimates, called *non-linear* least squares. While non-linear least squares is computationally more demanding than linear least squares, the key point, as with linear least squares, is that parameter estimates are sought that minimise the sum of squares of residuals, *SSR*. Even where transformation of data allows for fitting by linear least squares to proceed, this may not be the best approach to adopt, as the transformation affects the error distribution in the dependent variable. As a consequence of the transformation the assumption that the errors in the dependent variable are normally distributed is likely not to be valid. This leads to bias in the estimates of the parameters.[1] The adverse effect of transformation can be avoided by fitting equations using non-linear least squares, as no transformation of raw data is required.

SSR may be considered to be a continuous function of the parameter estimates. A surface may be constructed, sometimes referred to as a hypersurface[2] in M dimensional space, where M is the number of parameters appearing in the equation to be fitted to data. As with linear least squares, parameter estimates which result in a minimisation of *SSR* are regarded as the best estimates of the parameters in the equation. Figure 8.1 shows a hypersurface which depends on estimates a and b. There is a minimum in *SSR* for $a \approx 7$. Examination of tabulated values shown in table 8.1 indicates that the minimum in *SSR* occurs for $b \approx 4$.

Fitting by non-linear least squares begins with reasonable approximations for the best estimates of the parameters. These approximations are often referred to as 'starting values'. The objective is to modify the starting values in an iterative fashion until a minimum is found in *SSR*. The computational intensiveness of the iteration process means that non-linear least squares can only realistically be carried out using a computer.

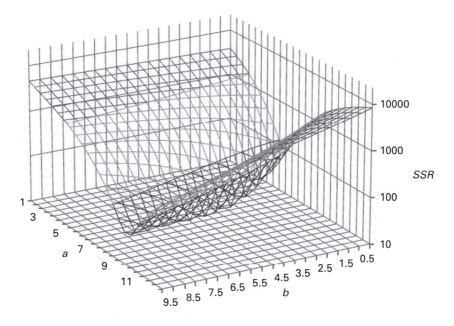

Figure 8.1. Variation of *SSR* with parameter estimates, a and b.

[1] See Motulski and Christopoulos (2005).
[2] See Bevington and Robinson (2002).

Table 8.1. *Values of SSR for a range of values of a and b. The highlighted cell contains the smallest value of SSR in the table; this indicates that best estimates of a and b are close to 7.0 and 4.0 respectively.*

		a				
		6.0	6.5	7.0	7.5	8.0
b	3.0	57.64611	21.8929	53.5454	152.6036	319.0675
	3.5	84.70766	25.42352	29.57799	97.17108	228.2028
	4.0	118.969	39.19182	19.26127	59.17733	158.94
	4.5	158.766	61.06023	19.93265	35.38321	107.4119
	5.0	202.8235	89.3755	29.5189	23.25373	70.57996
	5.5	250.1483	122.8372	46.378	20.77073	46.01537
	6.0	299.9563	160.4085	69.19152	26.30531	31.74987

There are many documented ways in which the values of a, b, c, etc., can be found which minimise *SSR*, including Grid Search (Bevington and Robinson, 2002), Gauss–Newton method (Nielsen-Kudsk, 1983) and the Levenberg-Marquardt algorithm (Bates and Watts, 1988).

Non-linear least squares is unnecessary when the derivatives of *SSR* with respect to the parameters are linear in those parameters. In such a situation, linear least squares offers a more efficient route to determining best estimates of the parameters (and the standard uncertainties in the best estimates). Nevertheless, a linear equation can be fitted to data using non-linear least squares. The answer obtained for best estimates of parameters and the standard uncertainties in the best estimates should agree, irrespective of whether a linear equation is fitted using linear or non-linear least squares.

Many computer-based statistical packages such as Origin[3] offer the facility to fit equations to data using non-linear least squares. A shortcoming of most packages is that the process by which the fitting takes place is hidden from the user, making it less easy to explore what is happening as fitting takes place. An alternative approach is to use a tool in Excel called Solver as this gives more control of the fitting process allowing for insights into what is happening as fitting by non-linear least squares proceeds.

[3] OriginLab Corporation, Northampton, Massachusetts, USA.

8.2   Excel's Solver add-in

Solver, first introduced in 1991, is one of many add-ins available in Excel.[4] Originally designed for business users, Solver is a powerful and flexible optimisation tool. Given a model, for example, representing the profitability of a commercial product, Excel's Solver is able to iteratively modify the parameters in that model and return the parameter estimates that would maximise the profitability. This is an example of optimisation, where parameters within a model or equation are adjusted until a desired outcome is achieved (which often involves maximising or minimising a particular quantity).

In the context of fitting equations to data, Solver is capable of finding the best estimates of parameters using least squares. It does this by iteratively altering the numerical values of variables contained in the cells of a spreadsheet until *SSR* is minimised. To solve non-linear problems, Solver uses Generalized Reduced Gradient (GRG2) code developed at the University of Texas and Cleveland State University.[5] Features of Solver are best described by reference to a particular example.

8.2.1   Example of use of Solver

Consider an experiment in which the air temperature in an enclosure (such as a living room) is measured as a function of time as heat passes into the enclosure, for example through a window. Table 8.2 contains the raw data. Figure 8.2 displays the same data in graphical form.

Through a consideration of the flow of heat into and out of an enclosure, a relationship may be derived for the air temperature, T, inside the enclosure as a function of time, t. The relationship can be expressed as

$$T = T_s + k[1 - \exp(\alpha t)], \tag{8.5}$$

where T_s, k and α are constants estimated using non-linear least squares. Using x and y as the independent and dependent variables respectively and a, b and c as best estimates of T_s, k and α respectively, equation 8.5 becomes

$$y = a + b[1 - \exp(cx)]. \tag{8.6}$$

To find best estimates, a, b and c, we proceed as follows.

(1) Enter the raw data from table 8.2 into columns A and B of an Excel worksheet as shown in sheet 8.1.

---

[4] See Fylstra, Lasdon, Watson and Waren (1998)

[5] See Excel's online Help. See also Smith and Lasdon (1992).

Table 8.2. *Variation of air temperature in an enclosure with time.*

Time (minutes)	Temperature (°C)
2	26.1
4	26.8
6	27.9
8	28.6
10	28.5
12	29.3
14	29.8
16	29.9
18	30.1
20	30.4
22	30.6
24	30.7

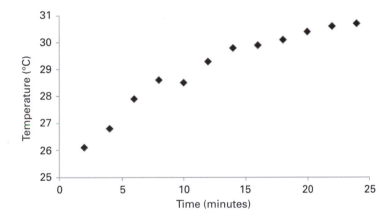

Figure 8.2. Temperature variation with time inside an enclosure.

(2) Type =**\$B\$15+\$B\$16*(1-EXP(\$B\$17*A2))** into cell C2 as shown in sheet 8.1. Cells B15 to B17 contain the starting values for *a*, *b* and *c* respectively.

(3) Use the cursor to highlight cells C2 to C13.

(4) Click on Home Ribbon, then choose Editing>Fill>Down.[6]

Sheet 8.2 shows the calculated values in the C column. As the squares of the residuals, $(y - \hat{y})^2$, are required, these are calculated in column D.

---

[6] As in chapter 2 we abbreviate these steps to Home>Editing>Fill>Down.

Sheet 8.1. *Temperature (y) and time (x) data from table 8.1 entered into a spreadsheet.*[7]

	A	B	C
	A	**B**	**C**
1	x (mins)	y (C)	ŷ (C)
2	2	26.1	= B15 + B16*(1-EXP(B17*A2))
3	4	26.8	
4	6	27.9	
5	8	28.6	
6	10	28.5	
7	12	29.3	
8	14	29.8	
9	16	29.9	
10	18	30.1	
11	20	30.4	
12	22	30.6	
13	24	30.7	
14			
15	a	1	
16	b	1	
17	c	1	

Sheet 8.2. *Calculation of sum of squares of residuals.*

	A	B	C	D
	A	**B**	**C**	**D**
1	x (mins)	y (C)	ŷ (C)	$(y-\hat{y})^2$ (C²)
2	2	26.1	−5.38906	991.560654
3	4	26.8	−52.5982	6304.066229
4	6	27.9	−401.429	184323.2129
5	8	28.6	−2978.96	9045405.045
6	10	28.5	−22024.5	486333300.3
7	12	29.3	−162753	26498009287
8	14	29.8	−1202602	1.44632E + 12
9	16	29.9	−8886109	7.89635E + 13
10	18	30.1	−6.6E + 07	4.31124E + 15
11	20	30.4	−4.9E + 08	2.35385E + 17
12	22	30.6	−3.6E + 09	1.28516E + 19
13	24	30.7	−2.6E + 10	7.01674E + 20
14			SSR =	7.14765E + 20
15	a	1		
16	b	1		
17	c	1		

The sum of the squares of residuals, *SSR*, is calculated in cell D14 by summing the contents of cells D2 through to D13. Starting values for *a*, *b* and *c*

---

[7] The estimated values of the dependent variable must be distinguished from values obtained through experiment. Estimated values are represented by the symbol, $\hat{y}$, and experimental values by the symbol, *y*.

are extremely poor, as the predicted values, $\hat{y}$, in column C of sheet 8.2 bear no resemblance to the experimental values in column B. As a consequence, *SSR* is very large. Choosing good starting values for parameter estimates is often crucial to the success of fitting equations using non-linear least squares and we will return to this issue in section 8.3.2.

SSR in cell D14 is reduced by altering the contents of cells B15 through to B17. Solver is able to adjust the parameter estimates in cells B15 to B17 until the number in cell D14 is minimised. To accomplish this, choose the Data Ribbon on Excel, go to the Analysis group and click on **?⇨ Solver** . If Solver does not appear, then it can be added in as follows.

(1) Click on the File tab,   **File**   , at the top left of the screen.

(2) Click on   **⊞ Options** .

(3) Click on   **Add-Ins**   .

(4) Select **Manage:** | Excel Add-ins   ✔ | **Go...**   then click on **Go**.

(5) The following the dialog box should appear.

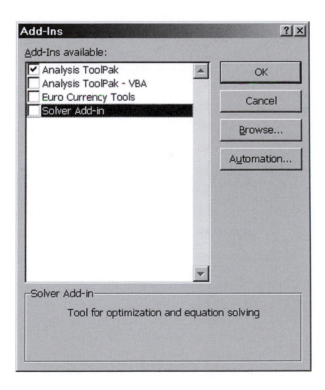

Solver is capable of adjusting cell contents such that the value in the target cell is maximised, minimised or reaches a specified value. For least squares analysis we require the content of the target cell to be minimised

We want to minimise the value in cell D14 so D14 becomes our 'Objective' cell.

Excel alters the values in this cell range in order to minimise the value in cell D14.

It is possible to constrain the values in one or more cells (for example a parameter estimate can be prevented from assuming a negative value, if a negative value is considered to be 'unphysical'). No constraints are applied in this example.

This button gives access to other options such as the maximum time allowed for fitting.

Unchecking this box allows for negative parameter estimates.

Several optimisation options are available in Solver. We will use only the 'GRG Nonlinear' method.

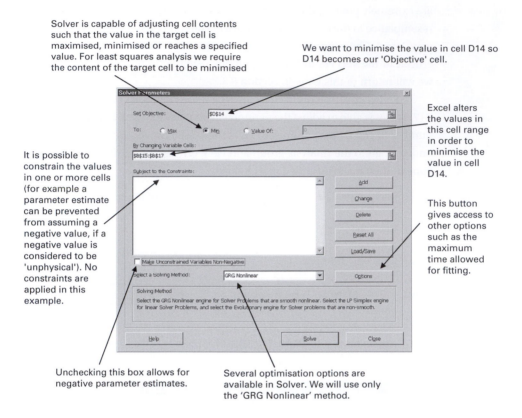

Figure 8.3. Solver dialog box with cell references inserted.

Tick the Solver Add-in box then press OK.

?⤸ Solver   should be visible on the right hand side of the Data Ribbon.

Click on Solver. The dialog box shown in figure 8.3 should appear. The annotations in figure 8.3 describe some of the features of Solver accessible through the dialog box.

After entering the information into the dialog box, click on the Solve button. After a few seconds Solver returns with the dialog box shown in figure 8.4. Click on OK.

Inspection of cells B15 to B17 in the spreadsheet indicates that Solver has adjusted the parameters. Sheet 8.3 shows the new parameters, predicted y values and SSR.

SSR in cell D14 in sheet 8.3 is almost 20 orders of magnitude smaller than that in cell D14 in sheet 8.2. However, all is not as satisfactory as it might seem. Consider the best line through the points shown in figure 8.5 which utilises the parameter estimates in cells B15 through to B17 of sheet 8.3.

Sheet 8.3. *Best values for a, b and c returned by Solver when starting values are poorly chosen.*[8]

	A	B	C	D
1	x (mins)	y (C)	$\hat{y}$ (C)	$(y - \hat{y})^2$ (C²)
2	2	26.1	24.750128	1.822154
3	4	26.8	27.996494	1.431597
4	6	27.9	29.10535	1.45287
5	8	28.6	29.484101	0.781635
6	10	28.5	29.613471	1.239817
7	12	29.3	29.657659	0.12792
8	14	29.8	29.672753	0.016192
9	16	29.9	29.677908	0.049325
10	18	30.1	29.679669	0.176678
11	20	30.4	29.68027	0.518011
12	22	30.6	29.680476	0.845525
13	24	30.7	29.680546	1.039286
14			SSR	9.501009
15	a	15.2458		
16	b	14.4347		
17	c	−0.5371		
18				

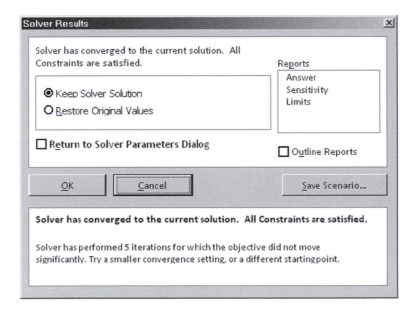

Figure 8.4. Solver dialog box indicating that fitting has been completed.

[8] If some of the Solver options have been modified then Solver may return slightly different parameter estimates to those shown in sheet 8.3. Solver options can be accessed by clicking on ⬚ Options ⬚ in the Solver Parameters dialogue box. For more details on Solver options, go to section 8.5.

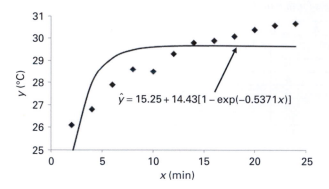

$$\hat{y} = 15.25 + 14.43[1 - \exp(-0.5371x)]$$

Figure 8.5. Graph of y versus x showing the line based on equation 8.6 where a, b and c have the values given in sheet 8.3.

As discussed in section 6.7, a plot of residuals (i.e. a plot of $(y_i - \hat{y}_i)$ versus x_i) is often used as an indicator of the 'goodness of fit' of an equation to data, with trends in the residuals indicating a poor fit.[9] However, no plot of residuals is required in this case to reach the conclusion that the line on the graph in figure 8.5 is **not** a good fit to the experimental data. Solver has found a minimum in SSR, but this is a *local* minimum[10] and the parameter estimates are of little worth. The source of the problem can be traced to the arbitrarily chosen starting values for a, b and c appearing in sheet 8.2 (i.e. $a = b = c = 1$). Working from these starting values, Solver has discovered a minimum in SSR. However, there is another combination of parameter estimates that will give a lower value for SSR.

Methods by which good starting values for parameter estimates may be obtained are considered in section 8.3.2. In the example under consideration here, we note (by reference to equation 8.6) that when $x = 0$, $y = a$. Drawing a line 'by eye' through the data in figure 8.2 indicates that, when $x = 0$, $y \approx 25.5$ °C. Starting values for b and c may be established by further preliminary analysis of data which we consider in section 8.3.2. Denoting starting values by a_0, b_0 and c_0, we find[11]

$a_0 = 25.5$, $b_0 = 5.5$ and $c_0 = -0.12$.

Inserting these values into sheet 8.2 and running Solver again gives the output shown in sheet 8.4 and in graphical form in figure 8.6.

---

[9] See Cleveland (1994) for a discussion of residuals.

[10] Local minima are discussed in section 8.3.1.

[11] All parameter estimates and standard uncertainties in parameters estimates in this example have units (for example the unit of c is min^{-1}, assuming time is measured in minutes). For convenience, units are omitted until the analysis is complete.

Sheet 8.4. *Best values for a, b and c returned by Solver when starting values for parameter estimates are good.*

	A	B	C	D
1	x (mins)	y (C)	ŷ (C)	$(y - ŷ)^2$ (C²)
2	2	26.1	26.072473	0.000758
3	4	26.8	26.977367	0.031459
4	6	27.9	27.727663	0.0297
5	8	28.6	28.349774	0.062613
6	10	28.5	28.8656	0.133663
7	12	29.3	29.293299	4.49E-05
8	14	29.8	29.647927	0.023126
9	16	29.9	29.941968	0.001761
10	18	30.1	30.185773	0.007357
11	20	30.4	30.387925	0.000146
12	22	30.6	30.55554	0.001977
13	24	30.7	30.694519	3E-05
14			SSR	0.292635
15	a	24.98113		
16	b	6.387886		
17	c	−0.09367		
18				

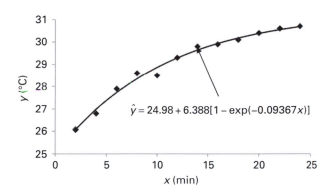

Figure 8.6. Graph of y versus x showing line and equation of line based on estimates *a*, *b* and *c* appearing in sheet 8.4.

The sum of squares of residuals in cell D14 of sheet 8.4 is less than that in cell D14 of sheet 8.3. This points to the parameter estimates obtained using Solver when good starting values are used being rather better than those obtained when the starting values are poorly chosen. More

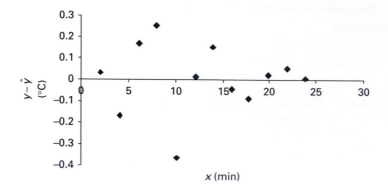

Figure 8.7. Plot of residuals based on the data and equation in figure 8.6.

Table 8.3. *Data for exercise A.*

x (rad)	y
0	1.77
0.1	2.28
0.2	3.83
0.3	5.47
0.4	7.50
0.5	8.60
0.6	8.59
0.7	10.56
0.8	10.11
0.9	10.36
1.0	10.15

importantly, the line fitted to the data in figure 8.6 (where the line is based upon the new best estimates of the parameters) is far superior to the line fitted to the same data as shown in figure 8.5. This is further reinforced by the plot of residuals shown in figure 8.7 which exhibits a random scatter about the x axis.

Exercise A

Consider the data in table 8.3.

Assuming the best line through the data in table 8.3 is of the form $y = a + b \sin cx$, use Solver to determine a, b and c. Try a range of starting values for a, b and c. Does Solver consistently converge to the same final values for a, b and c?

8.2.2 Limitations of Solver

Solver is able to efficiently solve for the best estimates of parameters in an equation, such as those appearing in equation 8.6. However, Solver does not provide standard uncertainties in the parameter estimates. Standard uncertainties in estimates are extremely important, as without them it is not possible to quote a coverage interval for the estimates and so we cannot decide if the estimates are good enough for any particular purpose to which they may be put.

The standard uncertainties in the parameter estimates a, b and c can be determined with the aid of the matrix of partial derivatives given by[12]

$$\mathbf{E} = \begin{bmatrix} \sum \left(\frac{\partial y_i}{\partial a} \right)^2 & \sum \left(\frac{\partial y_i}{\partial a} \frac{\partial y_i}{\partial b} \right) & \sum \left(\frac{\partial y_i}{\partial a} \frac{\partial y_i}{\partial c} \right) \\ \sum \left(\frac{\partial y_i}{\partial a} \frac{\partial y_i}{\partial b} \right) & \sum \left(\frac{\partial y_i}{\partial b} \right)^2 & \sum \left(\frac{\partial y_i}{\partial b} \frac{\partial y_i}{\partial c} \right) \\ \sum \left(\frac{\partial y_i}{\partial a} \frac{\partial y_i}{\partial c} \right) & \sum \left(\frac{\partial y_i}{\partial b} \frac{\partial y_i}{\partial c} \right) & \sum \left(\frac{\partial y_i}{\partial c} \right)^2 \end{bmatrix}. \tag{8.7}$$

Standard uncertainties in a, b and c are obtained using the diagonal elements of the $\mathbf{E}^{-1}$ matrix. Explicitly,

$$s_a = s\left(E_{11}^{-1} \right)^{1/2} \tag{8.8}$$

$$s_b = s\left(E_{22}^{-1} \right)^{1/2} \tag{8.9}$$

$$s_c = s\left(E_{33}^{-1} \right)^{1/2}, \tag{8.10}$$

---

[12] Note that this approach can be extended to any number of parameters. Kutner, Nachtsheim and Neter (2003) discuss the use of matrices to determine the variance in the parameter estimates.

where[13]

$$s \approx \left[\frac{1}{n-3} \sum (y_i - \hat{y}_i)^2 \right]^{1/2}. \qquad (8.11)$$

A convenient way to calculate the elements of the $\mathbf{E}$ matrix is to write

$$\mathbf{E} = \mathbf{D}^{\mathrm{T}}\mathbf{D}. \qquad (8.12)$$

$\mathbf{D}^{\mathrm{T}}$ is the transpose of the matrix, $\mathbf{D}$, where $\mathbf{D}$ is given by

$$\mathbf{D} = \begin{bmatrix} \dfrac{\partial y_1}{\partial a} & \dfrac{\partial y_1}{\partial b} & \dfrac{\partial y_1}{\partial c} \\[2mm] \dfrac{\partial y_2}{\partial a} & \dfrac{\partial y_2}{\partial b} & \dfrac{\partial y_2}{\partial c} \\[2mm] \dfrac{\partial y_i}{\partial a} & \dfrac{\partial y_i}{\partial b} & \dfrac{\partial y_i}{\partial c} \\[2mm] \dfrac{\partial y_n}{\partial a} & \dfrac{\partial y_n}{\partial b} & \dfrac{\partial y_n}{\partial c} \end{bmatrix}. \qquad (8.13)$$

The partial derivatives in equation 8.13 are evaluated on completion of fitting an equation using Solver, i.e. at values of a, b and c that minimise SSR. It is possible in some situations to determine the partial derivatives analytically. A flexible approach, and one that is generally more convenient, is to use the method of 'finite differences' to find $\frac{\partial y_1}{\partial a}$, $\frac{\partial y_2}{\partial a}$, etc. In general,

$$\left(\frac{\partial y_i}{\partial a} \right)_{b,c,x_i} \approx \frac{y[a(1+\delta), b, c, x_i] - y[a, b, c, x_i]}{a(1+\delta) - a}. \qquad (8.14)$$

As double precision arithmetic is used by Excel, the perturbation, δ, in equation 8.14 can be as small as $\delta = 10^{-6}$ or 10^{-7}.

Similarly, the partial derivatives, $\frac{\partial y_i}{\partial b}$ and $\frac{\partial y_i}{\partial c}$, are approximated by using

$$\left(\frac{\partial y_i}{\partial b} \right)_{a,c,x_i} \approx \frac{y[a, b(1+\delta), c, x_i] - y[a, b, c, x_i]}{b(1+\delta) - b} \qquad (8.15)$$

and

$$\left(\frac{\partial y_i}{\partial c} \right)_{a,b,x_i} \approx \frac{y[a, b, c(1+\delta), x_i] - y[a, b, c, x_i]}{c(1+\delta) - c}. \qquad (8.16)$$

---

[13] The expression $n-3$ in the denominator of the term in the square brackets of equation 8.11 appears because the estimate of the population standard deviation in the y values about the fitted line requires that the sum of squares of residuals be divided by the number of degrees of freedom. The number of degrees of freedom is the number of data points, n, minus the number of parameters, M, in the equation. In this example, $M = 3$.

8.2.3 Spreadsheet for determining standard uncertainties in parameter estimates

To illustrate the process of estimating standard uncertainties, we describe a step-by-step approach using Excel.

To find good approximations of the derivatives $\frac{\partial y_1}{\partial a}$ and $\frac{\partial y_2}{\partial a}$, it is necessary to perturb a slightly (say to $1.000001 \times a$) while leaving the parameter estimates b and c at their optimum values. Sheet 8.5 shows the optimum values, as obtained by Solver for a, b and c in cells G20 to G22. Cell H20 contains the value $1.000001 \times a$. Cell I21 contains the value $1.000001 \times b$ and cell J22 contains the value $1.000001 \times c$.

We use the modified parameter estimates to calculate the numerator in equation 8.14. The denominator in equation 8.14 may be determined by entering the formula = \$H\$20-\$G\$20 into a cell on the spreadsheet.

Sheet 8.5. *Modification of best estimates of parameters.*

	F	G	H	I	J
19		from solver	b,c constant	a,c constant	a,b constant
20	a	24.981126	24.9811512	24.98112622	24.98112622
21	b	6.3878859	6.387885919	6.387892307	6.387885919
22	c	−0.0936749	−0.093674867	−0.09367487	−0.093674961
23					

The partial derivative $\left(\frac{\partial y_1}{\partial a}\right)_{b,c,x_1}$ is calculated by entering the formula = (H2-C2)/(\$H\$20-\$G\$20) into cell L2 of sheet 8.6.[14] By using Home>Editing> Fill>Down, the formula may be copied into cells in the L column so that the partial derivative is calculated for every x_i. To obtain $\frac{\partial y_i}{\partial b}$, $\frac{\partial y_i}{\partial c}$, etc., this process is repeated for columns M and N, respectively, of sheet 8.6. The contents of cells L2 to N13 become the elements of the **D** matrix given by equation 8.13.

Excel's TRANSPOSE() function is used to transpose the **D** matrix. We proceed as follows.

- Highlight cells B24 to N26.
- In cell B24 type =**TRANSPOSE(L2:N13)**.
- While holding down the Ctrl and Shift keys press Enter to transpose the contents of cells L2 to N13 into cells B24 to M26.
- Multiply $\mathbf{D}^{\mathbf{T}}$ with **D** (using the MMULT matrix function in Excel) to give **E**, i.e.

[14] The values in the C column of the spreadsheet are shown in sheet 8.4.

$$\mathbf{E} = \mathbf{D}^{\mathrm{T}}\mathbf{D} = \begin{bmatrix} 12 & 7.65923 & 246.036 \\ 7.65923 & 5.49390 & 160.861 \\ 246.036 & 160.861 & 5251.55 \end{bmatrix}. \tag{8.17}$$

The MINVERSE() function in Excel is used to find the inverse of $\mathbf{E}$, i.e.

$$\mathbf{E}^{-1} = \begin{bmatrix} 2.239638 & -0.48571 & -0.09005 \\ -0.48571 & 1.870409 & -0.034537 \\ -0.09005 & -0.034537 & 0.005467 \end{bmatrix}. \tag{8.18}$$

Two more steps are required to calculate the standard uncertainties in the parameter estimates. The first is to calculate the square root of each diagonal element of the matrix $\mathbf{E}^{-1}$. The second is to calculate s using equation 8.11. Using the sum of squares of residuals appearing in cell D14 of sheet 8.4, we obtain

$$s = \left[\frac{1}{12 - 3} \times 0.2926 \right]^{1/2} = 0.1803.$$

It follows that

$$s_a = s(E_{11}^{-1})^{1/2} = 0.1803 \times (2.240)^{1/2} \qquad = 0.270 \tag{8.19}$$

$$s_b = s(E_{22}^{-1})^{1/2} = 0.1803 \times (1.870)^{1/2} \qquad = 0.247 \tag{8.20}$$

$$s_c = s(E_{33}^{-1})^{1/2} = 0.1803 \times (0.005467)^{1/2} \quad = 0.0133. \tag{8.21}$$

Sheet 8.6. *Calculation of partial derivatives.*

	H	I	J	K	L	M	N
1	$\hat{y}$ with b,c, constant	$\hat{y}$ with a,c constant	$\hat{y}$ with a,b, constant		$\partial y/\partial a$	$\partial y/\partial b$	$\partial y/\partial c$
2	26.07249786	26.07247397	26.07247387		1	0.170846	−10.5931
3	26.97739198	26.977369	26.97736864		1	0.312504	−17.5666
4	27.72768829	27.72766606	27.72766536		1	0.42996	−21.8481
5	28.34979926	28.34977764	28.34977654		1	0.527349	−24.1539
6	28.86562487	28.86560377	28.86560223		1	0.6081	−25.0341
7	29.29332358	29.29330291	29.29330093		1	0.675055	−24.9085
8	29.64795155	29.64793124	29.64792883		1	0.73057	−24.0952
9	29.94199265	29.94197263	29.94196981		1	0.776601	−22.8327
10	30.18579791	30.18577814	30.18577493		1	0.814768	−21.2983
11	30.38794995	30.38793038	30.38792681		1	0.846414	−19.6217
12	30.55556506	30.55554565	30.55554176		1	0.872654	−17.8964
13	30.69454375	30.69452448	30.69452029		1	0.894411	−16.1878

8.2.4 Coverage intervals for parameter estimates

We use parameter estimates and their respective standard uncertainties to quote a coverage interval for each parameter.[15] For the parameters appearing in equation 8.5,

$$T_s = a \pm t_{X\%,v} s_a \qquad\qquad (8.22)$$

$$k = b \pm t_{X\%,v} s_b \qquad\qquad (8.23)$$

$$\alpha = c \pm t_{X\%,v} s_c, \qquad\qquad (8.24)$$

where $t_{X\%,v}$ is the critical value of the t distribution for $X\%$ level of confidence with v degrees of freedom. Values of $t_{X\%,v}$ can be found in appendix 1. In this example $v = n - 3$ where n is the number of data points. In table 8.2 there are 12 points, so that $v = 9$. If we choose a level of confidence of 95% (the commonly chosen level),

$$t_{95\%,9} = 2.262.$$

Restoring the units of measurement and quoting a 95% coverage interval gives,

$$T_s = (24.98 \pm 2.262 \times 0.270)°C = (24.98 \pm 0.61)°C$$
$$k = (6.388 \pm 2.262 \times 0.247)°C = (6.39 \pm 0.56)°C$$
$$\alpha = (-0.09367 \pm 2.262 \times 0.0133) \text{ min}^{-1} = (-0.094 \pm 0.030) \text{ min}^{-1}.$$

Exercise B

The amount of heat entering an enclosure through a window may be reduced by applying a reflective coating to the window. An experiment is performed to establish the effect of a reflective coating on the rise in air temperature within the enclosure. The temperature within the enclosure as a function of time is shown in table 8.4.

Fit equation 8.6 to the data in table 8.4. Find a, b and c and their respective standard uncertainties.

---

[15] We could replace $t_{X\%,v}$ by the coverage factor, k, for v degrees of freedom with $X\%$ level of confidence (see section 5.9.1).

Table 8.4. *Data for exercise B.*

Time (minutes)	Temperature (°C)
2	24.9
4	25.3
6	25.4
8	25.8
10	26.0
12	26.3
14	26.4
16	26.6
18	26.5
20	26.8
22	27.0
24	26.9

8.3 More on fitting using non-linear least squares

There are several challenges to fitting equations using non-linear least squares, including:

(1) choosing an appropriate model to describe the relationship between x and y;

(2) avoiding local minima in *SSR*;

(3) establishing good starting values prior to fitting by non-linear least squares.

We consider (2) and (3) next.

8.3.1 Local minima in *SSR*

When data are noisy, or starting values are far from the best estimates, a non-linear least squares fitting routine can become 'trapped' in a local minimum.

To illustrate this situation, we draw on the analysis of data appearing in section 8.2.1. Equation 8.6 is fitted to the data in table 8.2 using the starting values given in sheet 8.2 and the best estimates, a, b and c are obtained for the parameters. For clarity, the relationship between only one parameter estimate (c) and *SSR* is considered. Solver finds a minimum in *SSR* when c is about −0.53 and terminates the fitting procedure. The variation of *SSR* with c is shown in figure 8.8.[16]

---

[16] For clarity, units of measurement are omitted from the graph.

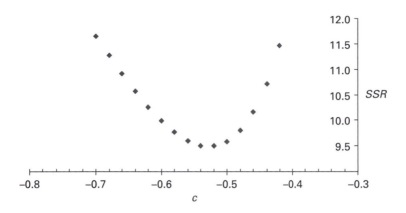

Figure 8.8. Variation of *SSR* with *c* near a local minimum for equation 8.6 is fitted to the data in table 8.1.

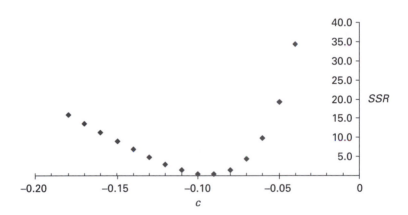

Figure 8.9. Variation of *SSR* with *c* near to the global minimum for equation 8.6 is fitted to the data in table 8.1.

The minimum in *SSR* in figure 8.8 is referred to as a *local* minimum as there is another combination of parameter estimates that will give a lower value for *SSR*. The lowest value of *SSR* obtainable corresponds to a *global* minimum. It is the global minimum that we seek in all fitting using least squares.

When starting values are used that are closer to the final values,[17] the non-linear fitting routine finds parameter estimates that produce a lower final value for *SSR*. Figure 8.9 shows the variation of *SSR* with *c* in the interval $(-0.18 < c < -0.04)$.

---

[17] See section 8.2.1.

A number of indicators can assist in identifying a local minimum, though there is no infallible way of deciding whether a local or global minimum has been discovered. A good starting point is to plot the raw data along with the fitted line (as illustrated in figure 8.5). A poor fit of the line to the data could indicate:

- a local minimum has been found;
- an inappropriate model has been fitted to the data.

When a local minimum in *SSR* is found, the standard uncertainties in the parameter estimates tend to be large. As an example, best estimates appearing in sheet 8.3 (resulting from being trapped in a local minimum), their respective standard uncertainties and the magnitude of the ratio of these quantities (expressed as a percentage) are:

$a = 15.25\,°C, s_a = 9.04\,°C$, so that $|s_a/a| \times 100\% = 59\%$;
$b = 14.43\,°C, s_b = 8.93\,°C$, so that $|s_b/b| \times 100\% = 62\%$;
$c = -0.5371\,\text{min}^{-1}, s_c = 0.276\,\text{min}^{-1}$, so that $|s_c/c| \times 100\% = 51\%$.

When the global minimum in *SSR* is found (see sheet 8.4), the best estimates of the parameters and standard uncertainties are:

$a = 24.98\,°C, s_a = 0.270\,°C$, so that $|s_a/a| \times 100\% = 1.1\%$;
$b = 6.388\,°C, s_b = 0.247\,°C$, so that $|s_b/b| \times 100\% = 3.9\%$;
$c = -0.09367\,\text{min}^{-1}, s_c = 0.0133\,\text{min}^{-1}$, so that $|s_c/c| \times 100\% = 14\%$.

There is merit in fitting the same equation to data several times, each time using different starting values for the parameter estimates. If, after fitting, there is consistency between the final values obtained for the best estimates, then it is likely that the global minimum has been identified.

8.3.2 Starting values

There are no general rules that may be applied in order to determine good starting values[18] for parameter estimates prior to fitting by non-linear least squares. Familiarity with the relationship being studied can assist in deciding what might be reasonable starting values for some of the parameters in an equation.

A useful approach to determining starting values is to begin by plotting the experimental data. Consider the data in figure 8.10, which has a smooth line drawn through the points 'by eye'.

If the relationship between x and y is given by equation 8.25, then we are able to estimate a and b by considering the data displayed in figure 8.10 and a

---

[18] Starting values are sometimes referred to as *initial estimates*.

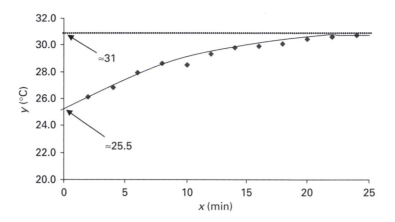

Figure 8.10. Line drawn 'by eye' through the data given in table 8.1.

'rough' line drawn through the data. Equation 8.25 predicts that $y = a$ when x is equal to zero:

$$y = a + b[1 - \exp(cx)]. \tag{8.25}$$

From figure 8.10 we see that when $x = 0$, $y \approx 25.5$ °C, so that $a \approx 25.5$ °C. When x is large (and assuming c is negative), then $\exp(cx)$ tends to zero so that, $y = a + b$. Inspection of the graph in figure 8.10 indicates that when x is large, $y \approx 31.0$ °C, i.e. $a + b \approx 31.0$ °C. It follows that $b \approx 5.5$ °C. If we write the starting values for a and b as a_0 and b_0 respectively, then $a_0 = 25.5$ °C and $b_0 = 5.5$ °C.

In order to determine a starting value for c, c_0, equation 8.25 is rearranged into the form[19]

$$\ln\left[1 - \left(\frac{y - a_0}{b_0}\right)\right] = c_0 x. \tag{8.26}$$

Equation 8.26 has the form of the equation of a straight line passing through the origin. It follows that plotting $\ln\left[1 - \left(\frac{y-a_0}{b_0}\right)\right]$ versus x should give a straight line with slope, c_0.

Figure 8.11 shows a plot of $\ln\left[1 - \left(\frac{y-a_0}{b_0}\right)\right]$ versus x. The best straight line and the equation of the line has been added using the Trendline option in Excel.[20] The slope of the line is approximately -0.12. The starting values may now be stated for this example, i.e.:

---

[19] Starting values, a_0 and b_0, are substituted into equation 8.26.

[20] For details on using Trendline, see section 2.7.1. Note that for a straight line forced through the origin, the slope would be -0.11 min^{-1}. Taking this to be c_0 rather than -0.12 min^{-1} would have no influence on the final fitted values.

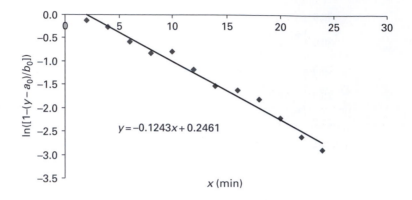

Figure 8.11. Line of best fit used to determine a starting value for c.

$$a_0 = 25.5°\text{C}; \; b_0 = 5.5°\text{C}; \; c_0 = -0.12\,\text{min}^{-1}.$$

These starting values were used in the successful fit of equation 8.6 to the data given in table 8.1 (the output of the fit is shown in sheet 8.4).

8.3.3 Starting values by curve stripping

Establishing starting values in some situations is quite difficult and may require a significant amount of pre-processing of data before employing fitting by non-linear least squares. For example, the fitting to data of an equation consisting of a sum of exponential terms, such as

$$y = a \exp bt + c \exp dt \tag{8.27}$$

or

$$y = a \exp bt + c \exp dt + e \exp ft, \tag{8.28}$$

where b, d and f are negative with $|b| > |d| > |f|$ and $t \geq 0$, is particularly challenging especially when data are noisy and/or the ratio of the parameters within the exponentials is less than approximately 3 (for example, when the ratio b/d in equation 8.27 is less than 3).[21] Fitting of equations such as equation 8.27 and equation 8.28 is quite common, for example the kinetics of drug transport through the human body is routinely modelled using 'compartmental analysis'.[22]

---

[21] See Kirkup and Sutherland (1988).

[22] See Jacquez (1996).

Compartmental analysis attempts to predict concentrations of drugs as a function of time (e.g. in blood or urine). The relationship between concentration and time is often well represented by a sum of exponential terms. In analytical chemistry, excited state lifetime measurements offer a means of identifying components in a mixture. The decay of phosphorescence with time that occurs after illumination of the mixture may be captured. The decay can be represented by a sum of exponential terms. Fitting a sum of exponentials by non-linear least squares allows for each component in the mixture to be discriminated.[23]

If an equation to be fitted to data consists of a sum of exponential terms, good starting values for parameter estimates are extremely important if local minima in *SSR* are to be avoided. It is also possible that, if starting values for the parameter estimates are too far from the optimum values, the *SSR* will increase during the iterative process to such an extent that the *SSR* exceeds the maximum floating point number that a spreadsheet (or other program) can handle. In this situation, fitting is terminated and an error message is returned by the spreadsheet.

Data in figure 8.12 have been gathered in an experiment in which the decay of photo-generated current in the wide band gap semiconductor cadmium sulphide (CdS) is measured as a function of time after photo-excitation of the semiconductor has ceased. There appears to be an exponential decay of the photocurrent with time. Theory indicates[24] that there may be more than one decay mechanism for photoconductivity. That, in turn, suggests that an equation of the form given by equation 8.27 or equation 8.28 may be appropriate.

If equation 8.27 is to be fitted to data, how are starting values for parameter estimates established? If b is large (and negative) then the contribution of the first term in equation 8.27 to y is small when t exceeds some value, which we will designate as t'. Equation 8.27 can now be written, for $t > t'$,

$$y \approx c \exp dt. \tag{8.29}$$

Equation 8.29 can be linearised by taking natural logarithms of both sides of the equation. The next step is to fit a straight line to the transformed data to find (approximate values) for c and d which we will designate as c_0 and d_0 respectively.

Now we revisit equation 8.27 and write, for $t < t'$,

$$y - c_0 \exp d_0 t = a \exp bt. \tag{8.30}$$

[23] See Demas (1983).

[24] See Bube (1960), chapter 6.

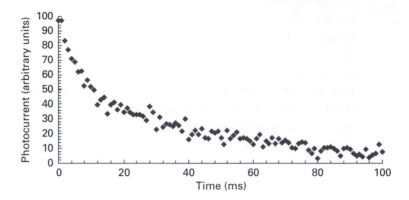

Figure 8.12. Photocurrent versus time data for cadmium sulphide.

Transforming equation 8.30 by taking natural logarithms of both sides of the equation then fitting a straight line to the transformed data will yield approximate values for a and b which can serve as starting values in a non-linear fit.

8.4 Weighted non-linear least squares

There are some situations in which the standard deviation of the residuals, $(y - \hat{y})$, is not constant (see section 6.7). These residuals are said to be heteroscedastic. Such a situation may be revealed by plotting[25] $(y - \hat{y})$ versus x. If residuals *are* heteroscedastic, then *weighted* fitting is required. The purpose of weighted fitting is to obtain best estimates of the parameters by forcing the line close to the data with least uncertainty, while giving much less weight to those data with larger uncertainty.

As described in section 6.10, the starting point for weighted fitting using least squares is to define a sum of squares of residuals that takes into account the standard deviation in the y values. We write

$$\chi^2 = \sum \left(\frac{y_i - \hat{y}_i}{s_i} \right)^2,$$ (8.31)

where χ^2 is the weighted sum of squares of residuals,[26] and s_i is the standard deviation in the ith y value. The purpose of the weighted fitting is to find best estimates of parameters that minimise χ^2 in equation 8.31.

---

[25] See section 6.7 for a discussion of residuals.

[26] $\sum \left(\frac{y_i - \hat{y}_i}{s_i} \right)^2$ follows a chi-squared distribution, hence the use of the symbol χ^2.

8.4.1 Weighted fitting using Solver

In order to establish best estimates of parameters using Solver when weighted fitting is performed, we use an approach similar to that described in section 8.2.3. For weighted fitting, an extra column in the spreadsheet containing the standard deviations s_i is required. It is possible that the absolute values of s_i are unknown and that only relative standard deviations are known. For example, equations 8.32 and 8.33 are sometimes used when weighted fitting is required,

$$s_i \propto \sqrt{y_i} \tag{8.32}$$

$$s_i \propto y_i. \tag{8.33}$$

Weighted fitting can be carried out so long as:

- the absolute standard deviations in values are known, or
- the relative standard deviations are known.

In order to accomplish weighted non-linear least squares, we proceed as follows.

(1) Fit the desired equation to data by calculating χ^2 as given by equation 8.31. Use Solver to modify parameter estimates so that χ^2 is minimised.
(2) Determine the elements in the **D** matrix, as described in section 8.2.2.
(3) Construct the weight matrix, **W**, in which the diagonal elements of the matrix contain the weights to be applied to the y values.
(4) Calculate the weighted standard deviation s_w, where s_w is given by

$$s_w = \left(\frac{\chi^2}{n - M} \right)^{1/2}, \tag{8.34}$$

and where χ^2 is given by equation 8.31, n is the number of data points and M is the number of parameters in the equation to be fitted to the data.
(5) Calculate the standard uncertainties in parameter estimates, given by[27]

$$\mathbf{s}(\mathbf{B}) = s_w \left[\left(\mathbf{D}^T \mathbf{W} \mathbf{D} \right)^{-1} \right]^{1/2}. \tag{8.35}$$

B is the matrix containing elements equal to the best estimates of the parameters, and s_w is the weighted standard deviation, given by equation 8.34.
(6) Calculate the coverage interval for each parameter appearing in the equation at a specified level of confidence (usually 95%).

To illustrate steps (1) to (6), we consider an example of weighted fitting using Solver.

---

[27] See Kutner, Nachtsheim and Neter (2003).

Table 8.5. *Current–voltage data for a germanium tunnel diode.*

V (mV)	I (mA)
10	4.94
20	6.67
30	10.57
40	10.11
50	10.44
60	12.90
70	10.87
80	9.73
90	7.03
100	5.61
110	3.80
120	2.36

8.4.2   Example of weighted fitting using Solver

The relationship between the current, I, through a tunnel diode[28] and the voltage, V, across the diode may be written[29]

$$I = AV(B - V)^2. \tag{8.36}$$

A and B are constants to be estimated using least squares. Table 8.5 shows current–voltage data for a germanium tunnel diode.

Equation 8.36 could be fitted to data using unweighted non-linear least squares (in the first instance it is prudent to carry out an unweighted fit, as the residuals may show marginal evidence of heteroscedasticity and so there is little point in performing a more complex analysis).

In this example we are going to assume that the error in the y quantity is proportional to the size of the y quantity, i.e. equation 8.33 is valid for these data.

The data in table 8.5 are entered into a spreadsheet as shown in sheet 8.7 and plotted in figure 8.13.

---

[28] A tunnel diode is a semiconductor device with characteristics that make it useful in microwave circuits.

[29] See Karlovsky (1962).

Sheet 8.7. *Data from table*
8.5 entered into a spreadsheet.

	A	B
	x(mV)	y(mA)
1	x(mV)	y(mA)
2	10	4.94
3	20	6.67
4	30	10.57
5	40	10.11
6	50	10.44
7	60	12.90
8	70	10.87
9	80	9.73
10	90	7.03
11	100	5.61
12	110	3.80
13	120	2.36
14		

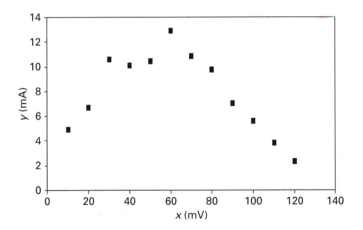

Figure 8.13. Current–voltage data for a germanium tunnel diode.

8.4.3 Best estimates of parameters using Solver

Consistent with symbols used in other least squares analyses, we rewrite
equation 8.36 as

$$y = ax(b - x)^2. \tag{8.37}$$

We can obtain a reasonable value for b, which we will use as a starting value, b_0,
by noting that equation 8.37 predicts that $y = 0$ when $x = b$. By inspection of

Sheet 8.8. *Fitted values and weighted sum of squares of residuals* before *optimisation occurs.*

	B	C	D
1	y (mA)	ŷ (mA)	$\left(\dfrac{y-\hat{y}}{y}\right)^2$
2	4.94	4.752	0.001448311
3	6.67	7.986	0.038927822
4	10.57	9.9	0.004017905
5	10.11	10.692	0.003313932
6	10.44	10.56	0.000132118
7	12.90	9.702	0.061457869
8	10.87	8.316	0.055205544
9	9.73	6.6	0.103481567
10	7.03	4.752	0.105001811
11	5.61	2.97	0.221453287
12	3.80	1.452	0.381793906
13	2.36	0.396	0.692562482
14		sum	1.668796555
15			
16			From solver
17		a	3.30E-05
18		b	130

figure 8.13 we find that when $y = 0$, $x \approx 130$ mV, so that $b_0 = 130$ mV. Equation 8.37 is rearranged to give

$$a = \frac{y}{x(b-x)^2}. \tag{8.38}$$

An approximate value for a (which we take to be the starting value, a_0) can be obtained by choosing any data pair from sheet 8.7 (say, $x = 50$ mV and $y = 10.44$ mA) and substituting these into equation 8.38 along with $b_0 = 130$ mV. This gives (to two significant figures) $a_0 = 3.3 \times 10^{-5}$ mA/(mV)^3.

Sheet 8.8 shows the cells containing the calculated values of current ($\hat{y}$) in column C based on equation 8.37. The parameter estimates are the starting values (3.3×10^{-5} and 130) in cells D17 and D18. Column D of sheet 8.8 contains the weighted sum of squares of residuals. The sum of these residuals appears in cell D14.

Running Solver using the default settings (see section 8.5.1) gives the output shown in sheet 8.9.

The weighted standard deviation is calculated using equation 8.34, i.e.

$$s_w = \left(\frac{\chi^2}{n-p}\right)^{1/2} = \left(\frac{0.0895702}{12-2}\right)^{1/2} = 0.09464. \tag{8.39}$$

Sheet 8.9. *Fitted values and weighted sum of squares of residuals* after *using Solver to minimise the value in cell D14 by changing the parameter estimates in cells D17 and D18.*[30]

	B	C	D
1	y (mA)	$\hat{y}$ (mA)	$\left(\dfrac{y-\hat{y}}{y}\right)^2$
2	4.94	4.4579265	0.009523
3	6.67	7.68209988	0.0230248
4	10.57	9.81027104	0.0051662
5	10.11	10.9801908	0.0074084
6	10.44	11.3296102	0.007261
7	12.90	10.99628	0.0217784
8	10.87	10.117951	0.0047867
9	9.73	8.83237427	0.0085107
10	7.03	7.2773006	0.0012375
11	5.61	5.59048087	1.211E−05
12	3.80	3.90966598	0.0008329
13	2.36	2.3726068	2.854E−05
14		sum	0.0895702
15			
16			From solver
17		a	2.2958E−05
18		b	149.34612

χ^2 (arrow pointing to cell D14 value 0.0895702)

8.4.4 Determining the **D** matrix

In order to determine the matrix of partial derivatives, we calculate

$$\left(\frac{\partial y_i}{\partial a}\right)_{b,x_i} \approx \frac{y[a(1+\delta), b, x_i] - y[a, b, x_i]}{a(1+\delta) - a} \tag{8.40}$$

and

$$\left(\frac{\partial y_i}{\partial b}\right)_{a,x_i} \approx \frac{y[a, b(1+\delta), x_i] - y[a, b, x_i]}{b(1+\delta) - b}, \tag{8.41}$$

where δ is chosen to be 10^{-6} (see section 8.2.2). Sheet 8.10 shows the values of the partial derivatives in the **D** matrix.

8.4.5 The weight matrix, **W**

The weight matrix is a square matrix with diagonal elements proportional to $\frac{1}{s_i^2}$ and other elements equal to zero.[31] In this example, s_i is taken to be equal to y_i, so that the diagonal matrix is as given in sheet 8.11.

---

[30] Note that prior to fitting, the Use Automatic Scaling box was checked in the Options dialogue box in Solver. It is recommended that this box be checked ahead of any fitting using Solver (see section 8.5.1 for details).

[31] For details on the weight matrix, see Kutner, Nachtsheim and Neter (2003).

Sheet 8.10. *Calculation of partial derivatives used in the **D** matrix.*

	E	F	G	H	I
1	$\hat{y}$ (b constant)	$\hat{y}$ (a constant)		$\partial y/\partial a$	$\partial l/\partial b$
2	4.457930955	4.457936052		194173.4121	0.063984
3	7.682107563	7.682117621		334608.3762	0.118784
4	9.810280846	9.810295588		427304.8921	0.1644
5	10.98020182	10.98022084		478262.9600	0.200834
6	11.32962152	11.32964425		493482.5800	0.228084
7	10.99629095	10.99631671		478963.7516	0.24615
8	10.11796114	10.11798911		440706.4753	0.255034
9	8.832383106	8.832412317		384710.7509	0.254733
10	7.277307872	7.277337222		316976.5784	0.24525
11	5.590486459	5.590514708		243503.9579	0226583
12	3.909669886	3.909695656		170292.8892	0.198732
13	2.372609174	2.372630951		103343.3725	0.161699
14					
15					
16	b constant	a constant			
17	2.295850E − 05	2.295848E − 05			
18	149.3461202	149.3462695			

Rows 2–13 are bracketed as **D**.

Sheet 8.11. *Weight matrix for tunnel diode analysis (while the weights are shown to only three decimal places, Excel retains all figures for the calculations).*

	C	D	E	F	G	H	I	J	K	L	M	N
24	0.041	0	0	0	0	0	0	0	0	0	0	0
25	0	0.022	0	0	0	0	0	0	0	0	0	0
26	0	0	0.009	0	0	0	0	0	0	0	0	0
27	0	0	0	0.010	0	0	0	0	0	0	0	0
28	0	0	0	0	0.009	0	0	0	0	0	0	0
29	0	0	0	0	0	0.006	0	0	0	0	0	0
30	0	0	0	0	0	0	0.008	0	0	0	0	0
31	0	0	0	0	0	0	0	0.011	0	0	0	0
32	0	0	0	0	0	0	0	0	0.020	0	0	0
33	0	0	0	0	0	0	0	0	0	0.032	0	0
34	0	0	0	0	0	0	0	0	0	0	0.069	0
35	0	0	0	0	0	0	0	0	0	0	0	0.180

Rows 24–35 are bracketed as **W**.

8.4.6 Calculation of $(\mathbf{D}^T\mathbf{W}\mathbf{D})^{-1}$

To obtain standard uncertainties in estimates a and b, we must determine $(\mathbf{D}^T\mathbf{W}\mathbf{D})^{-1}$. Sheet 8.12 shows the several steps required to determine $(\mathbf{D}^T\mathbf{W}\mathbf{D})^{-1}$. The steps consist of the following.

(a) Calculation of the matrix **WD**. The elements of this matrix are shown in cells C37 to D48. (**W** is multiplied with **D** using the MMULT() function in Excel).

Sheet 8.12. *Calculation of* $(\mathbf{D^TWD})^{-1}$.

	B	C	D	E	F	G	H
37	WD	7956.75278	0.0026219		$\mathbf{D^TWD}$	2.2597E + 10	15368.5701
38		7521.16542	0.00267			15368.5701	0.01347659
39		3824.61646	0.0014715				
40		4679.12273	0.0019649				
41		4527.62896	0.0020926		$(\mathbf{D^TWD})^{-1}$	1.9722E–10	–0.0002249
42		2878.21496	0.0014792			–0.0002249	330.689201
43		3729.84121	0.0021584				
44		4063.57839	0.0026907				
45		6413.81639	0.0049625				
46		7737.13727	0.0071995				
47		11793.1364	0.0137626				
48		18554.9003	0.0290323				

(b) Calculation of the matrix $\mathbf{D^TWD}$. The elements of this matrix are shown in cells G37 to H38.

(c) Inversion of the matrix, $\mathbf{D^TWD}$. The elements of the inverted matrix are shown in cells G41 to H42.

8.4.7 Bringing it all together

To calculate the standard uncertainties in a and b, the weighted standard deviation (given by equation 8.34) is multiplied by the square root of the diagonal elements of the $(\mathbf{D^TWD})^{-1}$ matrix, i.e.

$$s_a = s_w \left(1.9722 \times 10^{-10}\right)^{1/2}$$
$$= 0.094\,641 \times \left(1.9722 \times 10^{-10}\right)^{1/2} = 1.329 \times 10^{-6}$$

and

$$s_b = s_w (330.69)^{1/2} = 0.094\,641 \times (330.69)^{1/2} = 1.721.$$

It follows that the 95% coverage intervals for A and B are

$$A = a \pm t_{95\%,v} s_a \tag{8.42}$$

$$B = b \pm t_{95\%,v} s_b. \tag{8.43}$$

In this example, the number of degrees of freedom, $v = n - p = 12 - 2 = 10$. From table 2 in appendix 1,

$$t_{95\%,10} = 2.228.$$

It follows that (inserting units and rounding the expanded uncertainty to two significant figures)

$$A = (2.30 \pm 0.30) \times 10^{-5} \text{mA}/(\text{mV})^3$$
$$B = (149.3 \pm 3.8)\,\text{mV}.$$

Exercise C

Equation 8.36 may be transformed into a form suitable for fitting by linear least squares.

(i) Show that equation 8.36 can be rearranged into the form

$$\left(\frac{I}{V}\right)^{1/2} = A^{1/2}B - A^{1/2}V.$$ (8.44)

(ii) Plot a graph of $\left(\frac{I}{V}\right)^{1/2}$ versus V.

(iii) Use unweighted least squares to obtain best estimates of A and B and standard uncertainties in the best estimates.[32]

(iv) Why is it preferable to use non-linear least squares to estimate parameters rather than to linearise equation 8.36 followed by using linear least squares to find these estimates?

8.5 More on Solver

Solver uses the General Reduced Gradient (GRG) method for solving non-linear problems and this section is devoted to describing features of Solver that relate to this.[33]

Though optimisation can be carried out successfully with the default settings in the Solver Option dialog box, Solver possesses several options that can be adjusted by the user to assist in the optimisation process and we will describe those next.

The Solver dialog box, as shown in figure 8.3, offers the facility to constrain parameter estimates so that unrealistic estimates can be avoided. In any case the best estimates returned by Solver need to be compared with 'physical reality' before being accepted. Consider an example in which a parameter in an equation represents the speed of sound, v, in air. If, after fitting by least squares, the best estimate of v is $-212\,\text{m/s}$, it is fair to question whether this value is 'reasonable'. If it is not, then one course of action is to try new starting values for the parameter estimates. Alternatively, we could use the 'Subjects to Constraints box' in the Solver dialog box to constrain the estimate

---

[32] Care must be taken when calculating the uncertainty in the estimate of B as this requires use of both slope and the intercept and these are correlated.

[33] See Smith and Lasdon (1992). Solver includes two other methods for solving optimisation problems: one incorporates the Simplex method for solving linear problems, and the other 'Evolutionary Solver' which is capable of being used with any Excel formula or function. We will not consider those two methods in this book.

of v so that it cannot take on negative values. This cannot guarantee that a physically meaningful value will be found for v, only that the value will be non-negative.

If we are quite sure that a parameter will lie between certain limits, then specifying the constraints can reduce the fitting time. Perhaps even more useful, is that constraints can be used with an option new to Excel 2010 referred to as 'Multi-start' (described in the next section) which improves the probability of finding a global minimum in *SSR*. To use this option, bounds for all the parameters appearing in the equation to be fitted to data must be specified using the constraints box.

8.5.1 Solver Options

To view Solver options shown in figure 8.14, it is necessary to click on the Option button in the Solver dialog box. This dialog box may be used to modify, for example, the method by which the optimisation takes place. This, in turn, may provide for a better fit or reduce the fitting time over that obtained using the default settings.

We now consider some of the options in the Options dialog box.

Constraint Precision	Solver evaluates constraints within the tolerance specified within the Constraint Precision box. For example, suppose the constraint precision is set to 0.0001. If we specified that the parameter in cell \$D\$18 >=0, and after fitting the content of cell \$D\$18 was −0.000 015, then Solver would consider the constraint to be satisfied.
Use Automatic Scaling	In certain problems there may be many orders of magnitude difference between the data, the parameter estimates and the value in the cell which is referenced in the Set Objective box. This can lead to rounding problems owing to the finite precision arithmetic performed by Excel. If the 'Use Automatic Scaling' box is ticked, then Solver will scale values before carrying out optimisation (and 'unscale' the solution values before entering them into the spreadsheet). It is advisable to tick this box for all problems.
Show Iteration Results	Ticking this box causes Solver to pause after each iteration, allowing new estimates of parameters and the value in the cell referenced in the Set Objective box to be viewed. If parameter estimates are used to draw a line of best fit

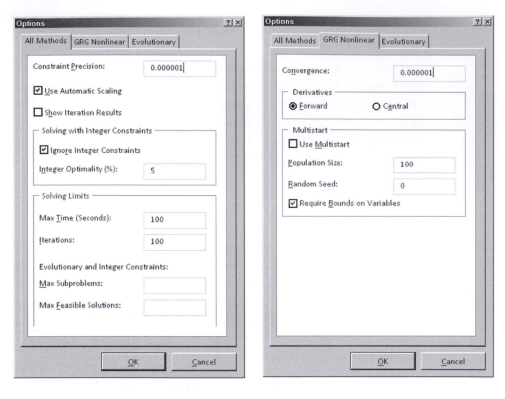

Figure 8.14. Solver Options dialog box illustrating default settings.

through the data, then the line will be updated after each iteration. Updating the fitted line on the graph after each iteration gives a valuable insight into the progress made by Solver to find best estimates of the parameters in an equation.

Max Time   This restricts the total time Solver spends searching for an optimum solution. Unless there are many data, the default value of 100 s is usually sufficient. If the maximum time is set too low, such that Solver has not completed its search, then a message is returned 'The maximum time limit was reached; continue anyway?'. Clicking on the Continue button will cause Solver to carry on searching for a solution.

Iterations   This is the maximum number of iterations that Solver will execute before terminating its search. The default value is 100, but this can be increased to a limit of 32767. Solver is likely to find an optimum solution before reaching such a limit or return a message that an optimum solution cannot be found. If the number of iterations is set too low, such that Solver has not completed its search, then a message will be returned 'The maximum iteration limit was reached; continue anyway?'. Clicking on the Continue button will cause Solver to carry on searching for a solution.

Convergence

As fitting proceeds, Solver compares the most recent solution (for our purposes this would be the value of SSR) with previous solutions. If the fractional reduction in the solution over five iterations is less than the value in the Convergence box, Solver reports that optimisation is complete. If this value is made very small (say, 10^{-6}) Solver will continue iterating (and hence take longer to complete) than if that number is larger (say, 10^{-2}).

Derivatives:
Forward or
Central

The partial derivatives of the function in the target cell with respect to the parameter estimates are found by the method of finite differences. It is possible to perturb the estimates 'forward' from a particular point or to perturb the estimates forward and backward from the point in order to obtain a better estimates of the partial derivatives. Both methods of determining the partial derivatives produce the same final results for the examples described in this chapter.

Multi-start

An important feature introduced in Excel 2010 that was not available in previous versions of Excel, is the ability of Solver to automatically choose many sets of starting values for the parameters. Excel performs a fit using each set of starting values in an effort to find the global minimum in SSR. It is necessary to specify the bounds of the parameter estimates (i.e. the maximum and minimum of each estimate) before Solver is able to execute its Multi-start feature. The bounds are entered by specifying the constraints for the parameters which is done in the main Solver Parameters dialog box. The wider the bounds that are specified, the longer it will take Solver to find the global minimum. Though this option does increase the probability of finding the global minimum, it does not guarantee it.

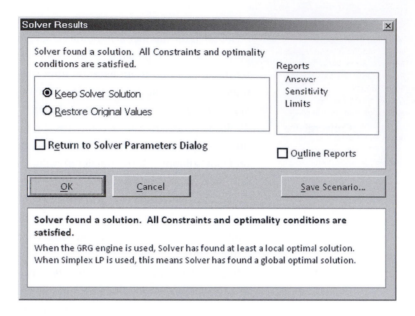

Figure 8.15. Solver Results dialog box.

8.5.2 Solver Results

Once Solver completes optimisation, it displays the Solver Results dialog box shown in figure 8.15.

Clicking on OK will retain the solution found by Solver (i.e. the starting parameters are permanently replaced by the final parameter estimates). At this stage Excel is able to present three reports: Answer, Sensitivity and Limits. Of the three reports, the Answer report is the most useful as it gives the starting values of the parameter estimates and the associated *SSR*. The report also displays the final parameter estimates and the final *SSR*, allowing for easy comparison with the original values. An Answer report is shown in figure 8.16.

8.6 Review

This chapter considered fitting equations to data using the technique of non-linear least squares. In particular, the use of the Solver tool within Excel was introduced and employed for non-linear least squares fitting.

In order to quote a coverage interval for parameters it is necessary to be able to determine standard uncertainties in the estimates made of any parameters appearing in an equation. As Solver does not provide standard

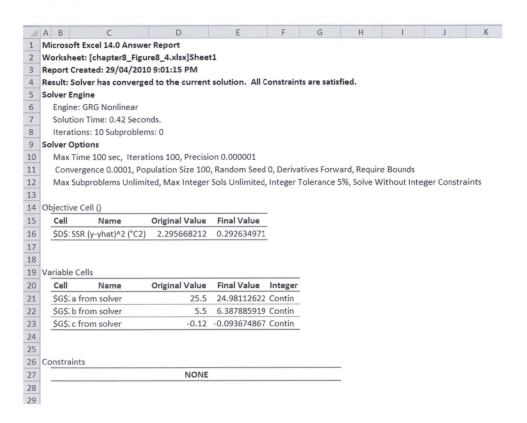

Figure 8.16. Answer report created by Excel.

uncertainties, this chapter describes a way in which these may be determined using Excel. An advantage of employing Excel is that aspects of fitting by non-linear least squares which are normally hidden from view when using a conventional computer based statistics package can be made visible with Excel. This approach leads to an insight into the process of fitting using non-linear-least squares which goes beyond the act of entering numbers into a statistics package and waiting for the fruits of the analysis to emerge.

Some general issues relating to fitting by non-linear least squares have been discussed, such as the existence of local minima in SSR and the means by which good starting values may be established in advance of fitting.

End of chapter problems

(1) Standard addition analysis is routinely used to establish the composition of a sample. In order to establish the concentration of Fe^{3+} in water, solutions

Table 8.6. *Data for problem (1).*

Concentration (ppm), x	Absorbance (arbitrary units), y
0	0.240
5.55	0.437
11.10	0.621
16.65	0.809
22.20	1.009

containing known concentrations of Fe^{3+} were added to water samples.[34] The absorbance of each solution, y, was determined for each concentration of added solution, x. The absorbance/concentration data are shown in table 8.6.

The relationship between absorbance, y, and concentration, x, of added solution may be written

$$y = B(x - x_C), \tag{8.45}$$

where B is the slope of the line of y versus x, and x_C is the intercept on the x axis which represents the concentration of Fe^{3+} in the water before additions are made.

Use non-linear least squares to fit equation 8.45 to the data in table 8.6. Determine

  (i) best estimates of B and x_C;
  (ii) standard uncertainties in best estimates of B and x_C.

(2) Another way to analyse the data in table 8.6 is to write

$$y = A + Bx. \tag{8.46}$$

Here A is the intercept on the y axis at $x = 0$, and B is the slope. The intercept on the x axis, x_C (found by setting $y = 0$ in equation 8.46), is given by

$$x_C = \frac{-A}{B}. \tag{8.47}$$

Use linear least squares to fit equation 8.46 to the data in table 8.6. Determine,

  (i) best estimates of A, B and x_C;
  (ii) standard uncertainties in the best estimates of A, B and x_C.

Note that the errors in the best estimate of slope and intercept in equation 8.47 are correlated and so the normal 'propagation of uncertainties' method is not valid when calculating x_C.

---

[34] This problem is adapted from Skoog and Leary (1992).

Table 8.7. *Data for problem (3).*

t (s)	$V(t)$ (mL)
145	4.0
314	7.6
638	12.2
901	15.6
1228	18.6
1691	21.6
2163	24.0
2464	24.8

(3) In a study of first order kinetics, the volume of titrant required, $V(t)$, to reach the end point of a reaction is measured as a function of time, t (see table 8.7). The data obtained are shown in table 8.7.[35]

The relationship between V and t can be written

$$V(t) = V_\infty - (V_\infty - V_0) \exp(-kt), \tag{8.48}$$

where k is the rate constant. V_∞ and V_0 are also constants.

(i) Find approximate values[36] for V_∞, V_0 and k.
(ii) Use Solver to establish best estimates of V_∞, V_0 and k.
(iii) Determine standard uncertainties in the estimates of V_∞, V_0 and k.
(iv) Determine the 95% coverage intervals for V_∞, V_0 and k.

(4) Table 8.8 contains data obtained from a simulation of a chemical reaction in which noise of constant variance has been added to the data.[37]

Assuming that the relationship between concentration, C, and time, t, can be written[38]

$$C = \frac{C_0}{1 + C_0 kt}, \tag{8.49}$$

where C_0 is the concentration at $t = 0$ and k is the second order rate constant.

(i) Find approximate values for C_0 and k.
(ii) Fit equation 8.49 to the data in table 8.8 to obtain best estimates for C_0 and k.
(iii) Determine standard uncertainties in the best estimates.

---

[35] See Denton (2000).

[36] Hint: draw a graph of $V(t)$ versus t. Equation 8.48 indicates that, as t gets large, $V(t)$ nears V_∞. V_0 is the value of $V(t)$ when $t = 0$ and this may be estimated from the graph. Using the estimates of V_∞ and V_0, equation 8.48 can be written in a form that allows for linearisation from which a starting value for k can be found (see section 8.3.2 for a similar problem).

[37] See Zielinski and Allendoerfer (1997).

[38] The assumption is made that a second order kinetics model can represent the reaction.

Table 8.8. *Simulated data drawn from Zielinski and Allendoerfer (1997).*

Time, t (s)	Concentration, C (mol/L)
0	0.01000
20000	0.00862
40000	0.00780
60000	0.00687
80000	0.00648
100000	0.00595
120000	0.00536
140000	0.00507
160000	0.00517
180000	0.00450
200000	0.00482
220000	0.00414
240000	0.00359
260000	0.00354
280000	0.00324
300000	0.00333
320000	0.00309
340000	0.00285
360000	0.00349
380000	0.00273
400000	0.00271

(5) Table 8.9 gives the temperature dependence of the energy gap of high purity crystalline silicon. The variation of energy gap with temperature can be represented by the equation

$$E_g(T) = E_g(0) - \frac{\alpha T^2}{\beta + T},$$
(8.50)

where $E_g(0)$ is the energy gap at absolute zero and α and β are constants.

Fit equation 8.50 to the data in table 8.9 to find best estimates of $E_g(0)$, α and β as well as standard uncertainties in the estimates. Use starting values, 1.1, 0.0004, and 600 respectively for estimates of $E_g(0)$, α and β .

(6) In an experiment to study phytoestrogens in Soya beans, an HPLC system was calibrated using known concentrations of the phytoestrogen, biochanin. Table 8.10 contains data of the area under the chromatograph absorption peak

Table 8.9. *Energy gap versus temperature data.*

T (K)	E_g (T) (eV)
20	1.1696
40	1.1686
60	1.1675
80	1.1657
100	1.1639
120	1.1608
140	1.1579
160	1.1546
180	1.1513
200	1.1474
220	1.1436
240	1.1392
260	1.1346
280	1.1294
300	1.1247
320	1.1196
340	1.1141
360	1.1087
380	1.1028
400	1.0970
420	1.0908
440	1.0849
460	1.0786
480	1.0723
500	1.0660
520	1.0595

as a function of biochanin concentration.[39] A comparison is to be made of two equations fitted to the data in table 8.10. The equations are

$$y = A + Bx \tag{8.51}$$

and

$$y = A + Bx^c. \tag{8.52}$$

[39] Some of these data were published by Kirkup and Mulholland (2004).

Table 8.10. *HPLC data for biochanin.*

Conc. (x) (mg/L)	Area (y) (arbitrary units)
0.158	0.121342
0.158	0.121109
0.315	0.403550
0.315	0.415226
0.315	0.399678
0.631	1.839583
0.631	1.835114
0.631	1.835915
1.261	3.840554
1.261	3.846146
1.261	3.825760
2.522	8.523561
2.522	8.539992
2.522	8.485319
5.045	16.80701
5.045	16.69860
5.045	16.68172
10.09	34.06871
10.09	33.91678
10.09	33.70727

Assuming an unweighted fit is appropriate, fit equations 8.51 and 8.52 to the data in table 8.10.

For each equation fitted to the data, calculate the

(i) best estimates of parameters;
(ii) standard uncertainties in estimates;
(iii) sum of squares of residuals (SSR);
(iv) corrected Akaike's information criterion.
(v) Draw graphs of residuals versus concentration for each equation fitted to the data.

Which equation better fits the data?

(7) The relationship between critical current, I_c, and temperature, T, for a high-temperature superconductor can be written

$$I_c = 1.74A\left(1 - \frac{T}{T_c}\right)^{1/2} \tanh\left[0.435B\frac{T_c}{T}\left(1 - \frac{T}{T_c}\right)^{1/2}\right], \tag{8.53}$$

Table 8.11. *Critical current versus temperature data for a high temperature superconductor with critical temperature of 90.1 K.*

T (K)	I_c (mA)
5	5212
10	5373
15	5203
20	4987
25	4686
30	4594
35	4245
40	4091
45	3861
50	3785
55	3533
60	3199
65	2903
70	2611
75	2279
80	1831
85	1098
90	29

where A and B are constants and T_c is the critical temperature of the super-conductor. The following data for critical current and temperature were obtained for a high-temperature superconductor with a T_c equal to 90.1 K.

(i) Draw a graph of I_c versus T.
(ii) Show that, as T tends to zero, equation 8.53 simplifies to $I_c = 1.74A$. Use this, and your graph, to estimate a starting value for A.
(iii) Estimate a starting value[40] for B.
(iv) Fit equation 8.53 to the data in table 8.11, to obtain best estimates for the parameters A and B.

---

[40] By choosing a data pair from table 8.11 and inserting the pair into equation 8.53 (as well as the estimate for A) that equation can be solved for B. This will give you an approximate value for B which is good enough to use with Solver.

Table 8.12. *Signal output from sensor as a function of electrical conductivity.*

σ (mS/cm)	V (volts)
1.504	6.77
2.370	7.24
4.088	7.61
7.465	7.92
10.764	8.06
13.987	8.14
14.781	8.15
17.132	8.19
24.658	8.27
31.700	8.31
38.256	8.34

(v) Determine the standard uncertainties in best estimates.

(vi) Determine the 95% coverage interval for A and B.

(8) A sensor developed to measure the electrical conductivity of salt solutions is calibrated using solutions of sodium chloride of known conductivity, σ. Table 8.12 contains data of signal output, V, of the sensor as a function of conductivity. Assume that the relationship between V and σ is

$$V = V_s + k[1 - \exp(\sigma^a)], \tag{8.54}$$

where V_s, k and a are constants.

Use unweighted non-linear least squares to determine best estimates of the constants and standard uncertainties in the best estimates.

(9) In a study of the propagation of an electromagnetic wave through a porous solid, the variation of relative permittivity, ε_r, of a solid was measured as a function of moisture content, v_w (expressed as a fraction). Table 8.13 contains the data obtained in the experiment.[41]

Assume the relationship between ε_r and v_w can be written

$$\varepsilon_r = v_w^2(\sqrt{\varepsilon_w} - \sqrt{\varepsilon_m})^2 + 2v_w(\sqrt{\varepsilon_w} - \sqrt{\varepsilon_m})\sqrt{\varepsilon_m} + \varepsilon_m, \tag{8.55}$$

where ε_w is relative permittivity of water, and ε_m is the relative permittivity of the (dry) porous material.

---

[41] Francois Malan 2002 (private communication).

Table 8.13. *Variation of relative permittivity with moisture content.*

v_w	ε_r
0.128	8.52
0.116	7.95
0.100	7.65
0.095	7.55
0.077	7.08
0.065	6.82
0.056	6.55
0.047	6.42
0.035	5.97
0.031	5.81
0.025	5.69
0.022	5.55
0.017	5.38
0.013	5.26
0.004	5.08

Use (unweighted) non-linear least squares to fit equation 8.55 to the data in table 8.13 and hence obtain best estimates of ε_w and ε_m and standard uncertainties in the best estimates.

(10) A common toughness test on steel is the Charpy V-notch test.[42] The energy that is absorbed by the specimen is measured during fracture over a range of temperatures. Table 8.14 shows the variation of absorbed energy, Y (in joules) as a function of temperature, T (in kelvins), for samples of heat-treated steel. An equation that is often fitted to the data is given by

$$Y = A + B\tanh\left[\frac{T - T_o}{C}\right], \tag{8.56}$$

where A, B, C and T_o are parameters that may be estimated using non-linear least squares.

  (i) Draw an x–y graph of the data in table 8.14.
  (ii) Determine approximate values for A, B, C and T_o.
  (iii) Use non-linear least squares to find best estimates of A, B, C and T_o.
  (iv) Determine the standard uncertainties in A, B, C and T_o.

---

[42] See Mathur, Needleman and Tvergaard (1994).

Table 8.14. *Variation of absorbed*
energy with temperature for heat
treated steel.

T (K)	Y (J)
270	11.8
280	6.6
290	16.3
300	17.5
310	22.7
320	22.7
330	32.7
340	66.2
350	90.7
360	98.8
370	114.8
380	113.4
390	113.1
400	110.1
410	114.8
420	108.4

(11) Unweighted least squares requires the minimisation of *SSR* given by

$$SSR = \sum (y_i - \hat{y}_i)^2. \tag{8.57}$$

A technique sometimes adopted when optimising parameters in optical design situations is to minimise *S4R*, where

$$S4R = \sum (y_i - \hat{y}_i)^4. \tag{8.58}$$

Carry out a simulation to compare parameter estimates obtained when equations 8.57 and 8.58 are used to fit an equation of the form $y = a + bx$ to simulated data. More specifically do the following.

(i)   Use the function $y = 2.1 - 0.4x$ to generate y values for $x = 1, 2, 3$, etc., up to $x = 20$.

(ii) Add normally distributed noise of mean equal to zero and standard deviation of 0.5 to the values generated in part (i).

(iii) Find best estimates of a and b by minimising *SSR* and *S4R* as given by equations 8.57 and 8.58. (Use Solver to minimise *SSR* and *S4R*.)

(iv) Repeat steps (ii) and (iii) until 50 sets of parameter estimates have been obtained using equations 8.57 and 8.58.

(v) Is there any significant difference between the parameter estimates obtained when minimising SSR and $S4R$?[43]

(vi) Is there any significant difference between the variance in the parameter estimates when minimising SSR and $S4R$?

---

[43] Return to parts (v) and (vi) of this question after reading chapter 9.

Chapter 9

Tests of significance

9.1 Introduction

What can reasonably be inferred from data gathered in an experiment? This simple question lies at the heart of experimentation, as an experiment can be judged by how much insight can be drawn from data. An experiment may have a broad or narrow focus, and may be designed to:

- challenge a relationship that has an established theoretical basis;
- critically examine a discovery that results from 'chance' observations;
- check for drift in an instrument;
- compare analysis of materials carried out in two or more laboratories.

Such general goals give way to specific questions that we hope can be answered by careful analysis of data gathered in well designed experiments. Questions that might be asked include:

- is there a linear relationship between quantities measured in an experiment;
- could the apparent correlation between variables have occurred 'by chance';
- does a new manufacturing process produce lenses with focal lengths that are less variable than the old manufacturing process;
- is there agreement between two methods used to determine the concentration of iron in a specimen;
- has the gain of an instrument changed since it was calibrated?

It is usually not possible to answer these questions with a definite 'yes' or definite 'no'. Though we hope data gathered during an experiment will provide evidence as to which reply to favour, we must be satisfied with answers expressed in terms of probability.

Consider a situation in which a manufacturer supplies an instrument containing an amplifier with a gain specified as 1000. Would it be reasonable to conclude that the instrument is faulty or needs recalibrating if the gain determined by a single measurement is 995? It is possible that random errors inherent in the measurement process, as revealed by making repeat measurements of the gain, would be sufficient to explain the discrepancy between the 'expected' value of gain of 1000 and the 'experimental' value of 995. What we would really like to know is whether, after taking into account the scatter in the values of the gain obtained through repeat measurements, the difference between the value we have reason to expect will occur and those actually obtained through experiment or observation is 'significant'.

This chapter introduces the notion of significance and how significance may be quantified so that it can be useful in the analysis of experimental data. This requires we make assumptions about which probability distribution is applicable to the data. While the t distribution is at the heart of many 'tests of significance', we introduce other distributions including the F distribution which describes the probability of obtaining a particular ratio of *variances* of two samples. With the aid of this distribution we will test whether there is a significant difference between the variability in data as exhibited by two samples.

9.2 Hypothesis testing

Formally, hypothesis testing involves comparing a hypothesised population parameter with the corresponding number (or statistic) determined from sample data. The testing process takes into account both the variability in the data as well as the number of values in the sample. Table 9.1 shows several population parameters and their corresponding sample statistics.[1]

For example, we can compare a hypothesised population mean,[2] μ_0, with a sample mean, $\bar{x}$, and address the following question.

> *What is the probability that a sample with mean, $\bar{x}$, has come from a population with mean, μ_0?*

Consider an example. An experimenter adds weights to one end of a wire suspended from a fixed point, causing the wire to extend. The experimenter wishes to be satisfied that the weights have been drawn from a population with

---

[1] As the number of values, n, in the sample tends to infinity then a sample statistic tends to its corresponding population parameter. So, for example, as $n \to \infty$, $\bar{x} \to \mu$.

[2] It is usual to denote hypothesised population parameters by attaching the subscript '0' to the symbol used for that parameter.

Table 9.1. *Some population parameters and their corresponding sample statistics.*

Population parameter	Sample statistic
mean, μ	mean, $\bar{x}$
standard deviation, σ	standard deviation, s
correlation coefficient, ρ	correlation coefficient, r
intercept, α (of a line through x–y data)	intercept, a
slope, β (of a line through x–y data)	slope, b

mean mass 50 g (as specified by the manufacturer). None of the weights will be 'exactly' 50 g, but it is reasonable to expect that, if many weights are tested, then the mean will be close to 50 g. The experimenter 'hypothesises' that the population from which the sample is drawn has a mean of 50 g. This is a conservative hypothesis in the sense that, due to the manufacturer's specification for the weights, the experimenter expects that the population mean *is* 50 g. Such a conservative hypothesis is referred to as the 'null hypothesis' and is represented symbolically by H_0. Formally, we write

H_0: μ = 50 g. This is read as 'the null hypothesis is that the population mean is 50 g'.

It is usual to state an alternate hypothesis, H_a, which is favoured if the probability of the null hypothesis being true is small. In this example we would write:

H_a: $\mu \neq$ 50 g. This is read as 'the alternate hypothesis is that the population mean is *not* equal to 50 g'.

The next stage is to decide what probability distribution is appropriate for the sample means. The central limit theorem[3] predicts that (so long as the sample size is large enough), the distribution of sample means is normal with a population mean, μ, and a standard uncertainty[4], $\sigma_{\bar{x}}$, where $\sigma_{\bar{x}} = \frac{\sigma}{\sqrt{n}}$, n is the number of values in the sample and σ is the population standard deviation. In the vast majority of situations s is not known so we use the approximation[5] $s_{\bar{x}} \approx \frac{s}{\sqrt{n}}$, where s is the best estimate of σ and $s_{\bar{x}}$ is the best estimate of $\sigma_{\bar{x}}$.

Next we decide under what circumstances we will reject the null hypothesis and so favour the alternate hypothesis. How small must be the probability, p (often referred to as the p-value), of the null hypothesis being true before we reject it? In principle, any value of p could be chosen, but conventionally a null hypothesis

[3] See section 3.8.

[4] Most statistic texts refer to $\frac{\sigma}{\sqrt{n}}$ as the 'standard error of the mean'. For consistency with earlier chapters we will continue to refer to $\frac{\sigma}{\sqrt{n}}$ as the 'standard uncertainty'.

[5] The approximation is good for n in excess of 30.

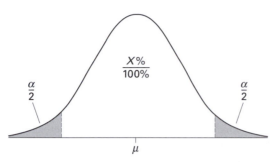

Figure 9.1. Relationship between level of significance, α, and level of confidence, $X\%$.

is rejected when the probability of it being true is less than 0.05. In some circum-stances[6] the null hypothesis is rejected when the p-value is less than 0.01.

The probabilities of 0.05 and 0.01 are arbitrary, but do represent 'bench marks' recognised and widely adopted by the scientific community when hypothesis testing is carried out. If the probability of the null hypothesis being true is less than 0.05, the difference between the hypothesised value for the population mean and the sample mean is regarded as *significant*.

The result of a hypothesis test may be that the null hypothesis is rejected at the level of significance[7], $\alpha = 0.05$. The 'α significance level' refers to the sum of the areas in the tails of the probability distribution shown in figure 9.1.[8] Note that α can be related directly to the level of confidence, $X\%$, as shown in figure 9.1.

As the total area under the curve in figure 9.1 equals 1, we write

$$\frac{X\%}{100\%} + \alpha = 1. \tag{9.1}$$

9.2.1   Distribution of the test statistic, z

If we consider a population with a hypothesised mean, μ_0, and a population standard deviation, σ, then samples consisting of n values should have means that are distributed normally about μ_0 such that the standard deviation of the means is $\sigma/\sqrt{n}$. A convenient way to express this is to use the continuous random variable, z, given by

$$z = \frac{\bar{x} - \mu_0}{\sigma/\sqrt{n}}. \tag{9.2}$$

The variable, z, is normally distributed with a mean of zero and a standard deviation of 1.

---

[6] See Peck, Olsen and Devore (2008), chapter 10.

[7] α is sometimes expressed as a percentage, so that a significance level of 0.05 may be stated as $\alpha = 5\%$.

[8] This is true for a 'two tailed' test. Section 9.2.3 discusses one and two tailed tests.

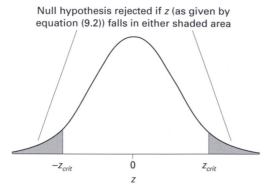

Null hypothesis rejected if z (as given by equation (9.2)) falls in either shaded area

$-z_{crit}$ 0 z_{crit}

z

Figure 9.2. Rejection regions (shaded) for the null hypothesis.

A value for z can be calculated using data gathered in an experiment. If data have been drawn from a population with mean μ_0 then large values of $|z|$ are unlikely. We can determine a 'critical' value of z, z_{crit} for each level of significance, α. If $|z| > z_{crit}$, the probability of the null hypothesis being true (i.e. that the sample has been drawn from a population with a mean μ_0) is *less than* the probability corresponding to the significance level α, and the null hypothesis is rejected. This is shown pictorially in figure 9.2.

By contrast, if $|z| < z_{crit}$, then the test has not revealed any significant difference (at the chosen α) between the hypothesised population mean, μ_0, and the sample mean and there is not sufficient evidence to reject the null hypothesis. Values of z_{crit} for particular significance levels may be determined using table 1 in appendix 1.

As always, it is not possible to determine the population standard deviation, σ, and we must 'make do' with the estimate of the population standard deviation, s, found using equation 1.16.

Example 1

The masses of a sample of 40 weights of nominal value 50 g are measured with an electronic balance. The values obtained are shown in table 9.2.

The experimenter assumes that the masses have been drawn from a population with a mean of 50.00 g. Is this assumption reasonable?

ANSWER

We perform a test to determine whether there is a significant difference between the hypothesised population mean and the mean of the values in table 9.2. Table 9.3 describes the steps in the test.

Table 9.2. *Masses of 40 weights.*

weights (g)									
50.06	50.02	50.18	50.05	50.05	50.12	50.05	50.07	50.20	49.98
50.13	50.05	49.99	49.99	49.94	50.20	50.03	49.97	50.09	49.77
50.10	49.93	49.99	50.01	50.12	50.16	50.12	50.22	50.13	50.09
50.22	50.09	50.07	50.02	50.05	50.14	50.10	50.18	50.08	50.02

Table 9.3. *Step by step description of the hypothesis test for example 1.*

Step	Details/comment
Decide the purpose of the test.	To compare a sample mean with a hypothesised population mean.
State the null hypothesis.	$H_0: \mu = 50.00$ g.
State the alternate hypothesis	$H_a : \mu \neq 50.00$ g.
Choose the significance level of the test, α.	No indication is given in the question of the level of significance at which we should test the null hypothesis, so we choose the most commonly used level, i.e. $\alpha = 0.05$.
Determine the mean, $\bar{x}$, and standard deviation, s, of the values in table 9.2.	$\bar{x} = 50.0695$ g, $s = 0.08950$ g.
Calculate the value of the test statistic, z, using equation 9.2.	$z = \dfrac{\bar{x}-\mu_0}{s/\sqrt{n}} = \dfrac{50.0695-50.00}{0.089\,50/\sqrt{40}} = 4.91.$
Determine the critical value of the test statistic, z_{crit}, based on the chosen significance level.	When the area in one tail of the standard normal distribution is 0.025 ($= \alpha/2$), the magnitude of the z value, found using table 1 in appendix 1, $= 1.96$. Hence $z_{crit} = 1.96$.
Compare $|z|$ and z_{crit}.	$|z| > z_{crit}$.
Make a decision.	As $|z| > z_{crit}$ we reject the null hypothesis. There is a significant difference between the sample mean and the hypothesised population mean. Specifically, the probability is less than 0.05 that a sample mean would differ from the hypothesised mean by 0.0695 g or more.

Table 9.4. *Voltages across bandgap reference diodes (measured in volts).*

1.255	1.259	1.257	1.256	1.260	1.262	1.259	1.257	1.260
1.260	1.261	1.263	1.258	1.261	1.253	1.259	1.257	1.254
1.251	1.257	1.252	1.258	1.253	1.248	1.265	1.255	1.256
1.258	1.265	1.257	1.257	1.264	1.262	1.256	1.258	1.265

Exercise A

Bandgap reference diodes are used extensively in electronic measuring equipment as they provide highly stable voltages against which other voltages can be compared.[9]

One manufacturer's specification indicates that the nominal voltage across the diodes should be 1.260 V. A sample of 36 diodes is tested and the voltage across each diode is shown in table 9.4.

Using the data in table 9.4, test whether the diodes have been drawn from a population with a mean of 1.260 V. Reject the null hypothesis at $\alpha = 0.05$.

9.2.2 Using Excel to compare sample mean and hypothesised population mean

We can calculate the probability that a particular sample mean would occur for a hypothesised population mean using Excel. The steps in the calculation of the probability are as follows.

(i) Calculate the mean, $\bar{x}$, the standard deviation, s, of the data. s is taken as the best estimate of σ.

(ii) For a hypothesised mean, μ_0, calculate the value of the z-statistic, z, using
$$z = \frac{\bar{x} - \mu_0}{s/\sqrt{n}}.$$

(iii) Use Excel's NORM.S.DIST() function to determine the area in the tail of the distribution[10] between $z = -\infty$ and $z = -|z|$.

(iv) Multiply the area by two to obtain the total area in both tails of the distribution.

(v) If a sample is drawn from a population with mean, μ_0, then the probability that the sample mean would be at least as far from the mean as $|z|$ is equal to the sum of the areas in the tails of the distribution.

---

[9] As examples, bandgap reference diodes are found in voltmeters and on interfacing cards used with personal computers.

[10] By calculating the area between $-\infty$ and $-|z|$ we are always choosing the tail to the left of $z = 0$.

Table 9.5. *Values for example 2.*

100.5	95.6	103.2	108.4
96.6	98.5	93.5	92.8
102.7	100.2	100.4	100.1
106.3	113.9	98.7	110.3
108.0	110.7	91.1	100.8
97.1	98.1	91.4	99.2
108.7	101.6	101.1	99.4
93.9	104.7	106.5	111.6
107.5	100.0	111.9	101.6

Example 2

Consider the values in table 9.5. Use Excel to determine the probability that the values in table 9.5 have been drawn from a population with a mean of 100.0.

ANSWER

The data in table 9.5 are shown in sheet 9.1.

Column F of sheet 9.1 contains the formulae required to determine the sample mean, standard deviation and so on. Sheet 9.2 shows the values returned in column F.

The number 0.067 983 returned in cell F6 is the probability of obtaining a sample mean at least as far from the population mean of 100.0 as 101.85. As this probability is greater than 0.05, we cannot reject a null hypothesis that the sample is drawn from a population with mean 100.0.

Sheet 9.1. *Determination of p value.*

	A	B	C	D	E	F
1	100.5	95.6	103.2	108.4	hypothesised mean	100.0
2	96.6	98.5	93.5	92.8	sample mean	= AVERAGE(A1:D9)
3	102.7	100.2	100.4	100.1	standard deviation, s	= STDEV.S(A1:D9)
4	106.3	113.9	98.7	110.3	$s/\sqrt{n}$	= F3/36^0.5
5	108.0	110.7	91.1	100.8	z-value	= (F2-F1)/F4
6	97.1	98.1	91.4	99.2	probability	= 2*NORM.S.DIST(-ABS(F5), TRUE)
7	108.7	101.6	101.1	99.4		
8	93.9	104.7	106.5	111.6		
9	107.5	100.0	111.9	101.6		

Sheet 9.2. *Numbers returned in column F.*

	E	F
1	hypothesised mean	100.0
2	sample mean	101.85
3	standard deviation, s	6.0818
4	$s/\sqrt{n}$	1.013633
5	z-value	1.825118
6	probability	0.067983
7		

Exercise B

Titanium metal is deposited on a glass substrate producing a thin film of nominal thickness 55 nm. The thickness of the film is measured at 30 points chosen at random across the film. The values obtained are shown in table 9.6.

Use Excel to determine whether there is a significant difference, at $\alpha = 0.05$, between the mean of the values in table 9.6 and a hypothesised population mean of 55 nm.

9.2.3   One tailed and two tailed tests of significance

In the test described in section 9.2.1, we are interested in whether the sample mean differs significantly from the hypothesised mean. That is, we are equally concerned whether the difference, $\bar{x} - \mu_0$, is positive or negative. Such a test is referred to as 'two tailed' and is appropriate if we have no preconceived notion as to whether to expect $\bar{x} - \mu_0$ to be positive or negative. There are situations in which, prior to carrying out the data analysis, we anticipate $\bar{x} - \mu_0$ to have a specific sign. For example, if we believe we have improved a process in which gold is recovered from sea water, then we would anticipate the new process to have a mean gold recovery greater than the old process. In this case we would express the null and alternate hypotheses as

Table 9.6. *Measured thickness (in nm) of titanium film.*

61	53	59	61	60	54	56	56	52	61
57	54	50	60	58	58	55	55	53	58
61	52	53	53	59	61	59	53	54	55

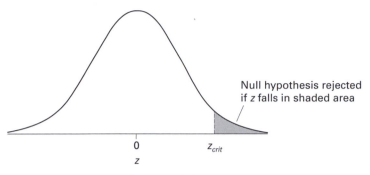

Null hypothesis rejected
if z falls in shaded area

0   z_{crit}
z

Figure 9.3. Rejection region for a one tailed test.

$$H_0 : \mu \leq \mu_0$$
$$H_a : \mu > \mu_0,$$

where μ_0 is the expected population mean based on the amount of gold recovered using the 'old' process.

By contrast, if a surface coating on a lens is designed to reduce the amount of reflected light, then we would write:

$$H_0 : \mu \geq \mu_0$$
$$H_a : \mu < \mu_0.$$

Here μ_0 would correspond to the expected reflection coefficient for the lens in the absence of a coating.

Tests in which the alternate hypothesis is $\mu > \mu_0$ or $\mu < \mu_0$ are referred to as one tailed tests. As the term 'one tailed test' implies, we consider areas only in one tail of the distribution as shown in figure 9.3.

If the significance level, $\alpha = 0.05$ is chosen for the test, then the shaded area in figure 9.3 would be equal to 0.05. Any value of z, determined using experimental data, that falls in the shaded region such that $z \geq z_{crit}$ would mean that the null hypothesis is rejected at the $\alpha = 0.05$ level of significance.

In order to determine z_{crit} for a one tailed test, we calculate $(1 - \alpha)$. This gives the area under the standard normal curve between $z = -\infty$ and $z = z_{crit}$. The z value corresponding to the area $(1 - \alpha)$ is found using table 1 in appendix 1.

Example 3

Determine the critical value of the z statistic for a one tailed test when the level of significance, $\alpha = 0.1$.

ANSWER

If $\alpha = 0.1$, then $(1 - \alpha) = 0.9$. Referring to table 1 in appendix 1, if the area under the standard normal curve between $z = -\infty$ and $z = z_{crit}$ is 0.9, then $z_{crit} = 1.28$.

Exercise C

Determine the critical value of the z statistic for a one tailed test when the level of significance α is equal to:

(i) 0.2;
(ii) 0.05;
(iii) 0.01;
(iv) 0.005.

9.2.4 Type I and Type II errors

In making the decision to reject or not reject a null hypothesis we must be aware that we will not always make the right decision.

Ideally,

- we reject a null hypothesis when the null hypothesis is *false*, or
- we do not reject a null hypothesis when the null hypothesis is *true*.

We can never be sure that the null hypothesis is true or false, for if we could there would be no need for hypothesis testing! This leads to two undesirable outcomes:

- we could reject a null hypothesis that is *true* (this is called a Type I error), or
- we could fail to reject a null hypothesis that is *false* (this is called a Type II error[11]).

By choosing a significance level of, say $\alpha = 0.05$, we limit the probability of a Type I error to 0.05. This means that if we repeat the test on many sets of similar data we will reject the null hypothesis when it should not have been rejected, on average, one time in twenty. Quantifying the probability of making a Type II error is more difficult but can be done in some situations.[12] Type I errors are usually regarded as more serious than Type II errors, as we wish to reduce the probability of rejecting a null hypothesis (usually chosen to be conservative) that is true. Choosing the level of significance to be, say, 0.05 fixes the probability of a Type I error so that it does not exceed 0.05.

9.3 Comparing $\bar{x}$ with μ_o when sample sizes are small

In order to compare a hypothesised population mean, μ_0, with the mean, $\bar{x}$, of a small sample, we follow the steps described in section 9.2.1 for large samples.

---

[11] Note Type I and Type II errors are *not* the same as the experimental errors discussed in chapter 5.
[12] See Devore (2007) for a discussion of Type II errors.

However we must take into account that when sample sizes are small, the distribution of means is not normal even if the values that make up the sample *are* normally distributed.

When the population standard deviation, σ, is known, then the variable

$$z = \frac{\bar{x} - \mu}{\sigma/\sqrt{n}} \tag{9.3}$$

is normally distributed. In practice, σ is rarely if ever known, and we must estimate σ with the data available. Note that s (given by equation 1.16) is a good estimate of σ so long as $n \geq 30$ and so equation 9.3 may still be used with s replacing σ. However, a sample size with $n \geq 30$ is a luxury that many experiments cannot afford due to cost or time considerations. If n is less than 30, s can no longer be regarded as a good estimate of σ, and we use the statistic, t, given by

$$t = \frac{\bar{x} - \mu}{s/\sqrt{n}}. \tag{9.4}$$

The distribution of t is symmetric with a characteristic 'bell' shape similar to the normal distribution.[13] An important difference between the normal and t distributions is that there is more area in the tails of the t distribution from, say $t = 2$ to $t = \infty$ than in the normal distribution between $z = 2$ to $z = \infty$. We can show this using Excel's T.DIST.RT() and NORM.S.DIST() functions.

The cumulative probability between $t = 2$ to $t = \infty$ can be obtained for any number of degrees of freedom using Excel's T.DIST.RT() function.[14] The area in the tail $t = 2$ to $t = \infty$ for $v = 10$ is found by typing =T.DIST.RT(2,10) into a cell on an Excel spreadsheet. Excel returns the number 0.03669. Similarly, the cumulative probability between $z = 2$ to $z = \infty$ can be obtained using Excel's NORM.S.DIST() function.[15] Specifically, the area in the tail between $z = 2$ to $z = \infty$ is found by typing = 1-NORM.S.DIST(2,1) into a cell in the spreadsheet. Excel returns the number 0.02275.

As a result of the increased area in the tail of the t distribution compared to the normal distribution, the critical t value, t_{crit} is larger for a given level of significance than the corresponding critical value for the standard normal distribution, z_{crit}. As the t distribution depends on the number of degrees of freedom, v, the critical value of the test statistic depends on v. Table 2 in appendix 1 gives the critical values of t for various levels of significance, α, for $1 \leq v \leq 1000$.

---

[13] See section 3.9.

[14] The T.DIST.RT() function returns the area in the right hand tail of the t distribution.

[15] See section 3.5.3 for a description of Excel's NORM.S.DIST() function.

Table 9.7. *Values of the acceleration due to gravity.*

$g \, (\text{m/s}^2)$	9.68	9.85	9.73	9.76	9.74

Example 4

Table 9.7 shows the value of the acceleration due to gravity, g, obtained through experiment.

Is the mean of the values in table 9.7 significantly different statistically from a hypothesised value of $g = 9.81 \, \text{m/s}^2$? Test the hypothesis at the $\alpha = 0.05$ level of significance.

ANSWER

H_0: $\mu = 9.81 \, \text{m/s}^2$;

H_a: $\mu \neq 9.81 \, \text{m/s}^2$ (i.e. this is a two tailed test).

From the data in table 9.7, $\bar{x} = 9.752 \, \text{m/s}^2$, $s = 0.062\,21 \, \text{m/s}^2$, $s/\sqrt{n} = 0.02782 \, \text{m/s}^2$

Using equation 9.4 with $\mu = \mu_0 = 9.81 \, \text{m/s}^2$ gives, $t = \frac{9.752 - 9.81}{0.027\,82} = -2.085$.

The critical value of the test statistic, t_{crit}, with $\alpha = 0.05$ and $\nu = n - 1 = 4$ is found from table 2 in appendix 1 to be $t_{crit} = 2.776$.

As $|t| < t_{crit}$, we cannot reject the null hypothesis, that is, there is no significant difference at $\alpha = 0.05$ between the sample mean and the hypothesised population mean.

Exercise D

The unit cell is the building block of all crystals. An experimenter prepares several crystals of a ceramic material and compares the size of the unit cell to that published by another experimenter. One dimension of the unit cell is the lattice dimension, c. Table 9.8 contains values of the c dimension obtained by the experimenter from eight crystals.

At the $\alpha = 0.05$ level of significance, determine whether the sample mean of the values in table 9.8 differs significantly from the published value of c of 1.1693 nm.

9.4 Significance testing for least squares parameters

In chapter 6 we considered fitting the equation $y = a + bx$ to x–y data using the technique of least squares. When a or b is close to zero, we might wonder if we have chosen the best equation to fit to data. For example, if a is not significantly different from zero, we would be justified in omitting a and fitting the equation

Table 9.8. *The c dimension of a ceramic crystal.*

c (nm)	1.1685	1.1672	1.1695	1.1702
	1.1665	1.1675	1.1692	1.1683

$y = bx$ to data. Similarly, if b is not significantly different from zero, we could justify removing b from the fit. The consequence of this is that we would find the best value of the constant, a, that would fit the data (which would be the mean of the y values).

In order to test whether a or b (or the estimate of other parameters found using least squares) is significantly different from zero we need:

(i) the best estimate of the parameter found using least squares;
(ii) the standard uncertainty in the estimate;
(iii) to choose a significance level for the test.

As the number of x–y pairs in a least squares analysis is routinely less than 30, we assume that the distribution of parameter estimates follows a t distribution. For the intercept of a line through data we hypothesise:

H_0: the population intercept, $\alpha = 0$;
H_a: the population intercept, $\alpha \neq 0$.

The test statistic, t, in this situation is given by

$$t = \frac{a - \alpha_0}{s_a} \quad \text{so that for } \alpha_0 = 0,$$
$$t = \frac{a}{s_a}, \tag{9.5}$$

where a is the best estimate of the population intercept, α, and s_a is the standard uncertainty in a. The critical value of the test statistic, t_{crit}, must be established for a given level of significance and degrees of freedom. If $|t| > t_{crit}$ the null hypothesis is rejected, i.e. the sample intercept, a, is not consistent with a hypothesised intercept of zero.

A similar approach is used to test for the significance of other parameters. For example, for the hypothesised population slope, β_0, we calculate

$$t = \frac{b - \beta_0}{s_b} \quad \text{so that for } \beta_0 = 0,$$
$$t = \frac{b}{s_b}, \tag{9.6}$$

where b is the best estimate of the population slope, β, and s_b is the standard uncertainty in b.

Table 9.9. *The x–y values for example 5.*

x	y
0.1	−0.9
0.2	−2.1
0.3	−3.6
0.4	−4.7
0.5	−5.9
0.6	−6.8
0.7	−7.8
0.8	−9.1

Example 5

Consider the x–y data in table 9.9.

   (i)   Using the values in table 9.9, find the intercept, a, and slope, b, of the best straight line through the points using unweighted least squares.
   (ii) Calculate the standard uncertainties in a and b.
   (iii) Determine, at the 0.05 significance level, whether the intercept and slope are significantly different from zero.

ANSWER

Solving for the intercept using unweighted least squares, gives, $a = 0.06786$ and the standard uncertainty in intercept, $s_a = 0.1453$. Similarly, the slope, $b = -11.51$ and the standard uncertainty in the slope, $s_b = 0.2878$.

   The degrees of freedom, $v = n - 2 = 8 - 2 = 6$.

   Test of intercept:

$H_0 : \alpha = 0;$
$H_a : \alpha \neq 0.$

Value of the test statistic, t, is: $t = \frac{a}{s_a} = \frac{0.067\,86}{0.1453} = 0.4670.$

   Using the table 2 in appendix 1 gives, for a two tailed test[16], $t_{crit} = t_{0.05,6} = 2.447$.

   As $|t| < t_{crit}$ we cannot reject the null hypothesis, i.e. the intercept is not significantly different from zero.

---

[16] A short hand way of writing the critical value of t which corresponds to the significance level, α, for v degrees of freedom is $t_{\alpha,v}$.

Test of slope:

$H_0 : \beta = 0;$
$H_a : \beta \neq 0.$

Value of the test statistic t is: $t = \frac{b}{s_b} = \frac{-11.51}{0.2878} = -40.00$

$t_{crit} = t_{0.05,6} = 2.447.$

As $|t| > t_{crit}$ we reject the null hypothesis, i.e. the slope is significantly different from zero.

As the intercept is not significantly different from zero, we have justification for fitting a line to data using least squares in which the line is 'forced' through zero (i.e. $a = 0$). A convenient way to do this is to use Excel's LINEST() function as discussed in section 6.3. Doing this gives the slope, $b = -11.39$ and the standard uncertainty in the slope, $s_b = 0.1231$.

Exercise E

Table 9.10 shows the variation in voltage across a germanium diode which was measured as the temperature of the diode increased from 250 K to 360 K.

Assuming there is a linear relationship between voltage and temperature, find the intercept and slope of the best line through the data in table 9.10. Test whether the intercept and slope differ from zero at the $\alpha = 0.05$ significance level.

9.5 Comparison of the means of two samples

As variability occurs in measured values, two samples will seldom have the same mean even if they are drawn from the same population. There are situations in which we would like to compare two means and establish whether they are significantly different.

For example, if a specimen of water from a river is divided and sent to two laboratories for analysis of lead content, it would be reasonable to anticipate that the difference between the means of the values obtained for lead content by each laboratory would not be statistically significant. If a significant difference *is* found then this might be traced to shortcomings in the analysis procedure of one of the laboratories or perhaps inconsistencies in storing or handling the specimens of river water. Other circumstances in which we might wish to compare two sets of data include where:

- the purity of a standard reference material is compared by two independent laboratories;
- the moisture content of a chemical is measured before and after a period of storage;

Table 9.10. *Variation of voltage with temperature for a germanium diode.*

T (K)	V (V)
250	0.440
260	0.550
270	0.469
280	0.486
290	0.508
300	0.494
310	0.450
320	0.451
330	0.434
340	0.385
350	0.458
360	0.451

- the oxygen content of a ceramic is determined before and after heating the ceramic in an oxygen rich environment.

In situations in which we want to know whether there is a significant difference in the mean of two samples in which each sample consists of less than 30 values, we use a t test. An assumption is made that the difference between, say, the mean of sample 1 and the mean of sample 2 follows a t distribution. The t statistic in this situation is given by

$$t = \frac{\bar{x}_1 - \bar{x}_2}{s_{\bar{x}_1 - \bar{x}_2}},$$
(9.7)

where $\bar{x}_1$ is the mean of sample 1, $\bar{x}_2$ is the mean of sample 2 and $s_{\bar{x}_1 - \bar{x}_2}$ is the standard uncertainty of the difference between the sample means. In situations in which the standard deviations of the samples are not too dissimilar[17]

$$s_{\bar{x}_1 - \bar{x}_2} = s_p \left(\frac{1}{n_1} + \frac{1}{n_2} \right)^{1/2},$$
(9.8)

where n_1 and n_2 are the number of values in sample 1 and sample 2 respectively. If the sample sizes are the same, such that $n_1 = n_2 = n$, equation 9.8 becomes

$$s_{\bar{x}_1 - \bar{x}_2} = s_p \left(\frac{2}{n} \right)^{1/2}$$
(9.9)

[17] See McPherson (1990), chapter 12, for a discussion of equation 9.8.

where s_p is the pooled or combined standard deviation and is given by

$$s_p = \left(\frac{s_1^2(n_1 - 1) + s_2^2(n_2 - 1)}{n_1 + n_2 - 2}\right)^{1/2}, \tag{9.10}$$

where s_1^2 and s_2^2 are the estimated population variances of sample 1 and 2 respectively.

If $n_1 = n_2$, then equation 9.10 reduces to

$$s_p = \left(\frac{s_1^2 + s_2^2}{2}\right)^{1/2}. \tag{9.11}$$

Example 6

In order to compare the adhesive qualities of two types of adhesive tape, six samples of each type of tape were pulled from a clean glass slide by a constant force. The time for 5 cm of each tape to peel from the glass is shown in table 9.11.

Test at the $\alpha = 0.05$ level of significance whether there is any significant difference in the peel off times for tape 1 and tape 2.

ANSWER

To perform the hypothesis test we write

H_0: $\mu_1 = \mu_2$ (we hypothesise that the population means for both tapes are the same);
H_a: $\mu_1 \neq \mu_2$ (two tailed test).

Using the data in table 9.11

$$\bar{x}_1 = 120.0 \text{ s}, s_1 = 53.16 \text{ s}, \bar{x}_2 = 147.7 \text{ s}, s_2 = 61.92 \text{ s}.$$

Using equation 9.11, we have

$$s_p = \left(\frac{2825.6 + 3834.3}{2}\right)^{1/2} = 57.71 \text{ s}.$$

Substituting 57.71 s into equation 9.9 (with $n = 6$) gives

$$s_{\bar{x}_1 - \bar{x}_2} = 57.71 \times \left(\frac{2}{6}\right)^{1/2} = 33.32 \text{ s},$$

the value of the test statistic is found using equation 9.7, i.e.

$$t = \frac{\bar{x}_1 - \bar{x}_2}{s_{\bar{x}_1 - \bar{x}_2}} = \frac{120.0 - 147.7}{33.32} = -0.8304.$$

Here the number of degrees of freedom, $v = n_1 + n_2 - 2 = 10$. The critical value of the test statistic, t_{crit} for a two tailed test at $\alpha = 0.05$, found from table 2 in appendix 1 is

$$t_{crit} = t_{0.05, 10} = 2.228.$$

As $|t| < t_{crit}$ we cannot reject the null hypothesis, i.e. based on the data in table 9.11 there is no convincing evidence that the peel off time of tape 1 differs from that of tape 2.

Table 9.11. *Peel off times for two types of adhesive tape.*

Peel off times (s)	
Tape 1	Tape 2
65	176
128	125
87	95
145	255
210	147
85	88

Table 9.12. *Coefficients of friction for two contact areas.*

Coefficient of kinetic friction, μ_k	
Block 1 (contact area = 174 cm^2)	Block 2 (contact area = 47 cm^2)
0.388	0.298
0.379	0.315
0.364	0.303
0.376	0.300
0.386	0.290
0.373	0.287

Exercise F

As a block slides over a surface, friction acts to oppose the motion of the block. An experiment is devised to determine whether the friction depends on the area of contact between a block and a wooden surface. Table 9.12 shows the coefficient of kinetic friction, μ_k, for two blocks with different contact areas between block and surface. Test at the $\alpha = 0.05$ level of significance whether μ_k is independent of contact area.

9.5.1 Excel's T.TEST()

A convenient way to establish whether there is a significant difference between the means of two samples is to use Excel's T.TEST() function.[18] The syntax of the function is

---

[18] In versions of Excel prior to Excel 2010, this function was written as TTEST().

T.TEST(array1,array2,tails,type)

where array1 and array2 are arrays of cells which contain the values of sample 1 and sample 2 respectively. The argument 'tails' has the value 1 for a one tailed test and the value 2 for a two tailed test; 'type' refers to which t-test should be performed. For the two sample test considered in section 9.5, type is equal to 2.

The number returned by the function is the probability that the sample means would differ by at least $\bar{x}_1 - \bar{x}_2$, when $\mu_1 = \mu_2$.

Example 7

Is there any significant difference in the peel off times between tape 1 and tape 2 in table 9.11?

ANSWER

We begin by stating the null and alternate hypotheses

H_0: $\mu_1 = \mu_2$;
H_a: $\mu_1 \neq \mu_2$.

Sheet 9.3 shows the data entered from table 9.11 along with the T.TEST() function.

Sheet 9.3. *Use of Excel's T.TEST() function.*

	A	B	C
1	Peel off times (s)		
2	tape1	tape2	
3	65	176	
4	128	125	
5	87	95	
6	145	255	
7	210	147	
8	85	88	
9			
10	= T.TEST(A3:A8,B3:B8,2,2)		

When the Enter key is pressed, the number 0.425681 is returned into cell A10. If the null hypothesis is correct, then the probability that the sample means $\bar{x}_1$ and $\bar{x}_2$ will differ by at least as much as $\bar{x}_1 - \bar{x}_2$ is ≈ 0.43 (or 43%). This probability is so large that we infer that the values are consistent with the null hypothesis and so we cannot reject that hypothesis.

Table 9.13. *Lead content of river water.*

Lead content (ppb)	Sample A	42	50	45	48	52	42	53
	Sample B	48	58	55	49	56	47	58

Exercise G

Table 9.13 shows values of the lead content of water drawn from two locations on a river.

Use Excel's T.TEST() function to test the null hypothesis that the mean lead content of both samples is the same.

9.5.2   The *t* test for paired samples

Another important type of test is that in which the contents of data sets are naturally linked or 'paired'. For example, suppose two chemical processes used to extract copper from ore are to be compared. Assuming that the processes are equally efficient at extracting the copper, the amount obtained from any particular batch of ore should be independent of the process used, so long as account is taken of random errors.

Other situations in which a paired *t* test might be considered include comparing:

- two methods for analysing the chemical composition of a range of materials;
- heart rate before and after administering caffeine to a group of people;
- the analysis of the same materials as carried out by two independent laboratories;
- the emf of batteries before and after a period of storage.

By calculating the mean of the differences, $\bar{d}$, of the paired values, and the standard deviation of the differences, $s_{\bar{d}}$, the *t* test statistic can be determined using

$$t = \frac{\bar{d} - \delta_0}{s_{\bar{d}}/\sqrt{n}}, \tag{9.12}$$

where δ_0 is the hypothesised population mean of the differences between the paired values and n is the number of paired values. As δ_0 is most often taken to be zero, we re-write equation 9.12 as

$$t = \frac{\bar{d}}{s_{\bar{d}}/\sqrt{n}}. \tag{9.13}$$

Table 9.14. *Copper yields from two processes.*

Batch	Process A Yield (%)	Process B Yield (%)	d_i = Process A Yield – Process B Yield (%)
1	32.5	29.6	2.9
2	30.5	31.2	−0.7
3	29.6	29.7	−0.1
4	38.4	37.1	1.3
5	32.8	31.3	1.5
6	34.0	34.5	−0.5
7	32.1	31.0	1.1
8	40.3	38.7	1.6

Example 8

Table 9.14 shows the amount of copper extracted from ore using two processes. Test at $\alpha = 0.05$ whether there is a significant difference in the yield of copper between the two processes.

ANSWER

Though it is possible to calculate the mean yield for process A and compare that with the mean of the yield for process B, the variability between batches, (say, due to the batches being obtained from different locations) encourages us to consider a comparison between yields on a batch by batch basis.

To perform the hypothesis test we write,

H$_0$: $\delta = 0$ (the null hypothesis is that the population mean of the differences between the paired values is zero);

H$_a$: $\delta \neq 0$.

Using the data in table 9.14, we find $\bar{d} = 0.8875\%$, $s_{\bar{d}} = 1.229\%$. Substituting these numbers into equation 9.13 (and noting the number of pairs, $n = 8$), we have

$$t = 2.042.$$

The number of degrees of freedom, $v = n - 1 = 7$. For a two tailed test, the critical value of the test statistic, t_{crit}, for $v = 7$ and $\alpha = 0.05$, found from table 2 in appendix 1, is

$$t_{crit} = t_{0.05,7} = 2.365.$$

As $|t| < t_{crit}$ we cannot reject the null hypothesis, i.e. based on the data in table 9.14 there is no difference in the efficiency of the extraction methods at the 0.05 level of significance.

Table 9.15. *Emfs of eight alkaline batteries.*

Battery number	Emf at $t = 0$ (V)	Emf at $t = 3$ months (V)
1	9.49	9.48
2	9.44	9.42
3	9.46	9.46
4	9.47	9.46
5	9.44	9.41
6	9.43	9.40
7	9.39	9.40
8	9.46	9.44

Exercise H

The emfs of a batch of 9 V batteries were measured before and after a storage period of 3 months. Table 9.15 shows the emfs of eight alkaline batteries before and after the storage period of three months.

Test at the $\alpha = 0.05$ level of significance whether the emfs of the batteries have changed over the storage period.

9.5.3 Excel's T.TEST() for paired samples

Excel's T.TEST() function may be used to determine the probability of obtaining a mean of differences between paired values, $\bar{d}$, given that the null hypothesis is that the population mean difference between paired values is $\delta_0 = 0$. The syntax of the function is

T.TEST(array1,array2,tails,type)

where array1 and array2 point to cells which contain the paired values of sample 1 and sample 2 respectively. The argument 'tails' is equal to 1 for a one tailed test and equal to 2 for a two tailed test. For the paired t test, the argument 'type' is equal to 1.

Exercise I

Consider the data given in table 9.14. Use the T.TEST() function to determine the probability that the mean difference in yields, $\bar{d}$, would be 0.8875% or larger. Assume that the null and alternate hypotheses given in example 8 still apply.

We can extend the comparison of means to establish whether there is a significant difference between the means of three or more samples. The approach adopted is termed analysis of variance (or ANOVA for short). We will consider ANOVA in section 9.8, but as this technique requires the comparison of variances we will consider this first, and in particular, the *F* test.

9.6 Comparing variances using the *F* test

The tests of significance we have discussed so far have mainly considered population and sample means. Another important population parameter estimated from sample data is the standard deviation, σ. There are situations in which we wish to know if two sets of data have come from populations with the same standard deviation or variance. These situations include where:

- the variability of values returned by two analytical techniques is compared;
- the control of a variable (such as temperature) has improved due to the introduction of new technology (for example, an advanced PID controller[19]);
- variability in human reaction time is compared before and after the intake of alcohol;
- the electrical noise picked up by a sensitive electronic circuit is compared when the circuit is (a) shielded, (b) unshielded;
- the precision of two instruments is compared.

For example, suppose we make ten measurements of the iron concentration in a specimen of stainless steel and calculate the standard deviation, s_1, of the sample. Ten measurements of iron concentration are made on another specimen of steel and the standard deviation of the values is s_2. By how much can s_1 differ from s_2 before the difference is regarded as significant? Before we are able to answer that question we consider a probability distribution that describes the probability of obtaining a particular ratio of *variances* of two samples, $\frac{s_1^2}{s_2^2}$. This is referred to as the *F* distribution.

9.6.1 The *F* distribution

A test that is routinely used to compare the variability of two samples is the '*F* test'. This test is based on the *F* distribution, first introduced by (and named in honour of) R. A. Fisher, a pioneer in the development of data analysis techniques and experimental design.[20]

---

[19] PID stands for Proportional, Integral and Derivative.
[20] For an account of the life of R. A. Fisher, see Box (1978).

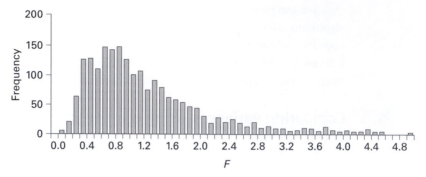

Figure 9.4. Histogram of 1000 simulated F values.

If we take two random samples of size n_1 and n_2, from the same normally distributed population, then the ratio

$$F = \frac{s_1^2}{s_2^2} \tag{9.14}$$

is a statistic which lies between 0 and $+\infty$, s_1^2 is the estimate of the population variance of sample 1, and s_2^2 is the estimate of the population variance of sample 2.

An indication of the shape of the distribution can be obtained by simulating many experiments in which the population variances estimates s_1^2 and s_2^2 are determined for samples containing n_1 and n_2 values respectively, where the samples are drawn from the same population. Figure 9.4 shows a histogram based on such a simulation in which $n_1 = n_2 = 10$ values were generated 1000 times using the normal distribution option of the Random Number Generator in Excel.[21]

Figure 9.4 indicates that, in contrast to the normal and t distributions, the distribution of the F statistic is not symmetrical. The shape of the distribution depends on the number of degrees of freedom of each sample. The probability density function for the F distribution can be written[22]

$$f(x) = K(v_1, v_2) \left(\frac{v_1}{v_2}\right)^{v_1/2} \frac{x^{\frac{v_1-2}{2}}}{\left(1 + \frac{v_1 x}{v_2}\right)^{\frac{v_1+v_2}{2}}} \tag{9.15}$$

where $x = \frac{s_1^2}{s_2^2}$, and $K(v_1, v_2)$ is a constant which depends on the degrees of freedom, v_1 and v_2.

---

[21] The population mean, $\mu = 32$, and the population standard deviation, $\sigma = 3$, were arbitrarily chosen for the simulation.

[22] See Graham (1993) for a discussion of equation 9.15.

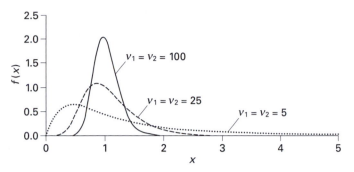

Figure 9.5. *F* distributions for several combinations of degrees of freedom, *v*.

$K(v_1, v_2)$ is chosen to ensure that the total area under the $f(x)$ curve is equal to one. Figure 9.5 shows curves of $f(x)$ versus x for $v_1 = v_2 = 5$, $v_1 = v_2 = 25$ and $v_1 = v_2 = 100$.

9.6.2 The *F* test

In order to determine whether the variances of two sets of data could have come from populations with the same population variance, we begin by stating the null hypothesis that the population variances for the two sets of data are the same, i.e. $\sigma_1^2 = \sigma_2^2$. The alternate hypothesis (for a two tailed test) would be $\sigma_1^2 \neq \sigma_2^2$. The next stage is to determine the critical value for the *F* statistic. If the *F* value determined using the data exceeds the critical value, then we have evidence that the two sets of data do not come from populations that have the same variance.

An added consideration when employing the *F* test is, due to the fact that the distribution is not symmetrical, there are two critical *F* values of unequal magnitude as indicated in figure 9.6. Figure 9.6 shows the upper and lower critical *F* values, F_U and F_L respectively, for a two tailed test at the significance level, α.

We cannot know σ_1^2 and σ_2^2, so we replace these by their estimates, s_1^2 and s_2^2 respectively. If $s_1^2 > s_2^2$ and the *F* value determined using the data exceeds F_U, then we reject the null hypothesis (for a given significance level, α). If $s_1^2 < s_2^2$ and the *F* value is less than F_L, then again we reject the null hypothesis. The difficulty of having two critical values is overcome by, when the ratio $\frac{s_1^2}{s_2^2}$ is calculated, the larger variance estimate is always placed in the numerator, so that $\frac{s_1^2}{s_2^2} > 1$. In doing this we need only consider the rejection region in the right hand tail of the *F* distribution. Critical values for *F* (in which the larger of the two variances is placed in the numerator) are given in table 3 in appendix 1 for various probabilities in the right hand tail of the distribution.

Table 9.16. *Capacitances of two samples.*

Sample 1 Capacitance (μF)	Sample 2 Capacitance (μF)
2.25	2.23
2.05	2.27
2.27	2.19
2.13	2.20
2.01	2.25
1.97	2.21

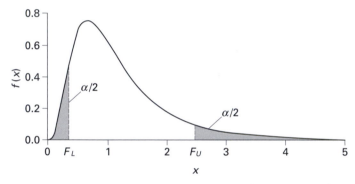

Figure 9.6. The F distribution showing the upper and lower critical values for a two tailed test at the α significance level.

Example 9

Two component manufacturers supply capacitors with nominal capacitance of 2.2 μF. Table 9.16 shows the capacitances of six capacitors supplied by each manufacturer.

Test at the $\alpha = 0.05$ level of significance whether there is any difference in the variances of capacitances of the components supplied by each manufacturer.

ANSWER

$H_0 : \sigma_1^2 = \sigma_2^2$ (the population variance of both samples is the same);
$H_a : \sigma_1^2 \neq \sigma_2^2$ (the population variance differs between samples).

Using the data in table 9.16, the variance of sample 1 is $s_1^2 = 1.575 \times 10^{-2} \mu F^2$ and the variance of sample 2 is $s_2^2 = 9.500 \times 10^{-4} \mu F^2$, hence $F = \frac{s_1^2}{s_2^2} = 16.58$. The number of degrees of freedom for each sample is one less than the number of values, i.e. $v_1 = v_2 = 5$.

The critical value, F_{crit} for F is obtained using table 3 in appendix 1. For $\alpha = 0.05$, the area in the right hand tail of the F distribution for a two tailed test is 0.025. For $v_1 = v_2 = 5$, $F_{crit} = 7.15$. As $F > F_{crit}$ this indicates that there is a significant difference between the variances of sample 1 and sample 2 and so we reject the null hypothesis.

Exercise J

In an experiment to measure the purity of reference quality morphine,[23] quantitative nuclear magnetic resonance (NMR) determined the purity as 99.920% with a standard deviation, *s*, of 0.052% (number of repeat measurements, *n* = 6). Using another technique called isotope dilution gas chromatography mass spectrometry (GCMS) the purity of the same material was determined as 99.879% with a standard deviation, *s*, of 0.035% (again the number of repeat measurements, *n* = 6).

Determine at the $\alpha = 0.05$ level of significance whether there is a significant difference in the variance of the values obtained using each analytical technique.

9.6.3 Excel's F.INV.RT() function

Excel's F.INV.RT() function can be used to determine the upper critical value from the *F* distribution, so avoiding the need to consult tables of critical values. The syntax of the function is[24]

$$\text{F.INV.RT}(\text{probability}, v_1, v_2)$$

where probability refers to the area in the right hand tail of the *F* distribution as shown in figure 9.6, v_1 is the number of degrees of freedom for sample 1 and v_2 is the number of degrees of freedom for sample 2.

Example 10

Determine the upper critical value of the *F* distribution when $v_1 = 5$, $v_2 = 7$ and a one tailed test is required to be carried out at the $\alpha = 0.05$ level of significance.

ANSWER

Using Excel, we type into a cell =F.INV.RT(0.05,5,7). After pressing the Enter key, Excel returns the number 3.971523.

---

[23] Agencies such as the National Institute of Standards and Technology (NIST) in the United States are able to supply reference materials and guarantee their elemental composition and purity. The materials can be as diverse as soil, powdered nickel oxides and stainless steel.

[24] The F.INV.RT() function in Excel 2010 is equivalent to the FINV() function found in earlier versions of Excel.

Exercise K

What would be the critical F value in example 10 if a two tailed test at $\alpha = 0.05$ level of significance were to be carried out?

9.6.4   Robustness of the *F* test

The F test produces satisfactory results so long as the distribution of the values in each sample is normal. However, if the values that make up a sample are not normally distributed or if the sample contains outliers, then a significance test which is believed to be being carried out at, say, $\alpha = 0.05$ could in fact be either above or below that significance level. That is, the F test is sensitive to the actual distribution of the values. As normality is difficult to establish when sample sizes are small, some workers are reluctant to bestow as much confidence in F tests compared to t tests (t tests can be shown to be much less sensitive to violations of normality of data than the F test[25]). A test that works well even when some of the assumptions upon which the test is founded are not valid is referred to as *robust*.

9.7   Comparing expected and observed frequencies using the χ^2 test

Up to this point we have made the assumption that distributions of real data are well described by a theoretical distribution, most notably (and most commonly) the normal distribution. The validity of our analyses would be strengthened if we could show that data are consistent with the distribution used to describe them. While a histogram is a good starting point for examining theoretical and actual distributions, there are tests we can perform to establish quantitatively if data are likely to have been drawn from some hypothesised distribution. One such test is the χ^2 test which employs the χ^2 distribution.

9.7.1   The χ^2 distribution

If a sample of data consisting of n values is drawn from a population in which values are normally distributed, then the statistic χ^2 is given by

---

[25] See Moore and McCabe (1989) for a discussion of the robustness of t and F tests.

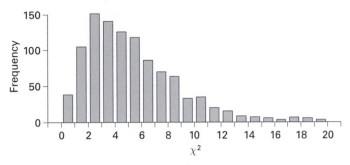

Figure 9.7. Histogram of 1000 simulated χ^2 values.

$$\chi^2 = \sum \left(\frac{x_i - \bar{x}}{\sigma} \right)^2, \tag{9.16}$$

where σ is the population standard deviation, and $\bar{x}$ is the mean of the n values. If many samples of the same size are taken from the population, then χ^2 follows a distribution which depends only on the number of degrees of freedom, v. As an example, figure 9.7 shows the result of a simulation in which values of χ^2 have been generated then displayed in the form of a histogram.

Each χ^2 value was determined by first generating five numbers using the normal distribution option of the Random Number Generator in Excel. The population mean of the distribution was chosen (arbitrarily) to have a mean of 50 and a standard deviation of 3. Equation 9.16 was used to determine χ^2 for each sample. This process was repeated 1000 times.

The distribution of χ^2 values is non-symmetrical with a single peak with mean = v and variance = $2v$. The probability density function for the χ^2 distribution with v degrees of freedom can be written[26]

$$f(x) = K(v)x^{\frac{v-2}{2}} \exp\left(\frac{-x}{2}\right), \tag{9.17}$$

where $x = \chi^2$. $K(v)$ is a constant which depends on the degrees of freedom. $K(v)$ is chosen to ensure that the total area under the $f(x)$ versus x curve is equal to one. Figure 9.8 shows the distribution function given by equation 9.17 for $v = 2$, 4 and 20. Note that, as v increases, the distribution becomes more symmetric and in fact tends to the normal distribution when v is very large.

---

[26] The χ^2 distribution is discussed by Kennedy and Neville (1986).

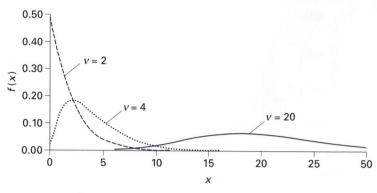

Figure 9.8. The χ^2 distribution for $v = 2$, 4 and 20.

9.7.2   The χ^2 test

An important application of the χ^2 distribution is as the basis of a test to establish how well the distribution of sample data gathered as part of an experiment compares with a theoretical distribution of data – this is sometimes called a test of the 'goodness of fit'. To compare a 'real' and theoretical distribution of data we begin by dividing the range of the data into conveniently sized categories or 'bins' and determine the frequency of values in each category. The categories should be chosen so that the frequency in each category is ≥ 5. This condition occasionally requires that categories containing small frequencies be combined.[27]

We generally refer to the actual frequency in the ith category as O_i (referred to as the 'observed' frequency). If an experiment is carried out many times, we would expect the mean frequency in the ith category to be equal to that predicted by the distribution being tested. We refer to the predicted or expected frequency based on the distribution as E_i. The difference between O_i and E_i is small if the hypothesised distribution describes the data well.

With reference to equation 9.16, it is possible to identify a correspondence between the observed frequency in the ith category, O_i, and the ith observed value of x, x_i. Similarly, the expected frequency can be regarded as corresponding to the mean, $\bar{x}$. The number of counts in any one category will vary from experiment to experiment but we discovered in section 4.3 that so long as events are random and independent then the distribution of counts in each category (should we perform the experiment many times) would follow the Poisson distribution. If this is the case, the standard deviation of the counts is approximately equal to $\sqrt{\bar{x}}$ or, using the symbolism introduced here, $\sigma_i \approx \sqrt{E_i}$.

---

[27] See Hamilton (1990) for a discussion on combining categories.

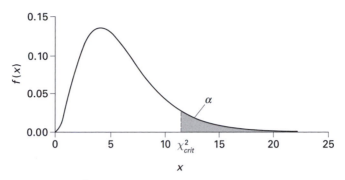

Figure 9.9. The χ^2 distribution showing the critical value, χ^2_{crit}, for the significance level, α.

Replacing x_i by O_i, $\bar{x}$ by E_i and σ_i by $\sqrt{E_i}$ in equation 9.16 gives

$$\chi^2 = \sum \frac{(O_i - E_i)^2}{E_i}. \tag{9.18}$$

If there is good agreement between observed and expected frequencies, so that $O_i \approx E_i$, χ^2 as given by equation 9.18 would be small. In fact, so long as χ^2 is approximately equal to the total number of categories, we can show that there is no evidence of significant difference between the actual and predicted frequencies. To put this more formally, if we know the number of degrees of freedom, then we can find the critical values for the χ^2 statistic, χ^2_{crit}, for any chosen level of significance, α. Figure 9.9 shows χ^2_{crit} for a significance level, α.

We can find χ^2_{crit} by using table 4 in appendix 1. A typical null hypothesis would be that the observed data come from a population that can be described by a particular theoretical distribution, such as the binomial, Poisson or normal distributions. If χ^2, as calculated using equation 9.16, exceeds χ^2_{crit}, we reject the null hypothesis.

9.7.3 Is the fit too good?

It is possible for the fit of a theoretical distribution to experimental data to be too good, i.e. that the observed and expected frequencies in each category match so well that this is unlikely to have happened 'by chance'. The critical value for χ^2 in this situation is found by considering the area corresponding the level of significance confined to the left hand tail of the χ^2 distribution. In this case if $\chi^2 < \chi^2_{crit}$ we reject the null hypothesis that the observed and expected distributions could match so well 'by chance'. If such a null hypothesis is rejected then it is possible that data gathered in the experiment, or the experimenter's methods, would be regarded with some concern as it is likely to be difficult to explain why $O_i = E_i$ for all i when random error is present.

As we seldom attempt to find out whether the fit between experimental data and hypothesised probability distribution is too good, we use a one tailed test in which the area corresponding to the level of significance is confined to the right hand tail of the χ^2 distribution, as shown in figure 9.9.

9.7.4 Degrees of freedom in χ^2 test

In order to determine χ^2_{crit} we need to know the number of degrees of freedom, v. Suppose the frequency of observations, O_i, are distributed in k categories or bins and that the sum of observations

$$\sum_{i=1}^{i=k} O_i = n, \tag{9.19}$$

where n is the total number of observations. One constraint on the expected number of observations, E_i, in each of the k categories is that

$$\sum_{i=1}^{i=k} E_i = n. \tag{9.20}$$

For each constraint that we impose upon the expected frequencies, we reduce the number of degrees of freedom by one, so that if the only constraint were given by equation 9.20, then[28]

$$v = k - 1. \tag{9.21}$$

The number of degrees of freedom is further reduced if the sample data must be used to estimate one or more parameters. As an example, in order to apply the χ^2 test to data that are assumed to be Poisson distributed, we usually need to estimate the mean of the distribution, μ, using the sample data. As the constraint expressed by equation 9.20 still applies, the number of degrees of freedom for a χ^2 test applied to a hypothesised Poisson distribution would be $v = k - 2$. Table 9.17 shows the number of degrees of freedom for χ^2 tests applied to various hypothesised distributions.

Example 11
Table 9.18 shows the frequency of occurrence of random numbers generated between 0 and 1 by the random number generator on a pocket calculator. The category width has been chosen as 0.1.

---

[28] If there are k categories, then the frequency in each of $k-1$ categories can take on any value. Since the total number of observations is fixed, then the kth category is constrained to have a frequency equal to the (total number of observations − sum of frequencies in the $k-1$ categories).

Table 9.17. *Degrees of freedom for χ^2 tests.*

Null hypothesis	Possible parameters estimated from sample data	Degrees of freedom, v
Data follow a uniform distribution	None	$k - 1$
Data follow a binomial distribution	The probability of a success on a single trial, p	$k - 2$
Data follow a Poisson distribution	Population mean, μ	$k - 2$
Data follow a normal distribution	Population mean, μ, and population standard deviation, σ	$k - 3$

Use a χ^2 test at the 0.05 level of significance to establish whether the values in table 9.18 are consistent with a number generator that produces random numbers distributed uniformly between 0 and 1.

ANSWER

H_0: data have been drawn from a population that is uniformly distributed between 0 and 1;
H_a: data have been drawn from a population that is not uniformly distributed between 0 and 1.

Table 9.18 contains the observed frequencies in each category. If the random number generator produces numbers evenly distributed between 0 and 1, then out of 100 random numbers we would expect, on average, ten numbers to lie between 0 and 0.1, ten to lie between 0.1 and 0.2 and so on. Table 9.19 shows the observed frequencies, expected frequencies and the terms in the summation given in equation 9.18.

Summing the last column of table 9.19 gives $\chi^2 = \sum \frac{(O_i - E_i)^2}{E_i} = 6.2$.

In order to find χ^2_{crit}, we use table 4 in appendix 1 with $v = k - 1 = 10 - 1 = 9$.

Then $\chi^2_{crit} = \chi^2_{0.05,9} = 16.92$.

As $\chi^2 < \chi^2_{crit}$, we cannot reject the null hypothesis, that is, the numbers are consistent with having been drawn from a population that is uniformly distributed between 0 and 1.

Table 9.18. *Distribution of numbers produced by uniform random number generator.*

Interval	Frequency
$0.0 \leq x < 0.1$	12
$0.1 \leq x < 0.2$	12
$0.2 \leq x < 0.3$	9
$0.3 \leq x < 0.4$	13
$0.4 \leq x < 0.5$	7
$0.5 \leq x < 0.6$	9
$0.6 \leq x < 0.7$	5
$0.7 \leq x < 0.8$	12
$0.8 \leq x < 0.9$	9
$0.9 \leq x < 1.0$	12

Table 9.19. *Observed and expected frequencies for example 11.*

Interval	Observed frequency, O_i	Expected frequency, E_i	$\dfrac{(O_i - E_i)^2}{E_i}$
$0.0 \leq x < 0.1$	12	10	0.4
$0.1 \leq x < 0.2$	12	10	0.4
$0.2 \leq x < 0.3$	9	10	0.1
$0.3 \leq x < 0.4$	13	10	0.9
$0.4 \leq x < 0.5$	7	10	0.9
$0.5 \leq x < 0.6$	9	10	0.1
$0.6 \leq x < 0.7$	5	10	2.5
$0.7 \leq x < 0.8$	12	10	0.4
$0.8 \leq x < 0.9$	9	10	0.1
$0.9 \leq x < 1.0$	12	10	0.4

Exercise L

The number of cosmic rays detected in 50 consecutive one minute intervals is shown in table 9.20.

(i) Assuming that the counts follow a Poisson distribution, determine the expected frequencies for 0, 1, 2, etc., counts. Note that the mean number of counts, determined using the data in table 9.20, is 1.86.

(ii) Use a χ^2 test at the 0.05 level of significance to determine whether the distribution of counts is consistent with a Poisson distribution that has a mean of 1.86.

Table 9.20. *Frequency of occurrence of cosmic rays in 50, one minute intervals.*

Counts	Frequency
0	10
1	13
2	7
3	15
4 or more	5

9.7.5   Excel's CHISQ.INV.RT() function

Critical values of the χ^2 statistic may be obtained using Excel's CHISQ.INV.RT() function instead of using table 4 in appendix 1. The syntax of the function is[29]

CHISQ.INV.RT(probability, v)

where probability refers to the area in the right hand tail of the distribution as shown by the shaded area in figure 9.9, and v is the number of degrees of freedom.

Example 12

Calculate the critical value of the χ^2 statistic when the area in the right hand tail of the χ^2 distribution is 0.1 and the number of degrees of freedom, $v = 3$.

ANSWER

Using Excel, type into a cell **=CHI.INV.RT(0.1,3)**. After pressing the Enter key, Excel returns the number 6.251389.

Exercise M

Use Excel to determine the critical value of the χ^2 statistic for

   (i) $\alpha = 0.005$, $v = 2$;
  (ii) $\alpha = 0.01$, $v = 3$;
 (iii) $\alpha = 0.05$, $v = 5$;
 (iv) $\alpha = 0.1$, $v = 10$.

[29] The CHISQ.INV.RT() function in Excel 2010 is equivalent to the CHIINV() function found in earlier versions of Excel.

9.8   Analysis of variance

Significance testing can be extended beyond comparing the means of two samples to comparing the means of many samples. One such test is based upon the analysis of variance or 'ANOVA' for short.[30] We consider 'one way' ANOVA in which a single factor or characteristic may vary from one sample to the next. For example, ANOVA might be used to compare the consistency of analysis of a specimen of blood divided into three parts and sent to three forensic laboratories. In this case the forensic laboratory is the factor that differs from sample to sample.

The null and alternate hypotheses in a one-way ANOVA is stated simply as:

H_0: The population means of all samples are equal;
H_a: The population means of all samples are not equal.

Situations in which we might use ANOVA include comparing:

- the effect of three (or more) types of fuel additive on the efficiency of a car engine;
- influence of storage temperature on the adhesive properties of sticky tape;
- the concentration of iron in an ore as determined by several independent laboratories;
- the concentration of carbon dioxide from four volcanic vents.

Although the purpose of ANOVA is to determine whether there is a significant difference between the means of several samples of data, the method actually relies on the calculation of variance of data within samples and compares that with the variance of data determined by considering the variability between samples. If the variances are significantly different then we can infer that the samples are not all drawn from the same population.

We must be careful to use ANOVA in situations in which it is most appropriate. In the physical sciences there is often a known, assumed or hypothesised relationship between variables. If is possible to express that relationship in the form of an equation, then we may prefer to apply the technique of least squares in order to estimate parameters that appear in the equation, rather than use ANOVA. For example, it makes little sense to make ten repeat measurements of the viscosity of oil at each of five different temperatures and then to use ANOVA to establish whether there is a significant difference between the mean viscosity at each temperature. It is well established that the viscosity of oil is dependent on temperature and a more useful study would focus upon determining a model of temperature dependence of viscosity which best describes the experimental data.

---

[30] ANOVA is a versatile and powerful technique. See McPherson (1990) for details of the usual variations of ANOVA.

9.8.1 Principle of ANOVA

Analysis using one way ANOVA relies on using the data in two ways to obtain an estimate of the variance of the population. Firstly, the variance of the values in each sample[31] is estimated. By averaging the estimate of the variance for all samples (assuming that all samples consist of the same number of values) we get the best estimate of the 'within-sample variance'. We may write this as $s^2_{within\ samples}$. Suppose, for example, we wish to compare the means of four samples. The estimate of the population variance for each sample is s^2_1, s^2_2, s^2_3 and s^2_4. The best estimate of the within-sample variance would be[32]

$$s^2_{within\ samples} = \frac{s^2_1 + s^2_2 + s^2_3 + s^2_4}{4}.$$
(9.22)

More generally, we write, for K samples,

$$s^2_{within\ samples} = \frac{s^2_1 + s^2_2 + s^2_3 + \cdots s^2_K}{K}, \text{ or}$$

$$s^2_{within\ samples} = \frac{1}{K}\sum_{j=1}^{j=k} s^2_j.$$
(9.23)

Another way to estimate the population variance is to find the variance, $s^2_{\bar{x}}$, of the sample means given by

$$s^2_{\bar{x}} = \sum_{j=1}^{j=K} \frac{\left(\bar{x}_j - \bar{X}\right)^2}{K - 1},$$
(9.24)

where $\bar{x}_j$ is the mean of the jth sample, $\bar{X}$ is the mean of all the sample means (sometimes referred to as the 'grand mean').

Thus $s^2_{\bar{x}}$ is related to the estimate of the between-sample variance, $s^2_{between\ samples}$ by

$$s^2_{\bar{x}} = \frac{s^2_{between\ samples}}{N},$$
(9.25)

where N is the number of values in each sample. The larger the difference between the samples means, the larger will be $s^2_{between\ samples}$. Rearranging equation 9.25 we obtain

$$s^2_{between\ samples} = Ns^2_{\bar{x}}.$$
(9.26)

---

[31] Note that some texts and software packages use the word 'group' in place of 'sample' and refer to 'between group variance' and 'within group variance'. In this text we will consistently use the word 'sample'.

[32] For convenience we assume that each sample consists of the same number of values.

If the values in all samples have been drawn from the same population, it should not matter which method is used to estimate the population variance as each method should give approximately the same variance. However, if the difference between one or more means is large (and it is this we wish to test) then $s_{between\ samples}^2$ will be larger than $s_{within\ samples}^2$. To determine if there is a significant difference between the variances, we use a one tailed F test[33] where the F statistic is given by,

$$F = \frac{s_{between\ samples}^2}{s_{within\ samples}^2}.$$ (9.27)

If all values come from the same population, we expect F to be close to 1. F will be much greater than 1 if $s_{between\ samples}^2 \gg s_{within\ samples}^2$ and this will occur if one or more of the sample means is significantly different from the other means.

In order to demonstrate the application of ANOVA, consider a situation in which an experimenter wishes to determine whether the gas emerging from four volcanic vents is tapping the same source.

9.8.2 Example of ANOVA calculation

Table 9.21 shows the concentration of CO_2 for gas emerging from four volcanic vents.

We will use ANOVA to test the hypothesis that the gas from each vent comes from a common reservoir. We will carry out the test at $\alpha = 0.05$ level of significance.

Table 9.21. *Percentage of CO_2 from four vents.*

vent 1 (% CO_2)	vent 2 (% CO_2)	vent 3 (% CO_2)	vent 4 (% CO_2)
21	25	30	31
23	22	25	25
26	28	24	27
28	29	26	28
27	27	27	33
25	25	30	28
24	27	24	32

---

[33] The F test is described in section 9.6.2.

Table 9.22. *Mean and variance of CO_2 data in table 9.21.*

	vent 1	vent 2	vent 3	vent 4
mean, $\bar{x}_j$	24.857	26.143	26.571	29.143
variance, s_j^2	5.810	5.476	6.619	8.476

The null and alternate hypotheses are:

H$_0$: The populations means of all the samples are the same.

H$_a$: The populations means of all the samples are not the same.

The estimate of the population variance of the *j*th vent is given by

$$s_j^2 = \frac{\sum \left(x_i - \bar{x}_j\right)^2}{N-1}, \tag{9.28}$$

where N is the number of values in the *j*th vent and $\bar{x}_j$ is the mean of the values for the *j*th vent. Table 9.22 shows the sample means and the estimated variances for the data in table 9.21.

Using equation 9.23, the estimate of the within-sample population variance is

$$s_{within\ samples}^2 = \frac{5.810 + 5.476 + 6.619 + 8.476}{4} = 6.595.$$

The grand mean of the data in table 9.21 is,

$$\bar{X} = \frac{1}{K}\sum \bar{x}_j = \frac{24.857 + 26.143 + 26.571 + 29.143}{4} = 26.679.$$

Substituting for $\bar{x}_j$ and $\bar{X}$ in equation 9.24 gives,

$$s_{\bar{x}}^2 = \frac{(24.857 - 26.679)^2 + (26.143 - 26.679)^2 + (26.571 - 26.679)^2 + (29.143 - 26.679)^2}{4 - 1}$$
$$= 3.230.$$

We use equation 9.26 to give the estimate of the between-sample variance, i.e.

$$s_{between\ samples}^2 = Ns_{\bar{x}}^2 = 7 \times 3.230 = 22.61.$$

The F statistic given by equation 9.27 is

$$F = \frac{s_{between\ samples}^2}{s_{within samples}^2} = \frac{22.61}{6.595} = 3.428.$$

In order to establish whether the difference in variances is significant we must determine the critical value of the F statistic at whatever significance level is chosen.

When calculating the variance of each sample, the number of degrees of freedom is $N-1$ (one degree of freedom is lost due to the fact that the mean of the sample is used in the calculation of the variance). As K samples contribute to the calculation of the within-sample variance, the number of degrees of freedom is

$$v_{within\ samples} = K(N-1). \tag{9.29}$$

When calculating the between-sample variance, we used equation 9.26 in which the number of degrees of freedom is K-1, so that

$$v_{between\ samples} = K-1. \tag{9.30}$$

For the data in table 9.21, $K = 4$ and $N = 7$, this gives, $v_{within\ samples} = 24$ and $v_{between\ samples} = 3$. We now require the critical value of the F statistic for $\alpha = 0.05$ when the degrees of freedom in the numerator = 3, and that in the denominator = 24. Using table 3 in appendix 1,

$$F_{crit} = F_{3,24} = 3.01.$$

Comparing this with $F = 3.428$ as determined using the data, indicates that we should reject the null hypothesis at $\alpha = 0.05$. That is, the population means of all the samples are not the same.

Exercise N

Experimental studies have linked the size of the alpha wave generated by the human brain to the amount of light falling on the retina of the eye. In one study, the size of the alpha signal for nine people is measured at three light levels. Table 9.23 shows the size of the alpha signal at each light level for the nine people.

Using ANOVA, determine at the $\alpha = 0.05$ level of significance whether the magnitude of the alpha wave depends on light level.

Table 9.23. *Variation of size of alpha wave with light level.*

Light level	Magnitude of alpha wave (µV)								
High	32	35	40	35	33	37	39	34	37
Medium	36	39	29	33	38	36	32	39	40
Low	39	42	47	39	45	51	43	43	39

9.9 Review

Hypothesis testing assists in data analysis by forcing us to look closely at the data and address in a formal manner the question, "Is there really something interesting in the data or can apparent relationships or differences between distributions of data be explained by chance?". A hypothesis is formulated, for example that the data are consistent with a normal distribution, and we establish, if the hypothesis is true, the probability of obtaining the data set being considered.

While hypothesis testing is powerful we should not forget that the rejection of a null hypothesis should be seen in the context of the whole experiment being performed. A null hypothesis could be rejected when there is a very small (and in practice unimportant) difference between, say, a sample mean and a hypothesised population mean.[34]

Traditionally, hypothesis testing requires referring to statistical tables at the back of text books to find critical values of a particular test statistic. This tedious activity is assisted by the useful built in functions available in Excel. In the next chapter we will consider more built in data analysis features in Excel, which allow us, amongst other things, to perform statistical tests efficiently.

End of chapter problems

(1) The porosity, r, was measured for two samples of the ceramic $YBa_2Cu_3O_7$ prepared at different temperature. The porosity data are shown in table 9.24.

Determine at the $\alpha = 0.05$ level of significance whether there is any difference in the porosity of the two samples.

(2) Blood was taken from eight volunteers and sent to two laboratories. The urea concentration in the blood of each volunteer, as determined by both laboratories, is shown in table 9.25.

Determine at the $\alpha = 0.05$ level of significance whether there is any difference in the urea concentration determined by the laboratories.

Table 9.24. *Porosity values for two samples of $YBa_2Cu_3O_7$.*

Porosity, r	Sample 1	0.397	0.385	0.394	0.387	0.362	0.388
	Sample 2	0.387	0.361	0.377	0.352	0.363	0.387

[34] This occurs especially when sample sizes are large.

Table 9.25. *Values of urea concentration for eight volunteers as determined by two laboratories.*

	Volunteer	1	2	3	4	5	6	7	8
Urea concentration (mmol/L)	Laboratory A	4.1	2.3	8.4	7.4	7.5	3.4	3.9	6.0
	Laboratory B	4.0	2.4	7.9	7.3	7.3	3.0	3.8	5.5

Table 9.26. *Variation of the output voltage of six TEGs with orientation of the heat sink attached to the TEG.*

	Heat sink number	1	2	3	4	5	6
Output voltage from	Horizontal orientation	19.9	21.6	17.8	27.3	18.6	17.4
TEG (mV)	Vertical orientation	23.5	24.9	20.1	25.7	19.6	21.3

(3) A thermoelectric generator (TEG) generates a voltage when the two surfaces of the TEG are kept at different temperatures. To generate a large voltage, a large temperature difference is required between the surfaces. A researcher wishes to establish the effect of the orientation of the heat sinks to which the TEG is attached. The researcher believes that a vertically mounted heat sink will allow cool air to flow across the heat sink more effectively, keeping one surface cooler than if the heat sink were mounted horizontally, and hence generate a higher voltage from the TEG.

In an experiment, six TEGs are attached to six identical horizontally oriented heat sinks. The same six TEGs with heat sinks attached are then oriented vertically. Table 9.26 show the output voltage of the TEG for the heat sinks in the two orientations. Is there sufficient evidence to be able to conclude that the vertical orientation of the heat sink generates a higher output from the TEGs? Choose an appropriate test at the $\alpha = 0.05$ level of significance.

(4) The area under a peak of a chromatogram is measured for various concentrations of isooctane in a mixture of hydrocarbons. Table 9.27 shows calibration data of peak area as a function of isooctane concentration.

Assuming that peak area is linearly related to concentration, perform unweighted least squares to find the intercept and slope of the best line through the data. Is the intercept significantly different from zero at the $\alpha = 0.05$ level of significance?

Table 9.27. *Variation of peak area with concentration of isooctane.*

Concentration (moles)	Peak area (arbitrary units)
0.00342	2.10
0.00784	3.51
0.0102	5.11
0.0125	5.93
0.0168	8.06

Table 9.28. *Mass of barium carbonate in containers packed by two machines.*

Machine A, mass (g)	50.0	49.2	49.4	49.8	48.3	50.0	51.3	49.7	49.5
Machine B, mass (g)	51.9	48.8	52.0	52.3	51.0	49.6	49.2	49.1	52.4

Table 9.29. *Current gain of transistor taken from two batches.*

Current gain	Batch A	251	321	617	425	430	512	205	325	415
	Batch B	321	425	502	375	427	522	299	342	420

(5) Two machines, A and B, are used to weigh and pack containers with 50 g of barium carbonate. Table 9.28 shows the mass of nine containers from each machine.

Do the data in table 9.28 indicate that there is a difference in the variability in the mass of barium carbonate packed by each machine? Take the level of significance for the hypothesis test to be $\alpha = 0.05$.

(6) After a modification to the process used to manufacture silicon transistors, the supplier claims that the variability in the current gain of the transistors has been reduced. Table 9.29 shows the current gain for transistors from two batches. Batch A corresponds to transistors tested before the new process was introduced and batch B after.

Test the claim that, at $\alpha = 0.05$ level of significance, the variability in the current gain has reduced as a result of the new process.

Table 9.30. *100 values of time of fall of ball (in seconds).*

1.35	1.15	1.46	1.67	1.65	1.76	0.97	1.36	1.63	1.19
1.27	1.07	1.04	1.21	1.26	0.99	1.30	1.33	1.44	1.34
1.34	1.34	1.68	1.39	1.37	1.31	1.80	1.58	1.89	1.28
1.74	1.09	1.52	1.59	1.79	1.39	1.31	1.55	1.33	1.56
1.12	1.24	1.11	1.34	1.40	1.42	1.35	1.85	1.06	1.26
0.89	1.70	1.15	1.28	1.56	1.50	1.58	1.53	1.14	1.19
1.55	1.47	1.22	1.36	1.44	1.52	1.44	1.23	1.79	1.51
1.42	1.58	1.58	1.28	1.23	1.63	1.17	1.10	1.55	1.54
1.85	1.70	1.67	1.43	1.41	1.50	1.40	1.20	1.06	1.58
1.50	1.53	1.45	1.20	1.66	1.35	1.24	1.25	1.32	1.32

Table 9.31. *Kinetic energy of an arrow as determined by three methods.*

Kinetic energy (joules)		
D45	MH	TE
10.5	9.9	11.0
10.7	10.1	10.9
11.2	11.0	11.3
10.1	10.8	11.4
11.5	10.3	10.9
10.8	10.6	10.6
10.0	10.8	11.1

(7) The time for a ball to fall 10 m was 'hand timed' using a stopwatch. Table 9.30 shows values of time for 100 successive measurements of elapsed time. Use the chi-squared test to determine whether the data are normally distributed. Carry out the test at the $\alpha = 0.05$ level of significance.

(8) Three methods are devised to determine the kinetic energy of an arrow shot from a bow. These methods are based on:

   (i)   distance travelled by the arrow when shot at 45° to the horizontal (D45),

  (ii) maximum height attained by the arrow when shot vertically (MH),

 (iii) time elapsed for arrow to hit the ground after shot vertically (TE).

     Table 9.31 shows values of kinetic energy of the arrow based on the three methods.

Table 9.32. *Calcium concentration in blood as measured by three laboratories.*

Amount of calcium in blood (mmol/L)		
Laboratory A	Laboratory B	Laboratory C
2.23	2.35	2.31
2.26	2.28	2.33
2.21	2.29	2.29
2.25	2.28	2.27
2.20	2.27	2.33

Use one way ANOVA to establish, at the $\alpha = 0.05$ level of significance, whether the determination the kinetic energy depends on the method used.

(9) Table 9.32 shows the concentration of calcium in a specimen of human blood as analysed by three laboratories.

Use the ANOVA utility in Excel to determine whether there is any significant difference in the sample means between the three laboratories at $\alpha = 0.05$ (refer to section 10.3 for details on how to use Excel's ANOVA utility).

Chapter 10

Data Analysis tools in Excel and the Analysis ToolPak

10.1 Introduction

The process of analysing experimental data frequently involves many steps which begin with the tabulation and graphing of data. Numerical analysis of data may require simple but repetitive calculations such as the summing and averaging of values. Spreadsheet programs are designed to perform these tasks, and in previous chapters we considered how Excel's built in functions such as AVERAGE() and CORREL() can assist data analysis. While the functions in Excel are extremely useful, there is still some effort required to:

- enter data into the functions;
- format numbers returned by the functions so that they are easy to assimilate;
- plot suitable graphs;
- combine functions to perform more advanced analysis.

Excel contains numerous useful data analysis tools designed around the built in functions which will, as examples, fit an equation to data using least squares or compare the means of many samples using analysis of variance. Once installed, these tools can be found via Analysis Group on the Data Ribbon. The dialog box that appears when a tool is selected allows for the easy input of data. Once the tool is selected and applied to data, results are displayed in a Worksheet with explanatory labels and headings. As an added benefit, some tools offer automatic plotting of data as graphs or charts.

In this chapter we consider several of Excel's advanced data analysis tools which form part of the Analysis ToolPak add-in, paying particular attention to those tools which relate directly to principles and methods described in this book. The Histogram and Descriptive Statistics tools are described in sections 2.8.1 and 2.8.2 respectively and will not be discussed further in this

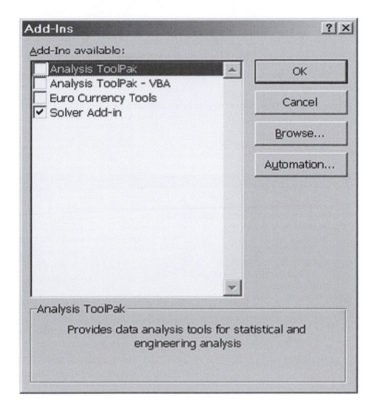

Figure 10.1. Add-in dialog box.

chapter. Tools which relate less closely to the material in this book are described
briefly with references given to where more information may be found.

10.2 Activating the Data Analysis tools

To activate the Data Analysis tools, click on the File tab. Next click on Options
then click on Add-ins. Near to the bottom of the dialog box, you should see

Click on Go.

Another dialog box similar to that in figure 10.1 should appear which
allows you to select one or more add-ins.

Check the box next to Analysis ToolPak, then click OK. Data Analysis
should now be added to the right hand side of the Data Ribbon as shown in
figure 10.2.

Figure 10.2. The Data Ribbon as it appears after installing the Analysis ToolPak.

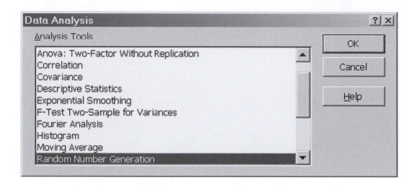

Figure 10.3. Data Analysis dialog box.

10.2.1 General features

Figure 10.3 shows the dialog box that appears after clicking on 🔲 Data Analysis which appears in the Analysis group of the Data Ribbon. The **?** or

Help   buttons in the dialog box can be used to search for information on any of the tools in the Analysis ToolPak.

Features of the tools in the Analysis ToolPak include the following.

(i) If data are changed after using a tool, the tool must be run again to update calculations, i.e. the output is not linked dynamically to the input data. Exceptions to this are the Moving Average and Exponential Smoothing tools.

(ii) Excel 'remembers' the cell ranges and numbers typed into each tool's dialog box. This is useful if you wish to run the same tool more than once. When Excel is exited the dialog boxes within all the tools are cleared.

The Data Analysis tools make extensive use of dialog boxes to allow for the easy entry of data, selection of options such as graph plotting and the input of numbers such as the level of significance, α.

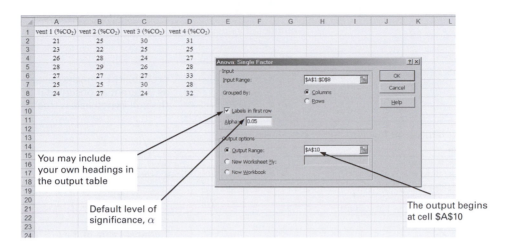

Figure 10.4. Example of the Anova: Single Factor tool.

10.3 Anova: Single Factor

One-way ANOVA (which Excel refers to as 'Single Factor' Anova) may be used to determine whether there is a significant difference between the means of two or more samples where each sample contains two or more values.[1]

As an example of using the 'Anova: Single Factor' tool in Excel consider figure 10.4, which shows data entered into an Excel Worksheet. The data appearing in the Worksheet are the percentages of CO_2 in gas emerging from four volcanic vents as discussed in section 9.8.2.

The data to be analysed are contained in cells A2 through to D8. This range is entered in the Input Range box as shown in figure 10.4. If we require the output of the ANOVA to appear on the same sheet as the data, we enter an appropriate cell reference into the Output Range box.[2] Choosing this cell as A10 and pressing OK returns the screen as shown (in part) in figure 10.5. Annotation has been added to figure 10.4 to clarify the labels and headings generated by Excel.

The value of the F statistic (3.428) is greater than the critical value of the F statistic, F_{crit} (3.009). We conclude that there is a significant difference (at $\alpha = 0.05$ level of significance) between the means of the samples appearing in figure 10.4. The same conclusion was reached when these data were analysed in

---

[1] Section 9.8 describes the principles of one-way ANOVA.

[2] Alternatively, the output from this (and all the other Data Analysis tools) can be made to appear in a new Worksheet or a new Workbook by clicking on New Worksheet Ply or New Workbook in the dialog box.

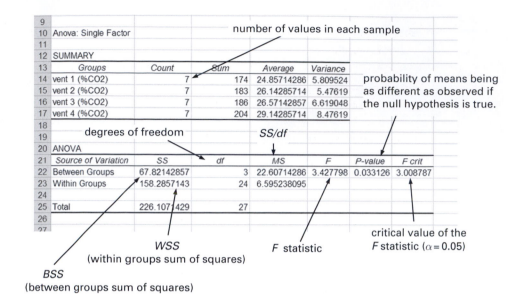

Figure 10.5. Output generated by the Anova: Single Factor tool.

section 9.8.2. However, by using the ANOVA tool in the Analysis ToolPak, the time and effort required to analyse the data is reduced considerably in comparison to the approach adopted in that section. Similar increases in efficiency are found when using other tools in the ToolPak, but one important matter should be borne in mind: if we require numbers returned by formulae to be updated as soon as the contents of cells are modified, it is better to avoid the tools in the ToolPak, instead creating the spreadsheet 'from scratch' using the built in functions, such as AVERAGE() and STDEV.S().

10.4   Correlation

The CORREL() function in Excel was introduced in section 6.6.1 to quantify the extent to which x–y data were correlated. The correlation between any two columns (or rows) containing data can be determined using the Correlation tool in the Analysis ToolPak. Figure 10.6 shows three samples of data entered into a worksheet.

Figure 10.7 shows the output of the Correlation tool as a matrix of numbers in which the correlation of every combination of pairs of columns is given.

The correlation matrix in figure 10.7 indicates that values in sample 1 and sample 2 are highly correlated, but the evidence of correlation between samples 1 and 3 and samples 2 and 3 is less convincing.

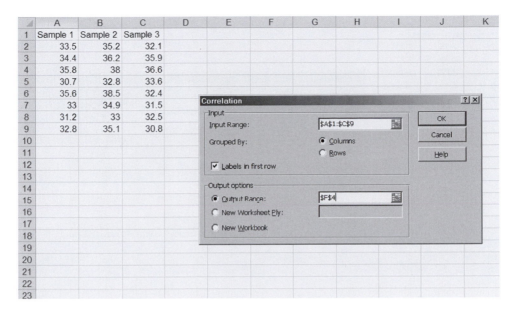

Figure 10.6. Worksheet showing the use of the Correlation tool.

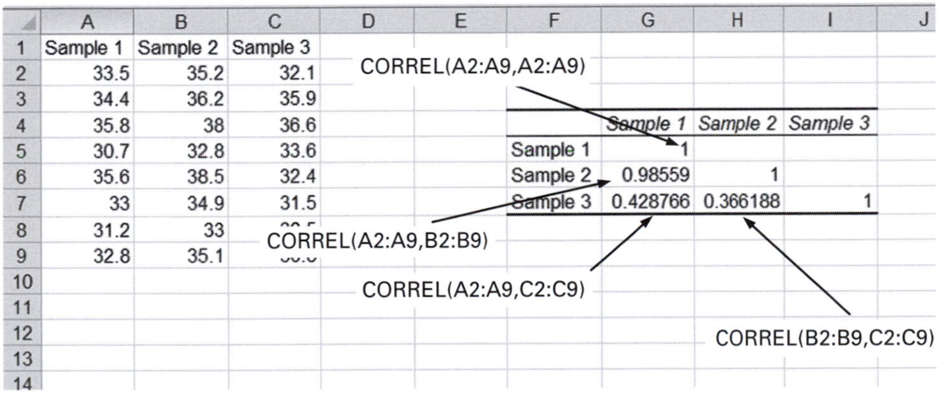

Figure 10.7. Output returned by the Correlation tool.

10.5 *F*-test two-sample for variances

The *F* test is used to determine whether two samples of data could have been drawn from populations with the same variance as discussed in section 9.6.2. The F-Test tool in Excel requires that the cell ranges containing the two samples be entered into the dialog box, as shown in figure 10.8. The data in figure 10.9

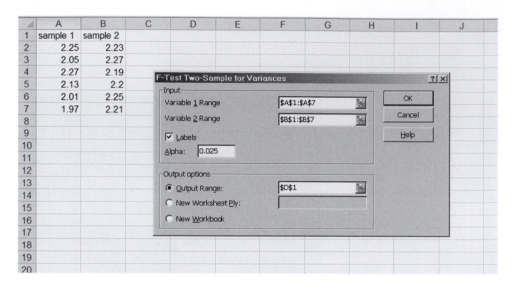

Figure 10.8. Worksheet showing the use of the F-Test tool.

	A	B	C	D	E	F	G
1	sample 1	sample 2		F-Test Two-Sample for Variances			
2	2.25	2.23					
3	2.05	2.27			sample 1	sample 2	
4	2.27	2.19		Mean	2.113333	2.225	
5	2.13	2.2		Variance	0.015747	0.00095	
6	2.01	2.25		Observations	6	6	
7	1.97	2.21		df	5	5	
8				F	16.57544		
9				P(F<=f) one-tail	0.003942		
10				F Critical one-tail	7.146382		
11							
12							

s_1^2 points to Variance sample 1, s_2^2 points to Variance sample 2. Ratio of sample variances, $F = \dfrac{s_1^2}{s_2^2}$

Figure 10.9. Output returned by Excel's F-Test tool.

refer to values of capacitors supplied by two manufacturers as described in example 9 of section 9.6.2.

The F-Test tool performs a one tailed test at the chosen significance level, α. When a two tailed F test is required, as in this example, we must enter $\alpha/2$ in the Alpha box in order to obtain the correct critical value of the test statistic (in this case $F_{crit} = 7.15$).

10.6 Random Number Generation

The Random Number Generation tool in Excel generates random numbers based on any of seven probability distributions, including the normal, Poisson

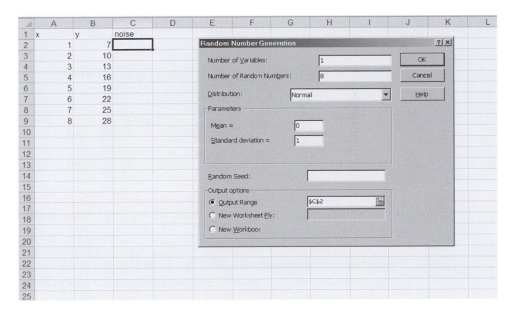

Figure 10.10. Use of the Random Number Generation tool.

and binomial distributions. Random numbers are extremely useful for simulating data. For example, the effectiveness of least squares in estimating parameters such as the slope and intercept can be investigated by adding 'noise' with a known distribution to otherwise 'noise free' data. To illustrate this, figure 10.10 shows x–y data in columns A and B of an Excel worksheet. The relationship between x and y is $y = 3x + 4$. Normally distributed random numbers are generated by choosing the normal distribution in the Random Number Generation dialog box. The default values for the mean and standard deviation of the parent distribution from which the numbers are generated are 0 and 1 respectively, but can be modified if required. The 'Number of Variables' in the Random Number Generation dialog box corresponds to the number of columns containing random numbers, and the 'Number of Random Numbers' corresponds to how many random numbers appear in each column.

The random numbers returned by Excel are shown in figure 10.11. Column D in the worksheet shows the normally distributed 'noise' added to the values in the column C.

10.7 Regression

The Regression tool in Excel estimates parameters using the method of linear least squares described in chapters 6 and 7. This tool is versatile and is able to

	A	B	C	D	E
1	x	y	noise	y+noise	
2	1	7	-0.10534	6.894658	
3	2	10	-0.55182	9.448181	
4	3	13	0.502416	13.50242	
5	4	16	-0.36598	15.63402	
6	5	19	0.121818	19.12182	
7	6	22	0.069094	22.06909	
8	7	25	-1.28691	23.71309	
9	8	28	1.062581	29.06258	
10					
11					

Figure 10.11. Normally distributed 'noise' in the C column. The cells in the D column show the noise added to y values appearing in column B.

- estimate parameters when the equation $y = a + bx$ is fitted to data, where y is the dependent variable, x is the independent variable. a and b are estimates of intercept and slope respectively, determined using least squares;
- estimate parameters when other equations that are linear in the parameters are to be fitted to data, such as $y = a + bx + cx^2$, where a, b and c are estimates of parameters obtained using least squares;
- determine standard uncertainty in parameter estimates (which Excel labels as 'standard error');
- calculate residuals and standardised residuals;
- plot measured and estimated values based on the line of best fit on an x–y graph;
- calculate the p value for each parameter estimate;
- determine coverage intervals for the parameters;
- determine the multiple correlation coefficient, R, the coefficient of multiple determination, R^2, and the adjusted coefficient of multiple determination, R^2_{ADJ} (see section 7.10.1 for more details);
- carry out analysis of variance.

As an example of using the Regression tool, consider the data in figure 10.12 which shows absorbance (y) versus concentration (x) data obtained during an experiment in which standard silver solutions were analysed by flame atomic absorption spectrometry. These data were described in example 2 in section 6.2.4. We use the Regression tool to fit the equation $y = a + bx$ to the data.

Features of the dialog box are described by the annotation on figure 10.12. Figure 10.13 shows the numbers returned by Excel when applying the Regression tool to the data in figure 10.12. Many useful statistics are shown in figure 10.13. For example, cells B27 and B28 contain the best estimates of intercept and slope respectively. Cells C27 and C28 contains the standard uncertainties in these

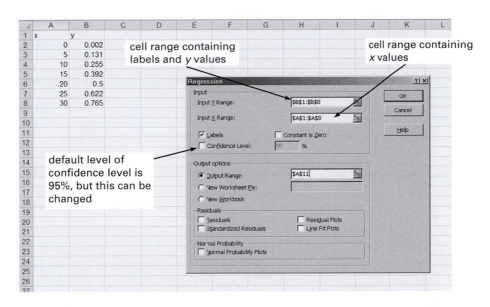

Figure 10.12. Worksheet showing the dialog box for the Regression tool in Excel.

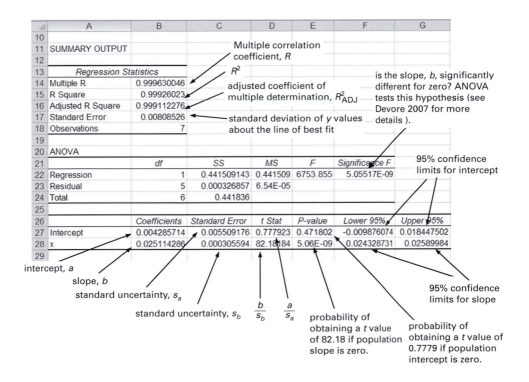

Figure 10.13. Annotated output from Excel's Regression tool.

estimates. Also worthy of special mention are the p values in cells E27 and E28. A p value of 0.472 for the intercept indicates that the intercept is not significantly different from zero. By contrast, the p value for the slope of 5.06×10^{-9} indicates that it is extremely unlikely that the 'true' slope is zero.

Note that the Regression tool in Excel is unable to perform weighted least squares.

10.7.1 Advanced linear least squares using Excel's Regression tool

It is possible to carry out advanced least squares using the Regression tool in Excel. As an example, consider fitting the equation $y = a + bx + cx^2$ to data as shown in figure 10.14 (these data are also analysed in example 5 in section 7.6).

Each term in the equation that contains the independent variable is allocated its own column in the spreadsheet. For example, column B of the spreadsheet in figure 10.14 contains x and column C contains x^2. To enter the x range (including a label in cell B1) into the dialog box we must highlight both the B and C columns, (or type \$B\$1:\$C\$12 into the Input X Range box). By choosing the output to begin at \$A\$14 on the same Worksheet, numbers are returned as shown in figure 10.15.

The three parameters, a, b and c estimated by the least squares technique appear in cells B30, B31 and B32 respectively, so that (to four significant figures) the equation representing the best line through the points can be written,

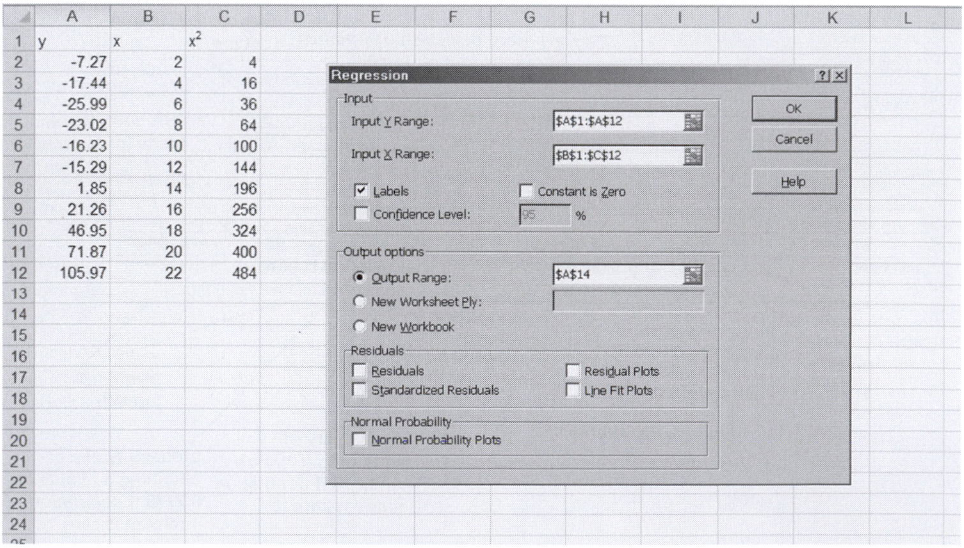

Figure 10.14. Fitting the equation $y = a + bx + cx^2$ to data using Excel's Regression tool.

◢	A	B	C	D	E	F	G
13							
14	SUMMARY OUTPUT						
15							
16	*Regression Statistics*						
17	Multiple R	0.998780654					
18	R Square	0.997562795					
19	Adjusted R	0.996953494					
20	Standard E	2.422563551					
21	Observatio	11					
22							
23	ANOVA						
24		*df*	*SS*	*MS*	*F*	*Significance F*	
25	Regression	2	19217.13338	9608.567	1637.224561	3.52832E-11	
26	Residual	8	46.95051329	5.868814			
27	Total	10	19264.08389				
28							
29		*Coefficients*	*Standard Error*	*t Stat*	*P-value*	*Lower 95%*	*Upper 95%*
30	Intercept	7.847272727	2.660478446	2.949572	0.018437751	1.712198428	13.98234703
31	x	-8.861643357	0.509492065	-17.3931	1.21727E-07	-10.03653417	-7.686752547
32	x2	0.605769231	0.020676239	29.29784	1.99539E-09	0.558089738	0.653448723

Figure 10.15. Output produced by Excel's Regression tool when fitting $y = a + bx + cx^2$ to data.

$$y = 7.847 - 8.862x + 0.6058x^2$$

These estimates, and the standard uncertainties in cells C30, C31 and C32, are consistent with those obtained in example 5 in section 7.6.

10.8 *t* tests

The Analysis ToolPak offers three types of *t* test:

- t-Test: Paired Two Sample for Means;
- t-Test: Two-Sample Assuming Equal Variances;
- t-Test: Two-Sample Assuming Unequal Variances.

The comparison of means of two samples (assuming equal variances) using a *t* test and the paired *t* test are discussed in sections 9.5 and 9.5.2 respectively. The comparison of means when samples have significantly different variances is not dealt with in this book.[3]

Figure 10.16 shows an Excel worksheet containing two samples of data and the dialog box that appears when the t-Test tool is selected: Two-Sample Assuming

---

[3] See Devore (2007) for a discussion of the *t* test when samples have unequal variances.

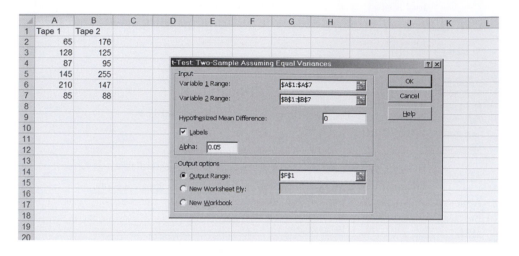

Figure 10.16. Peel off time data and t-Test dialog box.

	A	B	C	D	E	F	G	H	I
1	Tape 1	Tape 2				t-Test: Two-Sample Assuming Equal Variances			
2	65	176							
3	128	125					Tape 1	Tape 2	
4	87	95				Mean	120	147.6667	
5	145	255				Variance	2825.6	3834.267	
6	210	147				Observations	6	6	
7	85	88				Pooled Variance	3329.933		
8						Hypothesized Mean Difference	0		
9						df	10		
10						t Stat	-0.83042		
11						P(T<=t) one-tail	0.21284		
12						t Critical one-tail	1.812461		
13						P(T<=t) two-tail	0.425681		
14						t Critical two-tail	2.228139		
15									
16									

Combined variance of both samples

t statistic, found using $t = \dfrac{\bar{x}_1 - \bar{x}_2}{s_{\bar{x}_1 - \bar{x}_2}}$

Probability that the sample means would differ at least as much as observed, if the population means of both samples are the same

Figure 10.17. Annotated output from Excel's t-Test tool.

Equal Variances. The data are values of peel off time of two type of adhesive tape, as described in example 6 in section 9.5.

Applying the t-Test tool to data in figure 10.16 returns the numbers shown in figure 10.17.

10.9   Other tools

Excel possesses several other tools within the Analysis ToolPak which tend to be less frequently used by physical scientists. For example, the Moving Average tool is often used to smooth out variations in data observed over time, such as seasonal variations in commodity prices in sales and marketing. The remaining tools are outlined in the following sections, with reference to where more information may be obtained.

10.9.1 Anova: Two-Factor With Replication and Anova: Two-Factor Without Replication

In some situations an experimenter may wish to establish the influence of two factors on values obtained through experiment. For example, in a study to determine the amount of pollution by lead contamination of an estuary, both the geographical location as well as the time of day at which each sample is taken may affect the concentration of lead. Data categorised by geographical location and time of day can be studied using two-factor (also known as two-way) ANOVA. If only one value of lead is determined at each location at each time point then we would use the Anova: Two-Factor Without Replication tool in Excel. By contrast, if replicate measurements were made at each time point and location, then we would use the Anova: Two-Factor With Replication tool in Excel.[4]

10.9.2 Covariance

The covariance of two populations of data is defined as

$$Cov(x, y) = \frac{1}{n} \sum (x_i - \mu_x)(y_i - \mu_y). \tag{10.1}$$

If x_i is uncorrelated with y_i then the covariance will be close to zero as the product $(x_i - \mu_x)(y_i - \mu_y)$ will vary randomly in size and in sign. The Covariance tool in Excel determines the value of $Cov(x, y)$ of two populations of equal size tabulated in columns (or rows).

10.9.3 Exponential Smoothing

When quantities vary with time it is sometimes useful, especially if the data are noisy, to smooth the data. One way to smooth the data is to add the value at some

---

[4] Two-factor ANOVA is not dealt with in this text. See Devore (2007) for an introduction to this topic.

time point, $t+1$, to a fraction of the value obtained at the prior time point, t. Excel's Exponential Smoothing tool does this. The relationship Excel uses is

$$(\text{smoothed value})_{t+1} = \alpha \times (\text{actual value})_t + (1-\alpha) \times (\text{smoothed value})_t,$$

(10.2)

where α is called the smoothing constant and has a value between 0 and 1, and $(1-\alpha)$ is referred to by Excel as the 'damping factor'. Note that the output of this tool is dynamically linked to the input data, so that any change in the input values causes the output (including any graph created by the tool) to be updated. The Exponential Smoothing tool is discussed by Middleton (2003).

10.9.4   Fourier Analysis

The Fourier Analysis tool determines the frequency components of data using the Fast Fourier transform (FFT) technique. It is a powerful and widely used technique in signal processing to establish the frequency spectrum of time dependent signals. For a detailed discussion of the use Excel's Fourier Analysis tool, refer to Bloch (2003).

10.9.5   Moving Average

The Moving Average tool smoothes data by replacing the ith value in a column of values, x_i, by $x_{ismooth}$, where

$$x_{ismooth} = \frac{1}{N} \sum_{j=i-N+1}^{j=i} x_j,$$

(10.3)

where N is the number of values over which $x_{ismooth}$ is calculated.

The output of this tool is dynamically linked to the input data, so that any change in the input values causes the output (including any graph created by the tool) to be updated.

10.9.6   Rank and Percentile

The Rank and Percentile tool ranks values in a sample from largest to smallest. Each value is given a rank between 1 and n, where n is the number of values in the sample. The rank is also expressed as a percentage of the data set, such that the first ranked value has a percentile of 100% and that ranked last has percentile of 0%.

10.9.7 Sampling

The Sampling tool allows for the selection of values from a column or row in a worksheet. The selection may either be periodic (say every fifth value in a column of values) or random. When selecting at random from a group of values, a value may be absent, appear once or more than once (this is random selection 'with replacement'). Excel displays the selected values in a column.

10.10 Review

The task of analysing experimental data is considerably assisted by the use of spreadsheets or statistical packages. The Analysis ToolPak in Excel contains useful tools usually found in advanced statistical packages that can reduce the time to analyse and display data. For example, the Regression tool, with its ability to perform simple and advanced least squares, as well as plot useful graphs, is extremely useful for fitting equations to data. As is often the case with such utilities, we need to be careful that the ease with which an analysis may be carried out does not encourage us to suspend our 'critical faculties'. Questions to keep in mind when using the analysis utility are:

- has the appropriate analysis tool be chosen;
- have the raw data been entered correctly;
- are we able to interpret the output;
- does the output 'make sense'?

These words of caution aside, the Analysis ToolPak is a valuable and powerful aid to the analysis of experimental data.

Appendix 1

Statistical tables

(a) Table 1. *Cumulative distribution function for the standard normal distribution. The table gives the area under the standard normal probability curve between $z = -\infty$ and $z = z_1$.*
Example.

If $z_1 = -1.24$, then:

$P(-\infty < z < -1.24) = 0.10749$

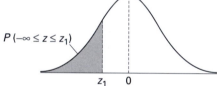

$P(-\infty \le z \le z_1)$

z_1 0

z_1	0.00	0.01	0.02	0.03	0.04	0.05	0.06	0.07	0.08	0.09
−4.00	0.00003	0.00003	0.00003	0.00003	0.00003	0.00003	0.00002	0.00002	0.00002	0.00002
−3.90	0.00005	0.00005	0.00004	0.00004	0.00004	0.00004	0.00004	0.00004	0.00003	0.00003
−3.80	0.00007	0.00007	0.00007	0.00006	0.00006	0.00006	0.00006	0.00005	0.00005	0.00005
−3.70	0.00011	0.00010	0.00010	0.00010	0.00009	0.00009	0.00008	0.00008	0.00008	0.00008
−3.60	0.00016	0.00015	0.00015	0.00014	0.00014	0.00013	0.00013	0.00012	0.00012	0.00011
−3.50	0.00023	0.00022	0.00022	0.00021	0.00020	0.00019	0.00019	0.00018	0.00017	0.00017
−3.40	0.00034	0.00032	0.00031	0.00030	0.00029	0.00028	0.00027	0.00026	0.00025	0.00024
−3.30	0.00048	0.00047	0.00045	0.00043	0.00042	0.00040	0.00039	0.00038	0.00036	0.00035
−3.20	0.00069	0.00066	0.00064	0.00062	0.00060	0.00058	0.00056	0.00054	0.00052	0.00050
−3.10	0.00097	0.00094	0.00090	0.00087	0.00084	0.00082	0.00079	0.00076	0.00074	0.00071
−3.00	0.00135	0.00131	0.00126	0.00122	0.00118	0.00114	0.00111	0.00107	0.00104	0.00100
−2.90	0.00187	0.00181	0.00175	0.00169	0.00164	0.00159	0.00154	0.00149	0.00144	0.00139
−2.80	0.00256	0.00248	0.00240	0.00233	0.00226	0.00219	0.00212	0.00205	0.00199	0.00193
−2.70	0.00347	0.00336	0.00326	0.00317	0.00307	0.00298	0.00289	0.00280	0.00272	0.00264
−2.60	0.00466	0.00453	0.00440	0.00427	0.00415	0.00402	0.00391	0.00379	0.00368	0.00357
−2.50	0.00621	0.00604	0.00587	0.00570	0.00554	0.00539	0.00523	0.00508	0.00494	0.00480

(a) Table 1. (*cont.*)

z_1	0.00	0.01	0.02	0.03	0.04	0.05	0.06	0.07	0.08	0.09
−2.40	0.00820	0.00798	0.00776	0.00755	0.00734	0.00714	0.00695	0.00676	0.00657	0.00639
−2.30	0.01072	0.01044	0.01017	0.00990	0.00964	0.00939	0.00914	0.00889	0.00866	0.00842
−2.20	0.01390	0.01355	0.01321	0.01287	0.01255	0.01222	0.01191	0.01160	0.01130	0.01101
−2.10	0.01786	0.01743	0.01700	0.01659	0.01618	0.01578	0.01539	0.01500	0.01463	0.01426
−2.00	0.02275	0.02222	0.02169	0.02118	0.02068	0.02018	0.01970	0.01923	0.01876	0.01831
−1.90	0.02872	0.02807	0.02743	0.02680	0.02619	0.02559	0.02500	0.02442	0.02385	0.02330
−1.80	0.03593	0.03515	0.03438	0.03362	0.03288	0.03216	0.03144	0.03074	0.03005	0.02938
−1.70	0.04457	0.04363	0.04272	0.04182	0.04093	0.04006	0.03920	0.03836	0.03754	0.03673
−1.60	0.05480	0.05370	0.05262	0.05155	0.05050	0.04947	0.04846	0.04746	0.04648	0.04551
−1.50	0.06681	0.06552	0.06426	0.06301	0.06178	0.06057	0.05938	0.05821	0.05705	0.05592
−1.40	0.08076	0.07927	0.07780	0.07636	0.07493	0.07353	0.07215	0.07078	0.06944	0.06811
−1.30	0.09680	0.09510	0.09342	0.09176	0.09012	0.08851	0.08692	0.08534	0.08379	0.08226
−1.20	0.11507	0.11314	0.11123	0.10935	0.10749	0.10565	0.10383	0.10204	0.10027	0.09853
−1.10	0.13567	0.13350	0.13136	0.12924	0.12714	0.12507	0.12302	0.12100	0.11900	0.11702
−1.00	0.15866	0.15625	0.15386	0.15151	0.14917	0.14686	0.14457	0.14231	0.14007	0.13786
−0.90	0.18406	0.18141	0.17879	0.17619	0.17361	0.17106	0.16853	0.16602	0.16354	0.16109
−0.80	0.21186	0.20897	0.20611	0.20327	0.20045	0.19766	0.19489	0.19215	0.18943	0.18673
−0.70	0.24196	0.23885	0.23576	0.23270	0.22965	0.22663	0.22363	0.22065	0.21770	0.21476
−0.60	0.27425	0.27093	0.26763	0.26435	0.26109	0.25785	0.25463	0.25143	0.24825	0.24510
−0.50	0.30854	0.30503	0.30153	0.29806	0.29460	0.29116	0.28774	0.28434	0.28096	0.27760
−0.40	0.34458	0.34090	0.33724	0.33360	0.32997	0.32636	0.32276	0.31918	0.31561	0.31207
−0.30	0.38209	0.37828	0.37448	0.37070	0.36693	0.36317	0.35942	0.35569	0.35197	0.34827
−0.20	0.42074	0.41683	0.41294	0.40905	0.40517	0.40129	0.39743	0.39358	0.38974	0.38591
−0.10	0.46017	0.45620	0.45224	0.44828	0.44433	0.44038	0.43644	0.43251	0.42858	0.42465
−0.00	0.50000	0.49601	0.49202	0.48803	0.48405	0.48006	0.47608	0.47210	0.46812	0.46414

(b) Table 1. *Cumulative distribution function for the standard normal distribution (continued).*
The table gives the area under the standard normal probability curve between $z = -\infty$ and $z = z_1$.
Example.

If $z_1 = 2.56$, then:

$P(-\infty < z < 2.56) = 0.99477$

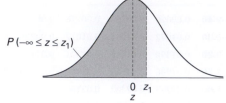

$P(-\infty \leq z \leq z_1)$

z_1	0.00	0.01	0.02	0.03	0.04	0.05	0.06	0.07	0.08	0.09
0.00	0.50000	0.50399	0.50798	0.51197	0.51595	0.51994	0.52392	0.52790	0.53188	0.53586
0.10	0.53983	0.54380	0.54776	0.55172	0.55567	0.55962	0.56356	0.56749	0.57142	0.57535
0.20	0.57926	0.58317	0.58706	0.59095	0.59483	0.59871	0.60257	0.60642	0.61026	0.61409
0.30	0.61791	0.62172	0.62552	0.62930	0.63307	0.63683	0.64058	0.64431	0.64803	0.65173
0.40	0.65542	0.65910	0.66276	0.66640	0.67003	0.67364	0.67724	0.68082	0.68439	0.68793
0.50	0.69146	0.69497	0.69847	0.70194	0.70540	0.70884	0.71226	0.71566	0.71904	0.72240
0.60	0.72575	0.72907	0.73237	0.73565	0.73891	0.74215	0.74537	0.74857	0.75175	0.75490
0.70	0.75804	0.76115	0.76424	0.76730	0.77035	0.77337	0.77637	0.77935	0.78230	0.78524
0.80	0.78814	0.79103	0.79389	0.79673	0.79955	0.80234	0.80511	0.80785	0.81057	0.81327
0.90	0.81594	0.81859	0.82121	0.82381	0.82639	0.82894	0.83147	0.83398	0.83646	0.83891
1.00	0.84134	0.84375	0.84614	0.84849	0.85083	0.85314	0.85543	0.85769	0.85993	0.86214
1.10	0.86433	0.86650	0.86864	0.87076	0.87286	0.87493	0.87698	0.87900	0.88100	0.88298
1.20	0.88493	0.88686	0.88877	0.89065	0.89251	0.89435	0.89617	0.89796	0.89973	0.90147
1.30	0.90320	0.90490	0.90658	0.90824	0.90988	0.91149	0.91308	0.91466	0.91621	0.91774
1.40	0.91924	0.92073	0.92220	0.92364	0.92507	0.92647	0.92785	0.92922	0.93056	0.93189
1.50	0.93319	0.93448	0.93574	0.93699	0.93822	0.93943	0.94062	0.94179	0.94295	0.94408
1.60	0.94520	0.94630	0.94738	0.94845	0.94950	0.95053	0.95154	0.95254	0.95352	0.95449
1.70	0.95543	0.95637	0.95728	0.95818	0.95907	0.95994	0.96080	0.96164	0.96246	0.96327
1.80	0.96407	0.96485	0.96562	0.96638	0.96712	0.96784	0.96856	0.96926	0.96995	0.97062
1.90	0.97128	0.97193	0.97257	0.97320	0.97381	0.97441	0.97500	0.97558	0.97615	0.97670
2.00	0.97725	0.97778	0.97831	0.97882	0.97932	0.97982	0.98030	0.98077	0.98124	0.98169
2.10	0.98214	0.98257	0.98300	0.98341	0.98382	0.98422	0.98461	0.98500	0.98537	0.98574
2.20	0.98610	0.98645	0.98679	0.98713	0.98745	0.98778	0.98809	0.98840	0.98870	0.98899
2.30	0.98928	0.98956	0.98983	0.99010	0.99036	0.99061	0.99086	0.99111	0.99134	0.99158
2.40	0.99180	0.99202	0.99224	0.99245	0.99266	0.99286	0.99305	0.99324	0.99343	0.99361
2.50	0.99379	0.99396	0.99413	0.99430	0.99446	0.99461	0.99477	0.99492	0.99506	0.99520
2.60	0.99534	0.99547	0.99560	0.99573	0.99585	0.99598	0.99609	0.99621	0.99632	0.99643
2.70	0.99653	0.99664	0.99674	0.99683	0.99693	0.99702	0.99711	0.99720	0.99728	0.99736
2.80	0.99744	0.99752	0.99760	0.99767	0.99774	0.99781	0.99788	0.99795	0.99801	0.99807
2.90	0.99813	0.99819	0.99825	0.99831	0.99836	0.99841	0.99846	0.99851	0.99856	0.99861

(b) Table 1. (*cont.*)

z_1	0.00	0.01	0.02	0.03	0.04	0.05	0.06	0.07	0.08	0.09
3.00	0.99865	0.99869	0.99874	0.99878	0.99882	0.99886	0.99889	0.99893	0.99896	0.99900
3.10	0.99903	0.99906	0.99910	0.99913	0.99916	0.99918	0.99921	0.99924	0.99926	0.99929
3.20	0.99931	0.99934	0.99936	0.99938	0.99940	0.99942	0.99944	0.99946	0.99948	0.99950
3.30	0.99952	0.99953	0.99955	0.99957	0.99958	0.99960	0.99961	0.99962	0.99964	0.99965
3.40	0.99966	0.99968	0.99969	0.99970	0.99971	0.99972	0.99973	0.99974	0.99975	0.99976
3.50	0.99977	0.99978	0.99978	0.99979	0.99980	0.99981	0.99981	0.99982	0.99983	0.99983
3.60	0.99984	0.99985	0.99985	0.99986	0.99986	0.99987	0.99987	0.99988	0.99988	0.99989
3.70	0.99989	0.99990	0.99990	0.99990	0.99991	0.99991	0.99992	0.99992	0.99992	0.99992
3.80	0.99993	0.99993	0.99993	0.99994	0.99994	0.99994	0.99994	0.99995	0.99995	0.99995
3.90	0.99995	0.99995	0.99996	0.99996	0.99996	0.99996	0.99996	0.99996	0.99997	0.99997
4.00	0.99997	0.99997	0.99997	0.99997	0.99997	0.99997	0.99998	0.99998	0.99998	0.99998

Table 2. *Critical values for the t distribution and coverage factor, k, for the X% coverage interval.*

Example 1.

For a 95% confidence interval, with $\nu = 6$, then

$t_{X\%,\nu} = t_{95\%,6} = 2.447$.

Example 2.

For a 95% coverage interval, with $\nu = 10$, then $k = 2.228$.

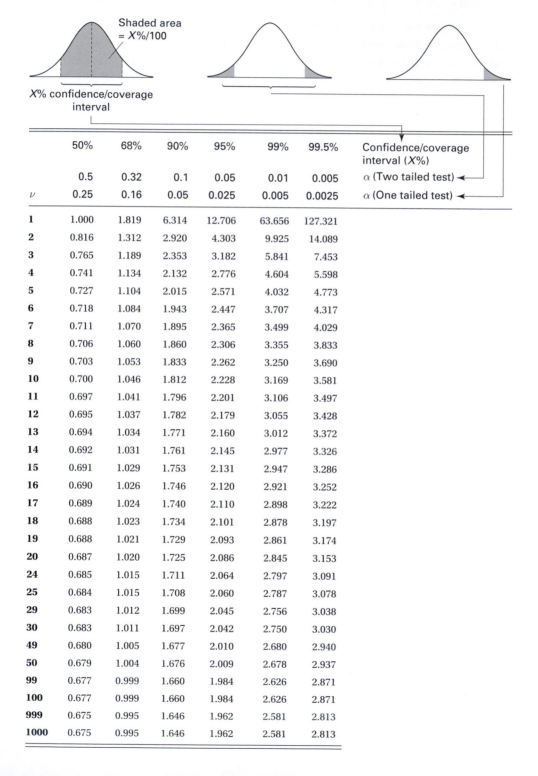

ν	50%	68%	90%	95%	99%	99.5%	Confidence/coverage interval (X%)
	0.5	0.32	0.1	0.05	0.01	0.005	α (Two tailed test)
	0.25	0.16	0.05	0.025	0.005	0.0025	α (One tailed test)
1	1.000	1.819	6.314	12.706	63.656	127.321	
2	0.816	1.312	2.920	4.303	9.925	14.089	
3	0.765	1.189	2.353	3.182	5.841	7.453	
4	0.741	1.134	2.132	2.776	4.604	5.598	
5	0.727	1.104	2.015	2.571	4.032	4.773	
6	0.718	1.084	1.943	2.447	3.707	4.317	
7	0.711	1.070	1.895	2.365	3.499	4.029	
8	0.706	1.060	1.860	2.306	3.355	3.833	
9	0.703	1.053	1.833	2.262	3.250	3.690	
10	0.700	1.046	1.812	2.228	3.169	3.581	
11	0.697	1.041	1.796	2.201	3.106	3.497	
12	0.695	1.037	1.782	2.179	3.055	3.428	
13	0.694	1.034	1.771	2.160	3.012	3.372	
14	0.692	1.031	1.761	2.145	2.977	3.326	
15	0.691	1.029	1.753	2.131	2.947	3.286	
16	0.690	1.026	1.746	2.120	2.921	3.252	
17	0.689	1.024	1.740	2.110	2.898	3.222	
18	0.688	1.023	1.734	2.101	2.878	3.197	
19	0.688	1.021	1.729	2.093	2.861	3.174	
20	0.687	1.020	1.725	2.086	2.845	3.153	
24	0.685	1.015	1.711	2.064	2.797	3.091	
25	0.684	1.015	1.708	2.060	2.787	3.078	
29	0.683	1.012	1.699	2.045	2.756	3.038	
30	0.683	1.011	1.697	2.042	2.750	3.030	
49	0.680	1.005	1.677	2.010	2.680	2.940	
50	0.679	1.004	1.676	2.009	2.678	2.937	
99	0.677	0.999	1.660	1.984	2.626	2.871	
100	0.677	0.999	1.660	1.984	2.626	2.871	
999	0.675	0.995	1.646	1.962	2.581	2.813	
1000	0.675	0.995	1.646	1.962	2.581	2.813	

Table 3. *Critical values of the F distribution for various probabilities, p, in the right hand tail of the distribution with v_1 degrees of freedom in the numerator, and v_2 degrees of freedom in the denominator.*

Example.

If $p = 0.1$, $v_1 = 6$, and $v_2 = 8$, then $F_{crit} = F_{0.1,6,8} = 2.67$.

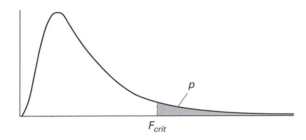

Numerator degrees of freedom, v_1

	p	1	2	3	4	5	6	7	8	10	12	15	20	50
1	0.1	39.86	49.50	53.59	55.83	57.24	58.20	58.91	59.44	60.19	60.71	61.22	61.74	62.69
	0.05	161.45	199.50	215.71	224.58	230.16	233.99	236.77	238.88	241.88	243.90	245.95	248.02	251.77
	0.025	647.79	799.48	864.15	899.60	921.83	937.11	948.20	956.64	968.63	976.72	984.87	993.08	1008.10
2	0.1	8.53	9.00	9.16	9.24	9.29	9.33	9.35	9.37	9.39	9.41	9.42	9.44	9.47
	0.05	18.51	19.00	19.16	19.25	19.30	19.33	19.35	19.37	19.40	19.41	19.43	19.45	19.48
	0.025	38.51	39.00	39.17	39.25	39.30	39.33	39.36	39.37	39.40	39.41	39.43	39.45	39.48
	0.01	98.50	99.00	99.16	99.25	99.30	99.33	99.36	99.38	99.40	99.42	99.43	99.45	99.48
	0.005	198.50	199.01	199.16	199.24	199.30	199.33	199.36	199.38	199.39	199.42	199.43	199.45	199.48
3	0.1	5.54	5.46	5.39	5.34	5.31	5.28	5.27	5.25	5.23	5.22	5.20	5.18	5.15
	0.05	10.13	9.55	9.28	9.12	9.01	8.94	8.89	8.85	8.79	8.74	8.70	8.66	8.58
	0.025	17.44	16.04	15.44	15.10	14.88	14.73	14.62	14.54	14.42	14.34	14.25	14.17	14.01
	0.01	34.12	30.82	29.46	28.71	28.24	27.91	27.67	27.49	27.23	27.05	26.87	26.69	26.35
	0.005	55.55	49.80	47.47	46.20	45.39	44.84	44.43	44.13	43.68	43.39	43.08	42.78	42.21
4	0.1	4.54	4.32	4.19	4.11	4.05	4.01	3.98	3.95	3.92	3.90	3.87	3.84	3.80
	0.05	7.71	6.94	6.59	6.39	6.26	6.16	6.09	6.04	5.96	5.91	5.86	5.80	5.70
	0.025	12.22	10.65	9.98	9.60	9.36	9.20	9.07	8.98	8.84	8.75	8.66	8.56	8.38
	0.01	21.20	18.00	16.69	15.98	15.52	15.21	14.98	14.80	14.55	14.37	14.20	14.02	13.69
	0.005	31.33	26.28	24.26	23.15	22.46	21.98	21.62	21.35	20.97	20.70	20.44	20.17	19.67
5	0.1	4.06	3.78	3.62	3.52	3.45	3.40	3.37	3.34	3.30	3.27	3.24	3.21	3.15
	0.05	6.61	5.79	5.41	5.19	5.05	4.95	4.88	4.82	4.74	4.68	4.62	4.56	4.44
	0.025	10.01	8.43	7.76	7.39	7.15	6.98	6.85	6.76	6.62	6.52	6.43	6.33	6.14
	0.01	16.26	13.27	12.06	11.39	10.97	10.67	10.46	10.29	10.05	9.89	9.72	9.55	9.24
	0.005	22.78	18.31	16.53	15.56	14.94	14.51	14.20	13.96	13.62	13.38	13.15	12.90	12.45
6	0.1	3.78	3.46	3.29	3.18	3.11	3.05	3.01	2.98	2.94	2.90	2.87	2.84	2.77
	0.05	5.99	5.14	4.76	4.53	4.39	4.28	4.21	4.15	4.06	4.00	3.94	3.87	3.75
	0.025	8.81	7.26	6.60	6.23	5.99	5.82	5.70	5.60	5.46	5.37	5.27	5.17	4.98
	0.01	13.75	10.92	9.78	9.15	8.75	8.47	8.26	8.10	7.87	7.72	7.56	7.40	7.09
	0.005	18.63	14.54	12.92	12.03	11.46	11.07	10.79	10.57	10.25	10.03	9.81	9.59	9.17

Denominator degrees of freedom, v_2

Table 3. (*cont.*)

Numerator degrees of freedom, v_1

	p	1	2	3	4	5	6	7	8	10	12	15	20	50
7	0.1	3.59	3.26	3.07	2.96	2.88	2.83	2.78	2.75	2.70	2.67	2.63	2.59	2.52
	0.05	5.59	4.74	4.35	4.12	3.97	3.87	3.79	3.73	3.64	3.57	3.51	3.44	3.32
	0.025	8.07	6.54	5.89	5.52	5.29	5.12	4.99	4.90	4.76	4.67	4.57	4.47	4.28
	0.01	12.25	9.55	8.45	7.85	7.46	7.19	6.99	6.84	6.62	6.47	6.31	6.16	5.86
	0.005	16.24	12.40	10.88	10.05	9.52	9.16	8.89	8.68	8.38	8.18	7.97	7.75	7.35
8	0.1	3.46	3.11	2.92	2.81	2.73	2.67	2.62	2.59	2.54	2.50	2.46	2.42	2.35
	0.05	5.32	4.46	4.07	3.84	3.69	3.58	3.50	3.44	3.35	3.28	3.22	3.15	3.02
	0.025	7.57	6.06	5.42	5.05	4.82	4.65	4.53	4.43	4.30	4.20	4.10	4.00	3.81
	0.01	11.26	8.65	7.59	7.01	6.63	6.37	6.18	6.03	5.81	5.67	5.52	5.36	5.07
	0.005	14.69	11.04	9.60	8.81	8.30	7.95	7.69	7.50	7.21	7.01	6.81	6.61	6.22
10	0.1	3.29	2.92	2.73	2.61	2.52	2.46	2.41	2.38	2.32	2.28	2.24	2.20	2.12
	0.05	4.96	4.10	3.71	3.48	3.33	3.22	3.14	3.07	2.98	2.91	2.85	2.77	2.64
	0.025	6.94	5.46	4.83	4.47	4.24	4.07	3.95	3.85	3.72	3.62	3.52	3.42	3.22
	0.01	10.04	7.56	6.55	5.99	5.64	5.39	5.20	5.06	4.85	4.71	4.56	4.41	4.12
	0.005	12.83	9.43	8.08	7.34	6.87	6.54	6.30	6.12	5.85	5.66	5.47	5.27	4.90
12	0.1	3.18	2.81	2.61	2.48	2.39	2.33	2.28	2.24	2.19	2.15	2.10	2.06	1.97
	0.05	4.75	3.89	3.49	3.26	3.11	3.00	2.91	2.85	2.75	2.69	2.62	2.54	2.40
	0.025	6.55	5.10	4.47	4.12	3.89	3.73	3.61	3.51	3.37	3.28	3.18	3.07	2.87
	0.01	9.33	6.93	5.95	5.41	5.06	4.82	4.64	4.50	4.30	4.16	4.01	3.86	3.57
	0.005	11.75	8.51	7.23	6.52	6.07	5.76	5.52	5.35	5.09	4.91	4.72	4.53	4.17
14	0.1	3.10	2.73	2.52	2.39	2.31	2.24	2.19	2.15	2.10	2.05	2.01	1.96	1.87
	0.05	4.60	3.74	3.34	3.11	2.96	2.85	2.76	2.70	2.60	2.53	2.46	2.39	2.24
	0.025	6.30	4.86	4.24	3.89	3.66	3.50	3.38	3.29	3.15	3.05	2.95	2.84	2.64
	0.01	8.86	6.51	5.56	5.04	4.69	4.46	4.28	4.14	3.94	3.80	3.66	3.51	3.22
	0.005	11.06	7.92	6.68	6.00	5.56	5.26	5.03	4.86	4.60	4.43	4.25	4.06	3.70
16	0.1	3.05	2.67	2.46	2.33	2.24	2.18	2.13	2.09	2.03	1.99	1.94	1.89	1.79
	0.05	4.49	3.63	3.24	3.01	2.85	2.74	2.66	2.59	2.49	2.42	2.35	2.28	2.12
	0.025	6.12	4.69	4.08	3.73	3.50	3.34	3.22	3.12	2.99	2.89	2.79	2.68	2.47
	0.01	8.53	6.23	5.29	4.77	4.44	4.20	4.03	3.89	3.69	3.55	3.41	3.26	2.97
	0.005	10.58	7.51	6.30	5.64	5.21	4.91	4.69	4.52	4.27	4.10	3.92	3.73	3.37
18	0.1	3.01	2.62	2.42	2.29	2.20	2.13	2.08	2.04	1.98	1.93	1.89	1.84	1.74
	0.05	4.41	3.55	3.16	2.93	2.77	2.66	2.58	2.51	2.41	2.34	2.27	2.19	2.04
	0.025	5.98	4.56	3.95	3.61	3.38	3.22	3.10	3.01	2.87	2.77	2.67	2.56	2.35
	0.01	8.29	6.01	5.09	4.58	4.25	4.01	3.84	3.71	3.51	3.37	3.23	3.08	2.78
	0.005	10.22	7.21	6.03	5.37	4.96	4.66	4.44	4.28	4.03	3.86	3.68	3.50	3.14
20	0.1	2.97	2.59	2.38	2.25	2.16	2.09	2.04	2.00	1.94	1.89	1.84	1.79	1.69
	0.05	4.35	3.49	3.10	2.87	2.71	2.60	2.51	2.45	2.35	2.28	2.20	2.12	1.97
	0.025	5.87	4.46	3.86	3.51	3.29	3.13	3.01	2.91	2.77	2.68	2.57	2.46	2.25
	0.01	8.10	5.85	4.94	4.43	4.10	3.87	3.70	3.56	3.37	3.23	3.09	2.94	2.64
	0.005	9.94	6.99	5.82	5.17	4.76	4.47	4.26	4.09	3.85	3.68	3.50	3.32	2.96

Denominator degrees of freedom, v_2

Table 3. (*cont.*)

Numerator degrees of freedom, v_1

	p	1	2	3	4	5	6	7	8	10	12	15	20	50
22	0.1	2.95	2.56	2.35	2.22	2.13	2.06	2.01	1.97	1.90	1.86	1.81	1.76	1.65
	0.05	4.30	3.44	3.05	2.82	2.66	2.55	2.46	2.40	2.30	2.23	2.15	2.07	1.91
	0.025	5.79	4.38	3.78	3.44	3.22	3.05	2.93	2.84	2.70	2.60	2.50	2.39	2.17
	0.01	7.95	5.72	4.82	4.31	3.99	3.76	3.59	3.45	3.26	3.12	2.98	2.83	2.53
	0.005	9.73	6.81	5.65	5.02	4.61	4.32	4.11	3.94	3.70	3.54	3.36	3.18	2.82
24	0.1	2.93	2.54	2.33	2.19	2.10	2.04	1.98	1.94	1.88	1.83	1.78	1.73	1.62
	0.05	4.26	3.40	3.01	2.78	2.62	2.51	2.42	2.36	2.25	2.18	2.11	2.03	1.86
	0.025	5.72	4.32	3.72	3.38	3.15	2.99	2.87	2.78	2.64	2.54	2.44	2.33	2.11
	0.01	7.82	5.61	4.72	4.22	3.90	3.67	3.50	3.36	3.17	3.03	2.89	2.74	2.44
	0.005	9.55	6.66	5.52	4.89	4.49	4.20	3.99	3.83	3.59	3.42	3.25	3.06	2.70
26	0.1	2.91	2.52	2.31	2.17	2.08	2.01	1.96	1.92	1.86	1.81	1.76	1.71	1.59
	0.05	4.23	3.37	2.98	2.74	2.59	2.47	2.39	2.32	2.22	2.15	2.07	1.99	1.82
	0.025	5.66	4.27	3.67	3.33	3.10	2.94	2.82	2.73	2.59	2.49	2.39	2.28	2.05
	0.01	7.72	5.53	4.64	4.14	3.82	3.59	3.42	3.29	3.09	2.96	2.81	2.66	2.36
	0.005	9.41	6.54	5.41	4.79	4.38	4.10	3.89	3.73	3.49	3.33	3.15	2.97	2.61
28	0.1	2.89	2.50	2.29	2.16	2.06	2.00	1.94	1.90	1.84	1.79	1.74	1.69	1.57
	0.05	4.20	3.34	2.95	2.71	2.56	2.45	2.36	2.29	2.19	2.12	2.04	1.96	1.79
	0.025	5.61	4.22	3.63	3.29	3.06	2.90	2.78	2.69	2.55	2.45	2.34	2.23	2.01
	0.01	7.64	5.45	4.57	4.07	3.75	3.53	3.36	3.23	3.03	2.90	2.75	2.60	2.30
	0.005	9.28	6.44	5.32	4.70	4.30	4.02	3.81	3.65	3.41	3.25	3.07	2.89	2.53
30	0.1	2.88	2.49	2.28	2.14	2.05	1.98	1.93	1.88	1.82	1.77	1.72	1.67	1.55
	0.05	4.17	3.32	2.92	2.69	2.53	2.42	2.33	2.27	2.16	2.09	2.01	1.93	1.76
	0.025	5.57	4.18	3.59	3.25	3.03	2.87	2.75	2.65	2.51	2.41	2.31	2.20	1.97
	0.01	7.56	5.39	4.51	4.02	3.70	3.47	3.30	3.17	2.98	2.84	2.70	2.55	2.25
	0.005	9.18	6.35	5.24	4.62	4.23	3.95	3.74	3.58	3.34	3.18	3.01	2.82	2.46
35	0.1	2.85	2.46	2.25	2.11	2.02	1.95	1.90	1.85	1.79	1.74	1.69	1.63	1.51
	0.05	4.12	3.27	2.87	2.64	2.49	2.37	2.29	2.22	2.11	2.04	1.96	1.88	1.70
	0.025	5.48	4.11	3.52	3.18	2.96	2.80	2.68	2.58	2.44	2.34	2.23	2.12	1.89
	0.01	7.42	5.27	4.40	3.91	3.59	3.37	3.20	3.07	2.88	2.74	2.60	2.44	2.14
	0.005	8.98	6.19	5.09	4.48	4.09	3.81	3.61	3.45	3.21	3.05	2.88	2.69	2.33
40	0.1	2.84	2.44	2.23	2.09	2.00	1.93	1.87	1.83	1.76	1.71	1.66	1.61	1.48
	0.05	4.08	3.23	2.84	2.61	2.45	2.34	2.25	2.18	2.08	2.00	1.92	1.84	1.66
	0.025	5.42	4.05	3.46	3.13	2.90	2.74	2.62	2.53	2.39	2.29	2.18	2.07	1.83
	0.01	7.31	5.18	4.31	3.83	3.51	3.29	3.12	2.99	2.80	2.66	2.52	2.37	2.06
	0.005	8.83	6.07	4.98	4.37	3.99	3.71	3.51	3.35	3.12	2.95	2.78	2.60	2.23
50	0.1	2.81	2.41	2.20	2.06	1.97	1.90	1.84	1.80	1.73	1.68	1.63	1.57	1.44
	0.05	4.03	3.18	2.79	2.56	2.40	2.29	2.20	2.13	2.03	1.95	1.87	1.78	1.60
	0.025	5.34	3.97	3.39	3.05	2.83	2.67	2.55	2.46	2.32	2.22	2.11	1.99	1.75
	0.01	7.17	5.06	4.20	3.72	3.41	3.19	3.02	2.89	2.70	2.56	2.42	2.27	1.95
	0.005	8.63	5.90	4.83	4.23	3.85	3.58	3.38	3.22	2.99	2.82	2.65	2.47	2.10

Denominator degrees of freedom, v_2

Table 4. *Critical values of the χ^2 distribution for various probabilities in the right hand tail of the distribution and degrees of freedom, v.*

Example.

If $p = 0.025$ and $v = 10$, then $\chi^2_{crit} = \chi^2_{0.025,10} = 20.483$

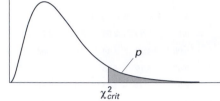

Degrees of freedom, v	Probability, *p*				
	0.1	0.05	0.025	0.01	0.005
1	2.706	3.841	5.024	6.635	7.879
2	4.605	5.991	7.378	9.210	10.597
3	6.251	7.815	9.348	11.345	12.838
4	7.779	9.488	11.143	13.277	14.860
5	9.236	11.070	12.832	15.086	16.750
6	10.645	12.592	14.449	16.812	18.548
7	12.017	14.067	16.013	18.475	20.278
8	13.362	15.507	17.535	20.090	21.955
9	14.684	16.919	19.023	21.666	23.589
10	15.987	18.307	20.483	23.209	25.188
11	17.275	19.675	21.920	24.725	26.757
12	18.549	21.026	23.337	26.217	28.3.00
13	19.812	22.362	24.736	27.688	29.819
14	21.064	23.685	26.119	29.141	31.319
15	22.307	24.996	27.488	30.578	32.801
16	23.542	26.296	28.845	32.000	34.267
17	24.769	27.587	30.191	33.409	35.718
18	25.989	28.869	31.526	34.805	37.156
19	27.204	30.144	32.852	36.191	38.582
20	28.412	31.410	34.170	37.566	39.997
25	34.382	37.652	40.646	44.314	46.928
30	40.256	43.773	46.979	50.892	53.672
35	46.059	49.802	53.203	57.342	60.275
40	51.805	55.758	59.342	63.691	66.766
45	57.505	61.656	65.410	69.957	73.166
50	63.167	67.505	71.420	76.154	79.490
60	74.397	79.082	83.298	88.379	91.952
70	85.527	90.531	95.023	100.425	104.215
80	96.578	101.879	106.629	112.329	116.321
90	107.565	113.145	118.136	124.116	128.299
100	118.498	124.342	129.561	135.807	140.170

Appendix 2

Propagation of uncertainties

If y depends on x and z, errors in values of x and z contribute to the error in the value of y. We are able to quote an uncertainty in y if we know the uncertainties in x and z. We derive an equation for the propagation of uncertainties under the assumption that there is no correlation between the errors in x and z, that is, neither the size nor the sign of the error in x is influenced by the size or sign of the error in z. We start with the expression for the variance in y, given by[1]

$$\sigma_y^2 = \frac{\sum (y_i - Y)^2}{n},$$ (A2.1)

where Y is the true value of the y quantity and n is the number of values. If we write $\delta y_i = (y_i - Y)$, then δy_i is the error in the ith value of y. Replacing $(y_i - Y)$ in A2.1 by δy_i, gives:

$$\sigma_y^2 = \frac{\sum (\delta y_i)^2}{n}.$$ (A2.2)

Assuming that y depends on x and z, i.e., $y = f(x,z)$, we use the approximation (valid for $\delta x_i \ll x_i$ and $\delta z_i \ll z_i$)

$$\delta y_i = \frac{\partial f}{\partial x} \delta x_i + \frac{\partial f}{\partial z} \delta z_i,$$ (A2.3)

where δx_i and δz_i are the errors in the ith values of x and z respectively.

Substituting A2.3 into A2.2 gives

$$\sigma_y^2 = \frac{\sum \left(\frac{\partial f}{\partial x} \delta x_i + \frac{\partial f}{\partial z} \delta z_i \right)^2}{n}.$$ (A2.4)

Expanding the brackets in A2.4,

[1] We assume for the moment that n is very large so that we can determine the population variance of y, σ_y.

453

$$\sigma_y^2 = \frac{\sum \left[\left(\frac{\partial f}{\partial x}\right)\delta x_i\right]^2}{n} + \frac{\sum \left[\left(\frac{\partial f}{\partial z}\right)\delta z_i\right]^2}{n} + \frac{\sum 2\frac{\partial f}{\partial x}\frac{\partial f}{\partial z}\delta x_i \delta z_i}{n},$$

(A2.5)

where $\frac{\partial f}{\partial x}$ and $\frac{\partial f}{\partial z}$ are determined at the best estimates of the true values x and z which are taken to be the means of the x and z values.

Concentrating on the last term in A2.5 for a moment, δx_i and δz_i take on both positive and negative values. If δx_i and δz_i are not correlated, then the summation of positive and negative terms should give a sum that tends to zero, particularly for large values of n. We therefore argue that the last term in equation A2.5 is negligible compared to the other terms and omit it.

The first two terms in equation A2.5 can be written as

$$\sigma_y^2 = \left(\frac{\partial f}{\partial x}\right)^2 \frac{\sum (\delta x_i)^2}{n} + \left(\frac{\partial f}{\partial z}\right)^2 \frac{\sum (\delta z_i)^2}{n},$$

(A2.6)

but $\frac{\sum (\delta x_i)^2}{n} = \sigma_x^2$ and $\frac{\sum (\delta z_i)^2}{n} = \sigma_z^2$, so that equation A2.6 becomes

$$\sigma_y^2 = \left(\frac{\partial f}{\partial x}\sigma_x\right)^2 + \left(\frac{\partial f}{\partial z}\sigma_z\right)^2$$

(A2.7)

or

$$\sigma_y = \left(\left(\frac{\partial f}{\partial x}\sigma_x\right)^2 + \left(\frac{\partial f}{\partial z}\sigma_z\right)^2\right)^{1/2}.$$

(A2.8)

Equation A2.7 can be expressed in terms of the standard uncertainties in y, x and z, i.e.

$$u_c^2(y) = \left(\frac{\partial f}{\partial x}u(x)\right)^2 + \left(\frac{\partial f}{\partial z}u(z)\right)^2,$$

(A2.9)

where $u_c(y)$ is the combined standard uncertainty in y and $u(x)$ and $u(z)$ are the standard uncertainties in x and z respectively.

Equation A2.9 can be extended to any number of quantities so long as the errors in those quantities are not correlated. For example, if y depends on a, b and c, i.e. $y = f(a,b,c)$, then

$$u_c^2(y) = \left(\frac{\partial f}{\partial a}u(a)\right)^2 + \left(\frac{\partial f}{\partial b}u(b)\right)^2 + \left(\frac{\partial f}{\partial c}u(c)\right)^2,$$

(A2.10)

where $u(a)$, $u(b)$ and $u(c)$ are the standard uncertainties in a, b and c respectively.

Appendix 3

Least squares and the principle of maximum likelihood

A3.1 Mean and weighted mean

We can use the principle of maximum likelihood to show that the best estimate of the population mean is the sample mean as given by equation 1.6. Using this approach we can also obtain the best estimate of the mean when the standard deviation is not constant. We derive an expression for what is usually termed the *weighted* mean. The argument is as follows.[1]

A distribution of values that occurs when repeat measurements are made is most likely to come from a population with a mean, μ, rather than from any other population. If the normal distribution is valid for the data, then the probability, P_i, of observing the value, x_i, is

$$P_i \propto \exp\left(-\frac{1}{2}\left[\frac{(x_i - \mu)}{\sigma_i}\right]^2\right), \tag{A3.1}$$

where μ is the population mean and σ_i is the standard deviation of the ith value. The greater is $x_i - \mu$, the smaller is the probability of observing the value x_i.

If we write the probability of observing n values when the population mean is μ, as $P(\mu)$, then so long as probabilities are independent,

$$P(\mu) = P_1 \times P_2 \times P_3 \times P_4 \cdots \times P_n, \tag{A3.2}$$

which may be written more succinctly as

$$P(\mu) = \prod P_i.$$

Substituting equation A3.1 into A3.2 gives,

---

[1] This follows the approach of Bevington and Robinson (2002).

455

$$P(\mu) \propto \exp\left(-\frac{1}{2}\left[\frac{(x_1 - \mu)}{\sigma_1}\right]^2\right) \times \exp\left(-\frac{1}{2}\left[\frac{(x_2 - \mu)}{\sigma_2}\right]^2\right) \cdots$$
$$\times \exp\left(-\frac{1}{2}\left[\frac{(x_n - \mu)}{\sigma_n}\right]^2\right).$$

(A3.3)

Equation A3.3 may be rewritten[2]

$$P(\mu) \propto \exp\left(-\frac{1}{2}\sum\left[\frac{(x_i - \mu)}{\sigma_i}\right]^2\right).$$

(A3.4)

For any *estimate*, X, of the population mean, we can calculate the probability, $P(X)$, of making a particular set of n measurements as

$P(X) = \Pi P_i$, where P_i is now given by

$$P_i \propto \exp\left(-\frac{1}{2}\left[\frac{(x_i - X)}{\sigma_i}\right]^2\right).$$

(A3.5)

It follows that

$$P(X) \propto \exp\left(-\frac{1}{2}\sum\left[\frac{(x_i - X)}{\sigma_i}\right]^2\right).$$

(A3.6)

We assume that the values, x_i, are more likely to have come from the distribution given by equation A3.1 than any other distribution. It follows that the probability given by equation A3.4 is the maximum attainable by equation A3.6. If we find the value of X that maximises $P(X)$, we will have found the best estimate of the population mean.

We introduce χ^2, which we term the 'weighted sum of squares' (and which is the χ^2 statistic discussed in chapter 9), where

$$\chi^2 = \sum\left[\frac{(x_i - X)}{\sigma_i}\right]^2.$$

(A3.7)

In order to maximise $P(X)$, it is necessary to find the value of X that *minimises* χ^2. This is done by differentiating χ^2 with respect to X, then setting the resulting equation equal to zero, i.e.

$$\frac{\partial \chi^2}{\partial X} = -2\sum\frac{1}{\sigma_i^2}(x_i - X) = 0$$

or

$$\sum\frac{x_i}{\sigma_i^2} = X\sum\frac{1}{\sigma_i^2},$$

it follows that

---

[2] Recalling $\exp(a) \times \exp(b) = \exp(a + b)$.

$$X = \frac{\sum \frac{x_i}{\sigma_i^2}}{\sum \frac{1}{\sigma_i^2}}.$$ (A3.8)

In preference to using X to represent the weighted mean we will use (for consistency with the way the mean is usually written) $\bar{x}_w$, so that,

$$\bar{x}_w = \frac{\sum \frac{x_i}{\sigma_i^2}}{\sum \frac{1}{\sigma_i^2}}$$ (A3.9)

As σ_i represents the population standard deviation, which we are only able to estimate, we replace σ_i in equation A3.9 by the estimate of the population standard deviation, s_i, so that:

$$\bar{x}_w = \frac{\sum \frac{x_i}{s_i^2}}{\sum \frac{1}{s_i^2}}$$ (A3.10)

Equation A3.10 gives the weighted mean. If the standard deviation for every value is the same such that $s_1 = s_2 = s$ then equation A3.10 becomes

$$\bar{x}_w = \frac{\frac{1}{s^2} \sum x_i}{\frac{1}{s^2} \sum 1} = \frac{\sum x_i}{n} = \bar{x}.$$ (A3.11)

If the weighted mean is found by combining several means, $\bar{x}_i$, each with its own standard uncertainty, u_i, then we can rewrite equation A3.10 as

$$\bar{x}_w = \frac{\sum \frac{\bar{x}_i}{u_i^2}}{\sum \frac{1}{u_i^2}}.$$ (A3.12)

Appendix 4 considers the standard uncertainty in the weighted mean.

A3.2 Best estimates of slope and intercept

Chapter 6 introduced the equations needed to calculate the 'best' straight line of the form $y = a + bx$ through a set of x–y data, where a is the intercept on the y axis at $x = 0$ and b is the slope of the line. Let us consider how the equations for a and b are derived.

Assume that sample observations are extracted from a population with parameters α and β, and that the true relationship between y and x is

$$y = \alpha + \beta x.$$ (A3.13)

We can never know the exact values of α and β, but we can find best estimates of these by using the principle of maximum likelihood.

For any given value of $x = x_i$, we can calculate the probability, P_i, of making a particular measurement of $y = y_i$. Assuming that the y observations are normally distributed and taking the true value of y at $x = x_i$ to be $y(x_i)$, then

$$P_i \propto \exp\left\{-\tfrac{1}{2}\left[\frac{y_i - y(x_i)}{\sigma_i}\right]^2\right\}. \tag{A3.14}$$

Here $y(x_i) = \alpha + \beta x_i$ and σ_i is the standard deviation of the observed y values.

The probability of making any observed set of n independent measurements, $P(\alpha,\beta)$ is the product of the individual probabilities,

$$P(\alpha,\beta) = \Pi P_i = P_1 \times P_2 \times P_3 \times P_4 \;\cdots\; \times P_n$$

$$\propto \exp\left\{-\tfrac{1}{2}\sum\left[\frac{y_i - y(x_i)}{\sigma_i}\right]^2\right\}. \tag{A3.15}$$

Similarly, if estimated values of the parameters, α and β, are a and b respectively, we can calculate the probability, $P(a,b)$, of making a particular set of n measurements.

$P(a,b) = \Pi P_i$, where P_i is now given by

$$P_i \propto \exp\left\{-\tfrac{1}{2}\left[\frac{y_i - \hat{y}_i}{\sigma_i}\right]^2\right\}, \tag{A3.16}$$

where $\hat{y}_i = a + bx_i$.

Therefore,

$$P(a,b) \propto \exp\left\{-\tfrac{1}{2}\sum\left[\frac{y_i - \hat{y}_i}{\sigma_i}\right]^2\right\}. \tag{A3.17}$$

We assume that the observed set of measurements is more likely to have come from the parent distribution given by equation A3.14 rather than any other distribution and therefore the probability given by equation A3.15 is the maximum probability attainable by equation A3.17. The best estimates of α and β are those values which maximise the probability given by equation A3.17. Maximising the exponential term *means* minimising the sum that appears within the exponential. Writing the summation in equation A3.17 as χ^2, we have

$$\chi^2 = \sum\left[\frac{y_i - \hat{y}_i}{\sigma_i}\right]^2. \tag{A3.18}$$

Equation A3.18 is at the heart of analysis by least squares. When χ^2 is minimised then we will have found the best estimates for the parameters which appear in the equation relating x to y.

If $\hat{y}_i = a + bx_i$, then equation A3.18 can be rewritten as

$$\chi^2 = \sum\left[\frac{y_i - a - bx_i}{\sigma_i}\right]^2. \tag{A3.19}$$

To find values for a and b which will minimise χ^2, we partially differentiate equation A3.19 with respect to a and b in turn, set the resulting equations to zero, then solve for a and b.

Note, in general, the equation relating x to y may be more complicated than simply $y = a + bx$, for example, $y = a + bx + cx^2$, where a, b and c are parameters. In such a situation we must differentiate with respect to each of the parameters in turn, set the equations to zero and solve for a, b and c.

Returning to equation A3.19 and differentiating, we get

$$\frac{\partial \chi^2}{\partial a} = -2 \sum \frac{1}{\sigma_i^2}(y_i - a - bx_i) = 0 \qquad (A3.20)$$

$$\frac{\partial \chi^2}{\partial b} = -2 \sum \frac{x_i}{\sigma_i^2}(y_i - a - bx_i) = 0, \qquad (A3.21)$$

where σ_i is the standard deviation associated with the ith value of y, y_i. In many cases, σ_i is a constant and can be replaced by σ. When σ_i is replaced by σ, we can write equations A3.20 and A3.21 as

$$\frac{-2}{\sigma^2} \sum (y_i - a - bx) = 0 \qquad (A3.22)$$

$$\frac{-2}{\sigma^2} \sum x_i(y_i - a - bx_i) = 0. \qquad (A3.23)$$

Equations A3.22 and A3.23 may be rearranged to give

$$na + b \sum x_i = \sum y_i \qquad (A3.24)$$

and

$$a \sum x_i + b \sum x_i^2 = \sum x_i y_i. \qquad (A3.25)$$

Manipulating equations A3.24 and A3.25 gives

$$a = \frac{\sum x_i^2 \sum y_i - \sum x_i \sum x_i y_i}{n \sum x_i^2 - (\sum x_i)^2} \qquad (A3.26)$$

$$b = \frac{n \sum x_i y_i - \sum x_i \sum y_i}{n \sum x_i^2 - (\sum x_i)^2}. \qquad (A3.27)$$

A3.3 The line of best fit passes through $\bar{x}, \bar{y}$

If we divide all the terms in equation A3.24 by the number of values, n, we obtain

$$a + b\frac{\sum x_i}{n} = \frac{\sum y_i}{n},$$

which can be written

$$\bar{y} = a + b\bar{x}. \qquad (A3.28)$$

Equation A3.28 indicates that a 'line of best fit' found using least squares passes through the point given by $\bar{x}, \bar{y}$.

A3.4 Weighting the fit

If σ_i is not constant (for example in a nuclear counting experiment, the standard deviation in the number of counts recorded, N, is not constant but is equal to $\sqrt{N}$), what do we do? We return to equations A3.20 and A3.21 and solve for a and b:

$$b \sum \frac{x_i}{\sigma_i^2} + a \sum \frac{1}{\sigma_i^2} = \sum \frac{y_i}{\sigma_i^2} \tag{A3.29}$$

$$b \sum \frac{x_i^2}{\sigma_i^2} + a \sum \frac{x_i}{\sigma_i^2} = \sum \frac{x_i y_i}{\sigma_i^2}. \tag{A3.30}$$

Solving for a and b gives

$$a = \frac{\sum \frac{x_i^2}{\sigma_i^2} \sum \frac{y_i}{\sigma_i^2} - \sum \frac{x_i}{\sigma_i^2} \sum \frac{x_i y_i}{\sigma_i^2}}{\Delta} \tag{A3.31}$$

$$b = \frac{\sum \frac{1}{\sigma_i^2} \sum \frac{x_i y_i}{\sigma_i^2} - \sum \frac{x_i}{\sigma_i^2} \sum \frac{y_i}{\sigma_i^2}}{\Delta}, \tag{A3.32}$$

where

$$\Delta = \sum \frac{1}{\sigma_i^2} \sum \frac{x_i^2}{\sigma_i^2} - \left(\sum \frac{x_i}{\sigma_i^2} \right)^2. \tag{A3.33}$$

Recognising that we cannot determine the population standard deviations, σ_i, we replace σ_i in equations A3.31 to A3.33 by the estimate of the population standard deviation, s_i, i.e.

$$a = \frac{\sum \frac{x_i^2}{s_i^2} \sum \frac{y_i}{s_i^2} - \sum \frac{x_i}{s_i^2} \sum \frac{x_i y_i}{s_i^2}}{\Delta} \tag{A3.34}$$

$$b = \frac{\sum \frac{1}{s_i^2} \sum \frac{x_i y_i}{s_i^2} - \sum \frac{x_i}{s_i^2} \sum \frac{y_i}{s_i^2}}{\Delta}, \tag{A3.35}$$

where

$$\Delta = \sum \frac{1}{s_i^2} \sum \frac{x_i^2}{s_i^2} - \left(\sum \frac{x_i}{s_i^2} \right)^2. \tag{A3.36}$$

Expressions for the standard uncertainties in a and b are considered in appendix 4.

Appendix 4

Standard uncertainties in mean, intercept and slope

A4.1 Standard uncertainty in the mean and weighted mean

If we can neglect the influence of systematic errors on values obtained through experiment, the best estimate of the true value of a quantity is the mean of values determined through repeat measurements. Using the values obtained through repeat measurements we are able to calculate the *standard error* of the mean. In this case the standard error of the mean *is* the standard uncertainty in the best estimate of the true value determined by a Type A evaluation (see section 5.7.2). For consistency with international guidelines for calculating and expressing uncertainty as discussed in chapter 5, we will refer to $s_{\bar{x}}$ in this appendix as the *standard uncertainty*.

In appendix 3 we showed that the weighted mean (that is, the mean when the standard deviation in x values is not constant) may be written as

$$\bar{x}_w = \frac{\sum \dfrac{x_i}{s_i^2}}{\sum \dfrac{1}{s_i^2}}, \tag{A4.1}$$

where x_i is the ith value, and s_i is the standard deviation in the ith value.

The 'unweighted' mean is written as usual as

$$\bar{x} = \frac{\sum x_i}{n}, \tag{A4.2}$$

where n is the number of values.

Now we consider the standard uncertainty in weighted and unweighted means. Assuming errors in values of x are independent, we write

$$s_{\bar{x}_w}^2 = \left(\frac{\partial \bar{x}_w}{\partial x_1} s_1\right)^2 + \left(\frac{\partial \bar{x}_w}{\partial x_2} s_2\right)^2 + \cdots \left(\frac{\partial \bar{x}_w}{\partial x_i} s_i\right)^2 + \cdots \left(\frac{\partial \bar{x}_w}{\partial x_n} s_n\right)^2 \tag{A4.3}$$

or

$$s_{\bar{x}_w}^2 = \sum \left(\frac{\partial \bar{x}_w}{\partial x_i} s_i \right)^2.$$ (A4.4)

From equation A4.1,

$$\frac{\partial \bar{x}_w}{\partial x_i} = \frac{\frac{1}{s_i^2}}{\sum \frac{1}{s_i^2}}.$$ (A4.5)

Substituting equation A4.5 into A4.4 gives

$$s_{\bar{x}_w}^2 = \sum \left(\frac{\frac{1}{s_i^2}}{\sum \frac{1}{s_i^2}} \right)^2 s_i^2 = \frac{\sum \frac{1}{s_i^2}}{\left(\sum \frac{1}{s_i^2} \right)^2} = \frac{1}{\sum \frac{1}{s_i^2}},$$

so that

$$s_{\bar{x}_w} = \left(\frac{1}{\sum \frac{1}{s_i^2}} \right)^{1/2}.$$ (A4.6)

Equation A4.6 gives the standard uncertainty for the *weighted* mean.

In situations in which $s_1 = s_2 = s_i = s$, then equation A4.6 reduces to

$$s_{\bar{x}} = \left(\frac{1}{\frac{1}{s^2} \sum 1} \right)^{1/2} = \left(\frac{s^2}{n} \right)^{1/2}$$

(note that $\sum 1 = n$), i.e.

$$s_{\bar{x}} = \frac{s}{\sqrt{n}}.$$ (A4.7)

If the weighted mean has been calculated by combining means (see appendix 3) where each mean, $\bar{x}_i$ has a standard uncertainty, $u_{\bar{x}_i}$, then equation A4.6 can be adapted to give the weighted standard uncertainty, i.e.

$$u_{\bar{x}_w} = \left(\frac{1}{\sum \frac{1}{u_{\bar{x}_i}^2}} \right)^{1/2}.$$ (A4.8)

A4.2 Standard uncertainty in intercept, *a*, and slope, *b*, for a straight line

The best line through linearly related data is written

$$y = a + bx.$$ (A4.9)

In order to estimate the standard uncertainties in *a* and *b*, we consider the propagation of uncertainties that arise due to the fact that each value of *y* has some uncertainty. The intercept, a, may be regarded as a function of the y values, i.e.

$a = f(y_1, y_2, y_3, y_4, y_5 \ldots y_n)$ where n is the number of data points.

Similarly, the slope, b, may be written

$b = f(y_1, y_2, y_3, y_4, y_5 \cdots y_n)$

(a and b can also be taken to be functions of x as well as y, but as we assume negligible error in x values, they do not contribute to the standard uncertainty in a and b).

The standard uncertainty in a is represented by s_a and is found from

$$s_a^2 = \sum s_i^2 \left(\frac{\partial a}{\partial y_i}\right)^2, \tag{A4.10}$$

where s_i is the standard deviation in the ith value.

Similarly, for s_b,

$$s_b^2 = \sum s_i^2 \left(\frac{\partial b}{\partial y_i}\right)^2. \tag{A4.11}$$

In an unweighted fit we take all the s_is to be the same and replace them by s, where[1]

$$s^2 = \frac{1}{n-2} \sum (\Delta y_i)^2 \tag{A4.12}$$

and n is the number of data points and $\Delta y_i = y_i - a - bx_i$

Now we must deal with the partial derivatives appearing in equations A4.10 and A4.11. For an unweighted fit, we have from section A3.2 in appendix 3,

$$a = \frac{\sum x_i^2 \sum y_i - \sum x_i \sum x_i y_i}{\Delta} \tag{A4.13}$$

$$b = \frac{n \sum x_i y_i - \sum x_i \sum y_i}{\Delta}, \tag{A4.14}$$

where

$$\Delta = n \sum x_i^2 - \left(\sum x_i\right)^2. \tag{A4.15}$$

Consider the jth value of y and its contribution to the uncertainty in a and b. We have

$$s_b^2 = \sum_{j=1}^{j=n} s_{b(y_j)}^2, \tag{A4.16}$$

---

[1] The sum of squares of the residuals is divided by $(n-2)$, where $(n-2)$ represents the number of degrees of freedom and is two less than the number of data points because a degree of freedom is 'lost' for every parameter that is estimated using the sample observations. Here there are two such parameter estimates, a and b.

where

$$s^2_{b(y_j)} = s^2 \left(\frac{\partial b}{\partial y_j} \right)^2 . \tag{A4.17}$$

A similar equation can be written for s^2_a (by replacing a by b in equations A4.16 and A4.17).

Return to equation A4.14 and differentiate[2] b with respect to y_j:

$$\frac{\partial b}{\partial y_j} = \frac{1}{\Delta} \left(nx_j - \sum x_i \right). \tag{A4.18}$$

Substituting into A4.17 gives

$$s^2_{b(y_j)} = \frac{s^2}{\Delta^2} \left(nx_j - \sum x_i \right)^2 . \tag{A4.19}$$

Substituting A4.18 into A4.16 gives

$$s^2_b = \frac{s^2}{\Delta^2} \left[\sum_{j=1}^{j=n} \left(n^2 x_j^2 - 2nx_j \sum x_i + \left(\sum x_i \right)^2 \right) \right] \tag{A4.20}$$

$$= \frac{s^2}{\Delta^2} \left[n^2 \sum x_i^2 - 2n \left(\sum x_i \right)^2 + n \left(\sum x_i \right)^2 \right] \tag{A4.21}$$

$$= \frac{s^2 n}{\Delta^2} \left[n \sum x_i^2 - \left(\sum x_i \right)^2 \right] . \tag{A4.22}$$

As $\Delta = n \sum x_i^2 - \left(\sum x_i \right)^2$, we have

$$s^2_b = \frac{s^2 n}{\Delta} \tag{A4.23}$$

or

$$s_b = \frac{s n^{1/2}}{\Delta^{1/2}} . \tag{A4.24}$$

Using the same arguments to find s^2_a,

$$\frac{\partial a}{\partial y_j} = \frac{1}{\Delta} \left(x_i^2 - x_j \sum x_i \right), \tag{A4.25}$$

so that

$$s^2_a = \frac{s^2 \sum x_i^2}{\Delta} \tag{A4.26}$$

---

[2] In order to be convinced that equation A4.18 is correct, return to equation A4.14 and write the summation out more fully, i.e.
$b = \frac{1}{\Delta} (n(x_1 y_1 + x_2 y_2 + x_3 y_3 \cdots) - (x_1 + x_2 + x_3 \cdots)(y_1 + y_2 + y_3 \cdots))$ then differentiate with respect to y_1, then y_2 and so on.

or

$$s_a = \frac{s\left(\sum x_i^2\right)^{1/2}}{\Delta^{1/2}}.$$ (A4.27)

If the standard deviations in the y values are not equal, we introduce s_i^2 explicitly into the equations for the standard uncertainty in a and b. s_a and s_b are now given by

$$s_a = \left(\frac{\sum \frac{x_i^2}{s_i^2}}{\Delta}\right)^{1/2}$$ (A4.28)

$$s_b = \left(\frac{\sum \frac{1}{s_i^2}}{\Delta}\right)^{1/2},$$ (A4.29)

where

$$\Delta = \sum \frac{1}{s_i^2} \sum \frac{x_i^2}{s_i^2} - \left(\sum \frac{x_i}{s_i^2}\right)^2.$$ (A4.30)

Appendix **5**

Introduction to matrices for least squares analysis

Applying least squares to the problem of finding the best straight line represented by the equation $y = a + bx$ through data creates two equations which must be solved for a and b (see equations A3.26 and A3.27 in appendix 3). The equations can be solved by the method of 'elimination and substitution' but this approach becomes increasingly cumbersome when equations to be fitted to data contain three or more parameters that must be estimated, such as

$$y = a + bx + cx \ln x + dx^2 \ln x. \tag{A5.1}$$

Fitting equation A5.1 to data using least squares creates four equations to be solved for a, b, c and d. The preferred method for dealing with fitting of equations to data where the equations consist of several parameters and/or independent variables is to use matrices. Matrices provide an efficient means of solving linear equations as well as offering compact and elegant notation. In this appendix we consider matrices and some of their basic properties, especially those useful in parameter estimation by linear least squares.[1]

A matrix consists of a rectangular array of elements. As examples,

$$\mathbf{A} = \begin{bmatrix} a_{11} & a_{12} & a_{13} \\ a_{21} & a_{22} & a_{23} \\ a_{31} & a_{32} & a_{33} \end{bmatrix}, \quad \mathbf{B} = \begin{bmatrix} b_1 \\ b_2 \\ b_3 \end{bmatrix}.$$

By convention, a matrix is represented by a bold symbol such as $\mathbf{A}$, and a_{11}, a_{12}, a_{13}, etc., are elements of the matrix $\mathbf{A}$.

If a matrix consists of more than one row and more than one column (such as matrix $\mathbf{A}$) it is customary to use a double subscript to identify a particular element within the matrix. A general element of a matrix is designated a_{ij}, where i refers to

---

[1] Kutner, Nachtsheim and Neter (2003) deal with the application of matrices to least squares.

the ith row and j refers to the jth column. Matrix **B** above consists of a single column of elements where each element can be identified unambiguously by a single subscript.

In general, a matrix consists of m rows and n columns and is usually referred to as a matrix of *dimension* $m \times n$ (note that the number of rows is specified first, then the number of columns). If $m = n$, as it does for matrix **A**, then the matrix is said to be 'square'. By contrast, **B** is a 3×1 matrix. A matrix consisting of a single column of elements is sometimes referred to as a column vector, or simply as a vector.

Many operations such as addition, subtraction and multiplication can be defined for matrices. For data analysis using least squares, it is often required to multiply two matrices together.

A5.1 Matrix multiplication

Consider matrices **A** and **B**, where

$$\mathbf{A} = \begin{bmatrix} 2 & 4 \\ 7 & 5 \end{bmatrix}, \mathbf{B} = \begin{bmatrix} 1 & 5 \\ 3 & 9 \end{bmatrix}.$$

If $\mathbf{AB} = \mathbf{P}$, where **A** can be written most generally, as $\mathbf{A} = \begin{bmatrix} a_{11} & a_{12} \\ a_{21} & a_{22} \end{bmatrix}$ and $\mathbf{B} = \begin{bmatrix} b_{11} & b_{12} \\ b_{21} & b_{22} \end{bmatrix}$ then the elements of **P**, $\begin{bmatrix} p_{11} & p_{12} \\ p_{21} & p_{22} \end{bmatrix}$, can be found by multiplying each row of **A**, by each column of **B**, in the following manner:

$$p_{11} = a_{11} \times b_{11} + a_{12} \times b_{21} = 2 \times 1 + 4 \times 3 = 14$$
$$p_{12} = a_{11} \times b_{12} + a_{12} \times b_{22} = 2 \times 5 + 4 \times 9 = 46$$
$$p_{21} = a_{21} \times b_{11} + a_{21} \times b_{21} = 7 \times 1 + 5 \times 3 = 22$$
$$p_{22} = a_{21} \times b_{12} + a_{22} \times b_{22} = 7 \times 5 + 5 \times 9 = 80$$

so that $\mathbf{P} = \begin{bmatrix} 14 & 46 \\ 22 & 80 \end{bmatrix}.$

If we reverse the order of **A** and **B**, we find that $\mathbf{BA} = \begin{bmatrix} 37 & 29 \\ 69 & 57 \end{bmatrix}$. In this example (and generally), $\mathbf{AB} \neq \mathbf{BA}$, so that the order in which the matrices are multiplied is important.

If matrix **A** consists of r_1 rows and c_1 columns and matrix **B** consists of r_2 rows and c_2 columns, then the product **AB** can only be formed if $c_1 = r_2$. If $c_1 = r_2$, then **AB** is a matrix with r_1 rows and c_2 columns.

product **AB** has dimension

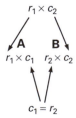

As an example,

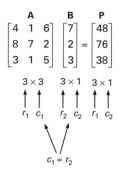

By contrast,

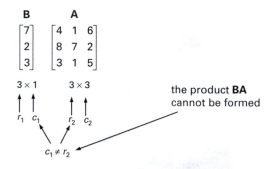

the product **BA** cannot be formed

A5.2 The identity matrix, I

Identity matrices play a role equivalent to the number 1 in conventional algebra (and are sometimes termed 'unit' matrices). When you multiply any number or variable by unity, the number or variable remains unchanged. Similarly, if you multiply a matrix, **A**, by its identity matrix, **I**, we find $\mathbf{AI} = \mathbf{A}$.

As an example, if we multiply the square matrix, $\mathbf{I} = \begin{bmatrix} 1 & 0 & 0 \\ 0 & 1 & 0 \\ 0 & 0 & 1 \end{bmatrix}$, by any other 3 × 3 matrix such as, $\mathbf{A} = \begin{bmatrix} 2 & 1 & -4 \\ 1 & 8 & 2 \\ 3 & 4 & 2 \end{bmatrix}$ we find that $\mathbf{IA} = \mathbf{AI} = \mathbf{A}$.

The matrix, $\begin{bmatrix} 1 & 0 & 0 \\ 0 & 1 & 0 \\ 0 & 0 & 1 \end{bmatrix}$ is the identity matrix of dimension 3.

A5.3 Inverse matrix, A⁻¹

If we multiply the matrix, $\mathbf{A} = \begin{bmatrix} 2 & 1 & -4 \\ 1 & 8 & 2 \\ 3 & 4 & 2 \end{bmatrix}$ by the matrix

$$\mathbf{A}^{-1} = \begin{bmatrix} 0.08 & -0.18 & 0.34 \\ 0.04 & 0.16 & -0.18 \\ -0.20 & -0.05 & 0.15 \end{bmatrix} \text{ we obtain the identity matrix, } \mathbf{I}. \text{ That is,}$$

$$\begin{bmatrix} 2 & 1 & -4 \\ 1 & 8 & 2 \\ 3 & 4 & 2 \end{bmatrix} \begin{bmatrix} 0.08 & -0.18 & 0.34 \\ 0.04 & 0.16 & -0.18 \\ -0.20 & -0.05 & 0.15 \end{bmatrix} = \begin{bmatrix} 1 & 0 & 0 \\ 0 & 1 & 0 \\ 0 & 0 & 1 \end{bmatrix},$$

where $\mathbf{A}^{-1}$ is referred to as the inverse matrix and in general,

$$\mathbf{A}\mathbf{A}^{-1} = \mathbf{A}^{-1}\mathbf{A} = \mathbf{I}.$$

It is not always possible to determine an inverse matrix, for example if two columns in a matrix have equal elements, such as,

$$\mathbf{A} = \begin{bmatrix} -4 & 1 & -4 \\ 2 & 8 & 2 \\ 2 & 4 & 2 \end{bmatrix},$$

then it is not possible to determine $\mathbf{A}^{-1}$ and $\mathbf{A}$ is said to be 'singular'. If $\mathbf{A}^{-1}$ can be found then $\mathbf{A}$ is said to be 'non-singular'. Matrix inversion is a challenging operation to perform 'by hand' even for small matrices. A computer package with matrix manipulation routines is almost mandatory if matrices larger than 3×3 are to be inverted.

A5.4 Using the inverse matrix to solve for parameter estimates in least squares

In chapters 6 and 7 we discovered that the application of the least squares technique leads to two or more simultaneous equations that must be solved to find best estimates for the parameters that appear in an equation that is to be fitted to data. The equations can be written in matrix form (see equation 7.11) as

$$\mathbf{AB} = \mathbf{P}, \tag{A5.2}$$

where it is the elements of matrix $\mathbf{B}$ which are the best estimates of the parameters. To isolate these elements we multiply both sides of equation A5.2 by $\mathbf{A}^{-1}$. This gives

$$\mathbf{A}^{-1}\mathbf{AB} = \mathbf{A}^{-1}\mathbf{P}, \tag{A5.3}$$

but $\mathbf{A}^{-1}\mathbf{A} = \mathbf{I}$, so equation A5.3 becomes,

$$\mathbf{IB} = \mathbf{A}^{-1}\mathbf{P}. \tag{A5.4}$$

Now $\mathbf{IB} = \mathbf{B}$, so we have

$$\mathbf{B} = \mathbf{A}^{-1}\mathbf{P}. \tag{A5.5}$$

Example

Suppose that after applying the method of least squares to experimental data, we obtain the following equations, which must be solved for a, b, c and d:

$$1.75a + 18.3b + 42.8c - 25.9d = 49.3$$
$$3.26a - 19.8b + 17.4c - 32.2d = 65.3$$
$$18.6a + 14.7b + 12.2c + 14.3d = -18.1$$
$$65.7a - 15.3b - 18.9c + 25.3d = 19.1.$$

We can write this in matrix form as

$$
\overset{\mathbf{A}}{
\begin{bmatrix}
1.75 & 18.3 & 42.8 & -25.9 \\
3.26 & -19.8 & 17.4 & -32.2 \\
18.6 & 14.7 & 12.2 & 14.3 \\
65.7 & -15.3 & -18.9 & 25.3
\end{bmatrix}}
\overset{\mathbf{B}}{
\begin{bmatrix}
a \\ b \\ c \\ d
\end{bmatrix}}
=
\overset{\mathbf{P}}{
\begin{bmatrix}
49.3 \\ 65.3 \\ -18.1 \\ 19.1
\end{bmatrix}}.
$$

The elements of $\mathbf{B}$ are found from $\mathbf{B} = \mathbf{A}^{-1}\,\mathbf{P}$. An efficient way to obtain $\mathbf{A}^{-1}$ is to use the MINVERSE() function in Excel as described in section 7.4.1. Using this function we find

$$
\mathbf{A}^{-1} =
\begin{bmatrix}
0.028909 & -0.01895 & -0.03514 & 0.025339 \\
0.072522 & -0.07832 & -0.0971 & 0.029445 \\
-0.05058 & 0.064565 & 0.114316 & -0.03422 \\
-0.069 & 0.050075 & 0.117925 & -0.03403
\end{bmatrix}
$$

$$
\mathbf{B} = \mathbf{A}^{-1}\mathbf{P} =
\begin{bmatrix}
1.307\,834 \\
0.780\,933 \\
-1.000\,25 \\
-2.916\,25
\end{bmatrix}.
$$

It follows that, to four significant figures,

$a = 1.308$, $b = 0.7809$, $c = -1.000$ and $d = -2.916$.

As a final step, it is useful to verify that $\mathbf{AB} = \mathbf{P}$.

Appendix 6

Useful formulae

If events are mutually exclusive	$P(\text{A or B}) = P(\text{A}) + P(\text{B})$
If events are independent	$P(\text{A and B}) = P(\text{A}) \times P(\text{B})$
Standard normal variable, z	$z = \frac{x - \mu}{\sigma}$
t variable	$t = \left(\frac{\bar{x} - \mu}{s/\sqrt{n}} \right)$
Mean	$\bar{x} = \frac{\sum x_i}{n}$
Weighted mean	$\bar{x}_w = \dfrac{\sum \frac{x_i}{s_i^2}}{\sum \frac{1}{s_i^2}}$
Standard deviation	$s = \left(\dfrac{\sum (x_i - \bar{x})^2}{n-1} \right)^{1/2}$
Standard error of mean	$s_{\bar{x}} = \dfrac{s}{\sqrt{n}}$
Approximation for $s_{\bar{x}}$, valid for $n < 12$	$s_{\bar{x}} \approx \dfrac{\text{range}}{n}$
Standard uncertainty in weighted mean	$u_{\bar{x}_w} = \left(\dfrac{1}{\sum \frac{1}{u_{x_1}^2}} \right)^{1/2}$
Level of significance, α, for $X\%$ confidence level	$\alpha = \dfrac{(100\% - X\%)}{100\%}$
Relative standard uncertainty	$\dfrac{u}{|\bar{x}|}$

Standard uncertainty
expressed as a percentage
$$\frac{u}{|\bar{x}|} \times 100\%$$

Uncertainty in y (y is function of a only)
$$u(y) = \left|\frac{df}{da}\right| u(a)$$

Combined standard uncertainty in y, when errors in a, b and c are not correlated

$$u_c(y) = \left(\left(\frac{\partial f}{\partial a} u(a)\right)^2 + \left(\frac{\partial f}{\partial b} u(b)\right)^2 + \left(\frac{\partial f}{\partial c} u(c)\right)^2 \right)^{1/2}$$

Welch–Satterthwaite formula
$$\nu_{\text{eff}} = \frac{u_c^4(y)}{\displaystyle\sum_{i=1}^{N} \frac{u_i^4(y)}{\nu_i}}$$

Equation of a straight line
$$y = a + bx$$

Residual
$$\Delta y_i = y_i - \hat{y}_i$$

Standardised residual
$$\Delta y_{is} = \frac{\Delta y_i}{s_i}$$

Unweighted sum of squares of residuals,
$$SSR = \sum (y_i - \hat{y}_i)^2$$

Weighted sum of squares of residuals
$$\chi^2 = \sum \left[\frac{y_i - \hat{y}_i}{s_i}\right]^2$$

Slope (unweighted least squares)
$$b = \frac{n \sum x_i y_i - \sum x_i \sum y_i}{n \sum x_i^2 - \left(\sum x_i\right)^2}$$

Intercept (unweighted least squares)
$$a = \frac{\sum x_1^2 \sum y_i - \sum x_i \sum x_i y_i}{n \sum x_i^2 - \left(\sum x_i\right)^2}$$

Standard deviation in y values (unweighted least squares)
$$s = \left[\frac{1}{n-2} \sum (y_i - \hat{y}_i)^2\right]^{1/2}$$

Standard uncertainty in slope (unweighted least squares)
$$s_b = \frac{s n^{1/2}}{\left[\sum x_i^2 - \left(\sum x_i\right)^2\right]^{1/2}}$$

Standard uncertainty in intercept (unweighted least squares)
$$s_a = \frac{s \left(\sum x_i^2\right)^{1/2}}{\left[\sum x_i^2 - \left(\sum x_i\right)^2\right]^{1/2}}$$

Unweighted correlation coefficient

$$r = \frac{n \sum x_i y_i - \sum x_i \sum y_i}{\left[n \sum x_i^2 - \left(\sum x_i\right)^2\right]^{1/2} \left[n \sum y_i^2 - \left(\sum y_i\right)^2\right]^{1/2}}$$

Estimate of y for $x = x_0$

$$\hat{y}_0 = \bar{y} + b(x_0 - \bar{x})$$

Standard uncertainty in estimate of y for $x = x_0$

$$s_{\hat{y}_0} = s \left(\frac{1}{n} + \frac{n(x_0 - \bar{x})^2}{n \sum x_i^2 - \left(\sum x_i\right)^2}\right)^{1/2}$$

Estimate of x for mean y, $\bar{y}_0$

$$\hat{x}_0 = \frac{\bar{y}_0 - a}{b}$$

Standard uncertainty in estimate of x for mean y, $\bar{y}_0$

$$s_{\hat{x}_0} = \frac{s}{b} \left[\frac{1}{m} + \frac{1}{n} + \frac{n(\bar{y}_0 - \bar{y})^2}{b^2 \left(n \sum x_i^2 - \left(\sum x_i\right)^2\right)}\right]^{1/2}$$

Slope (weighted least squares)

$$b = \frac{\sum \frac{1}{s_i^2} \sum \frac{x_i y_i}{s_i^2} - \sum \frac{x_i}{s_i^2} \sum \frac{y_i}{s_i^2}}{\Delta}$$

Intercept (weighted least squares)

$$a = \frac{\sum \frac{x_i^2}{s_i^2} \sum \frac{y_i}{s_i^2} - \sum \frac{x_i}{s_i^2} \sum \frac{x_i y_i}{s_i^2}}{\Delta}$$

Denominator for weighted least squares equations

$$\Delta = \sum \frac{1}{s_i^2} \sum \frac{x_i^2}{s_i^2} - \left(\sum \frac{x_i}{s_i^2}\right)^2$$

Standard uncertainty in intercept (weighted least squares, absolute s_i known)

$$s_a = \left[\frac{\sum \frac{x_i^2}{s_i^2} \sum \frac{x_i^2}{s_i^2}}{n\Delta}\right]^{1/2}$$

Standard uncertainty in slope (weighted least squares, absolute s_i known)

$$s_b = \left(\frac{\sum \frac{1}{s_i^2}}{\Delta}\right)^{1/2}$$

Standard uncertainty in intercept (weighted least squares, relative s_i known)

$$s_b = \left(\frac{\sum \frac{1}{s_i^2}}{\Delta}\right)^{1/2}$$

Standard uncertainty in slope (weighted least squares, relative s_i known)

$$s_b = \frac{s_w \sum \frac{1}{s_i^2}}{(n\Delta)^{1/2}}$$

Standard deviation in y values (weighted least squares)

$$s_w = \frac{\left(\dfrac{n}{n-2}\right)^{1/2}}{\sum \dfrac{1}{s_i^2}} \left[\sum \frac{1}{s_i^2} \sum \frac{y_i^2}{s_i^2} - \left(\sum \frac{y_i}{s_i^2} \right)^2 - \frac{\left(\sum \dfrac{1}{s_i^2} \sum \dfrac{x_i y_i}{s_i^2} - \sum \dfrac{x_i}{s_i^2} \sum \dfrac{y_i}{s_i^2} \right)^2}{\Delta} \right]^{1/2}$$

Weighted correlation coefficient

$$r_w = \frac{\sum \dfrac{1}{s_i^2} \sum \dfrac{x_i y_i}{s_i^2} - \sum \dfrac{x_i}{s_i^2} \sum \dfrac{y_i}{s_i^2}}{\left[\sum \dfrac{1}{s_i^2} \sum \dfrac{x_i^2}{s_i^2} - \left(\sum \dfrac{x_i}{s_i^2} \right)^2 \right]^{1/2} \left[\sum \dfrac{1}{s_i^2} \sum \dfrac{y_i^2}{s_i^2} - \left(\sum \dfrac{y_i}{s_i^2} \right)^2 \right]^{1/2}}$$

Adjusted coefficient of multiple determination

$$R_{ADJ}^2 = \frac{(n-1)^2 - (M-1)}{n-M}$$

Standard uncertainty in slope (weighted least squares, relative s_i known)

$$s_b = \frac{s_w \sum \dfrac{1}{s_i^2}}{(n\Delta)^{1/2}}$$

Akaike's information criterion

$$AIC = n \ln\left(\frac{SSR}{n}\right) + 2K$$

Corrected Akaike's information criterion

$$AIC_c = AIC + \frac{2K(K+1)}{n-K-1}$$

Answers to exercises and end of chapter problems

Chapter 1

Exercises

A $kg \cdot m^2 \cdot s^{-2} \cdot A^{-2}$.

B (1) (i) 13.8 zJ, (ii) 0.36 µs, (iii) 43.258 kW, (iv) 780 Mm/s.
(2) (i) 6.50×10^{-10} m, (ii) 3.7×10^{-11} C, (iii) 1.915×10^6 W, (iv) 1.25×10^{-4} s.

C (1) (i) three, (ii) three, (iii) two, (iv) four, (v) one, (vi) four, (vii) four, (vi) three.
(2)

Part	2 significant figures	3 significant figures	4 significant figures
(i)	7.8×10^5 m/s^2	7.76×10^5 m/s^2	7.757×10^5 m/s^2
(ii)	1.3×10^{-3} s	1.27×10^{-3} s	1.266×10^{-3} s
(iii)	-1.1×10^2 °C	-1.05×10^2 °C	-1.054×10^2 °C
(iv)	1.4×10^{-5} H	1.40×10^{-5} H	1.400×10^{-5} H
(v)	1.2×10^4 J	1.24×10^4 J	1.240×10^4 J
(vi)	1.0×10^{-7} m	1.02×10^{-7} m	1.016×10^{-7} m

D (i) Using the guidelines in section 1.5.1, the number of intervals, $N = \sqrt{n} = \sqrt{52} \approx 7$. Dividing range (which is 0.61 g) by 7 and rounding up, gives an interval width of 0.1 g. Now we can construct a grouped frequency distribution, as follows.

Interval (g)	Frequency
$49.8 < x \le 49.9$	4
$49.9 < x \le 50.0$	3
$50.0 < x \le 50.1$	22
$50.1 < x \le 50.2$	18
$50.2 < x \le 50.3$	4
$50.3 < x \le 50.4$	0
$50.4 < x \le 50.5$	1

E   Graph with semi-logarithmic scales is most appropriate.

F $\bar{x} = 102.04$ pF, median = 101.25 pF.

G   (1) Expanding equation 1.10 gives:

$$\sigma = \left[\frac{\sum x_i^2 - 2\bar{x}\sum x_i + \sum (\bar{x})^2}{n} \right]^{1/2}.$$

Now $\sum x_i = n\bar{x}$, and $\sum (\bar{x})^2 = n(\bar{x})^2$, so that

$$\sigma = \left[\frac{\sum x_i^2}{n} - \frac{2n(\bar{x})^2}{n} + \frac{n(\bar{x})^2}{n} \right]^{1/2}$$

hence equation 1.11 follows.
(2) (i) range = 0.22 cm, (ii) $\bar{x} = 4.163$ cm, (iii) median = 4.15 cm, (iv) variance = 3.9×10^{-3} cm^2, (v) standard deviation = 0.063 cm.

H $\bar{x} = 2.187$ s, standard deviation, $s = 0.75$ s.

I   (i) $s = 0.052$ s (using equation 1.16).
(ii) $s = 0.047$ s (using equation 1.19).
(iii)   Percentage difference $\approx 10\%$.

End of chapter problems

(1) (i) J/(kg·K), (ii) Pa or N/m^2, (iii) W/(m·K), (iv) N·s, (v) Ω·m, (vi) Wb, (vii) W/m^2.
(2) (i) m^2·s^{-2}·K^{-1}, (ii) m^{-1}·kg·s^{-2}, (iii) m·kg·s^{-3}·K^{-1}, (iv) m·kg·s^{-1}, (v) m^3·kg·s^{-3}·A^{-2}, (vi) m^2·kg·s^{-2}·A^{-1}, (vii) kg·s^{-3}.

(3) (i) K^{-1}, (ii) $kg^{-1} \cdot m^{-3} \cdot s^4 \cdot A^2$.

(4) (i) 5.7×10^{-5} s, (ii) 1.4×10^4 K, (iii) 1.4×10^3 m/s, (iv) 1.0×10^5 Pa, (v) 1.5×10^{-3} Ω.

(5) (ii) median lead content = 51.5 ppb.

(6) (ii) median retention time = 6.116 s, (iii) standard deviation, s = 0.081s.

(7) (ii) When the resistance is 5000 Ω, the humidity is approximately 65%.

(8) (i) $\bar{x}$ = 0.4915 μmol/mL, standard deviation, s = 0.019 μmol/mL.

(9) (i) mean = 120.4 m, median = 120 m, (ii) range = 32 m, (iii) s = 9.5 m, s^2 = 90 m^2.

(10) mean mass = 0.5481 g, $s^2 = 5.2 \times 10^{-6}$ g^2.

Chapter 2

In the interests of brevity, answers to exercises and end of chapter problems for this chapter show only relevant extracts from a Worksheet.

Exercises

A   (1)

t(s)	V(volts)	I(amps)	Q(coulombs)
0	3.98	3.32E-07	1.8706E-06
5	1.58	1.32E-07	7.426E-07
10	0.61	5.08E-08	2.867E-07
15	0.24	2E-08	1.128E-07
20	0.094	7.83E-09	4.418E-08
25	0.035	2.92E-09	1.645E-08
30	0.016	1.33E-09	7.52E-09
35	0.0063	5.25E-10	2.961E-09
40	0.0031	2.58E-10	1.457E-09
45	0.0017	1.42E-10	7.99E-10
50	0.0011	9.17E-11	5.17E-10
55	0.0007	5.83E-11	3.29E-10
60	0.0006	5E-11	2.82E-10

(2)

t(s)	V(volts)	I(amps)	Q(coulombs)	$I^{0.5}$(amps)$^{0.5}$
0	3.98	3.32E-07	1.87E-06	0.000576
5	1.58	1.32E-07	7.43E-07	0.000363
10	0.61	5.08E-08	2.87E-07	0.000225
15	0.24	2E-08	1.13E-07	0.000141
20	0.094	7.83E-09	4.42E-08	8.85E-05
25	0.035	2.92E-09	1.65E-08	5.4E-05
30	0.016	1.33E-09	7.52E-09	3.65E-05
35	0.0063	5.25E-10	2.96E-09	2.29E-05
40	0.0031	2.58E-10	1.46E-09	1.61E-05
45	0.0017	1.42E-10	7.99E-10	1.19E-05
50	0.0011	9.17E-11	5.17E-10	9.57E-06
55	0.0007	5.83E-11	3.29E-10	7.64E-06
60	0.0006	5E-11	2.82E-10	7.07E-06

(3)

V (V)	E (J)
0.2	1.1
0.4	4.4
0.6	9.9
0.8	17.6
1	27.5
1.2	39.6
1.4	53.9
1.6	70.4
1.8	89.1
2	110
2.2	133.1
2.4	158.4
2.6	185.9
2.8	215.6

B

20	6.5	130
30	7.2	216
40	8.5	340

C   (1) When $g = 9.81$,   (2) When $g = 1.6$,
     column B reads: column B reads:

t (s)		
0.638551		
0.903047		
1.106003		
1.277102		
1.427843		

t (s)		
1.581139		
2.236068		
2.738613		
3.162278		
3.535534		

D

T (K)	H (J/s)			
1000	3515.4		SB (W/(m^2·K^4))	5.67E-08
2000	56246.4		A (m^2)	0.062
3000	284747.4			
4000	899942.4			
5000	2197125			
6000	4555958			

E (1)

r_m (m)	6.35E-03
r_n (m)	6.72E-03
m	52
n	86
λ(m)	6.02E-07
$R_{numerator}$	4.84E-06
$R_{denominator}$	2.05E-05
R (m)	2.36E-01

(2)

d (m)	v_d (m/s)		
0.1	339.3848	v (m/s)	344
0.2	341.6924	f (Hz)	5
0.3	342.4616		
0.4	342.8462		
0.5	343.077		
0.6	343.2308		
0.7	343.3407		
0.8	343.4231		
0.9	343.4872		
1	343.5385		

F (1)

t(s)	V(volts)	I(amps)	ln(I)	LOG10(I)
0	3.98	3.32E-07	−14.9191	−6.4793
5	1.58	1.32E-07	−15.843	−6.88052
10	0.61	5.08E-08	−16.7947	−7.29385
15	0.24	2E-08	−17.7275	−7.69897
20	0.094	7.83E-09	−18.6649	−8.10605
25	0.035	2.92E-09	−19.6528	−8.53511
30	0.016	1.33E-09	−20.4356	−8.87506
35	0.0063	5.25E-10	−21.3676	−9.27984
40	0.0031	2.58E-10	−22.0768	−9.58782
45	0.0017	1.42E-10	−22.6775	−9.84873
50	0.0011	9.17E-11	−23.1129	−10.0378
55	0.0007	5.83E-11	−23.5648	−10.2341
60	0.0006	5E-11	−23.719	−10.301

(2)

h (m)	P (Pa)				
1.00E+03	8.92E+04			T (K)	273
2.00E+03	7.88E+04			Po (Pa)	1.01E+05
3.00E+03	6.96E+04				
4.00E+03	6.15E+04				
5.00E+03	5.43E+04				
6.00E+03	4.79E+04				
7.00E+03	4.23E+04				
8.00E+03	3.74E+04				
9.00E+03	3.30E+04				

(3)

x(m)	y(m)
-100	270.3
-80	262.9
-60	257.2
-40	253.2
-20	250.8
0	250.0
20	250.8
40	253.2
60	257.2
80	262.9
100	270.3

G

x	0	0.1	0.2	0.3	0.4	0.5	0.6	0.7	0.8	0.9	1
ASIN(x)	0	5.73917	11.53696	17.4576	23.57818	30	36.8699	44.427	53.1301	64.15807	90
ACOS(x)	90	84.26083	78.46304	72.5424	66.42182	60	53.1301	45.573	36.8699	25.84193	0
ATAN(x)	0	5.710593	11.30993	16.69924	21.80141	26.56505	30.96376	34.99202	38.65981	41.98721	45

H

	x	y	xy	x^2
	1.75	23	40.25	3.0625
	3.56	34	121.04	12.6736
	5.56	42	233.52	30.9136
	5.85	42	245.7	34.2225
	8.76	65	569.4	76.7376
	9.77	87	849.99	95.4529
sums	35.25	293	2059.9	253.0627

I

max	min	range
98	12	86

J

mean	median	mode
23.60417	25	27

K

Mean	6.74
Harmonic Mean	6.726957
Average Deviation	0.2376
Standard Deviation	0.29966

L   (2)

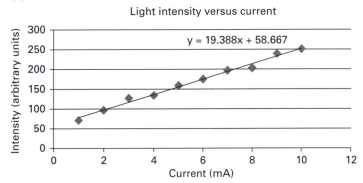

Light intensity versus current

$y = 19.388x + 58.667$

M   (3)

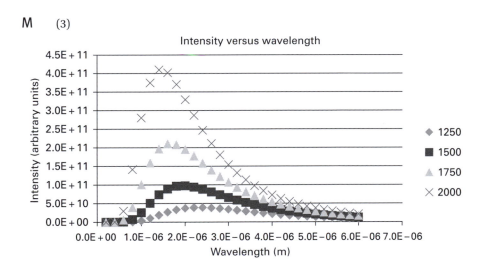

Intensity versus wavelength

End of chapter problems

(1) Note that each term appearing within the square root in equation 2.13 is evaluated in a separate column of the spreadsheet.

λ (m)	1st_term	2nd_term	v (m/s)			
0.01	0.015613	4.59E-02	0.247952		γ(N/m)	7.30E-02
0.02	0.031226	2.29E-02	0.232723		ρ(kg/m³)	1.00E+03
0.03	0.046839	1.53E-02	0.249256		g (m/s²)	9.81
0.04	0.062452	1.15E-02	0.271881			
0.05	0.078065	9.17E-03	0.295362			
0.06	0.093679	7.64E-03	0.318313			
0.07	0.109292	6.55E-03	0.340359			
0.08	0.124905	5.73E-03	0.361439			
0.09	0.140518	5.10E-03	0.381594			

(2) Parentheses required in the denominator of the formula in cell B2.

v (m/s)	m (kg)		
2.90E+08	1.39E-29	mo	9.11E-31
2.91E+08	1.54E-29	c_	3.00E+08
2.92E+08	1.73E-29		
2.93E+08	1.98E-29		
2.94E+08	2.30E-29		
2.95E+08	2.76E-29		
2.96E+08	3.44E-29		
2.97E+08	4.58E-29		
2.98E+08	6.86E-29		
2.99E+08	1.37E-28		

(3)   (i)

mean (%)	66.19
standard deviation (%)	5.482193
maximum (%)	79
minimum (%)	54
range (%)	25

(4)   (ii)

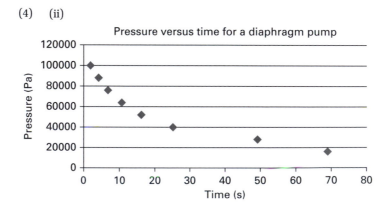

(5)

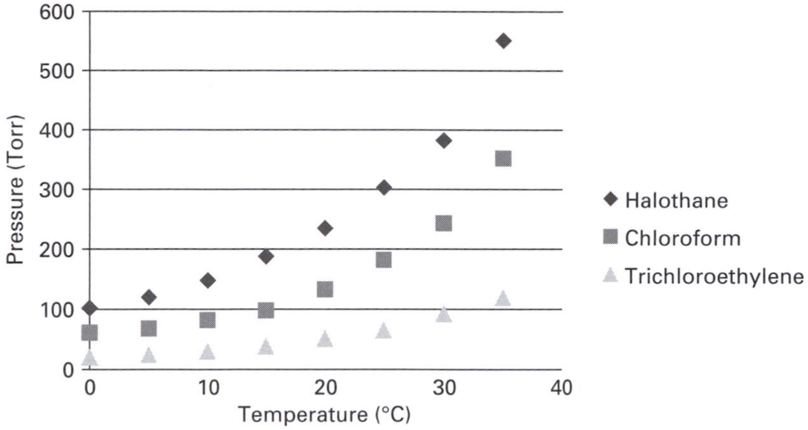

(6) (i) and (ii)

	$\frac{\rho}{\rho_P}$				
H (%)	f = 4	f = 3	f = 2.5	f = 2	f = 1.5
0	1	1	1	1	1
5	1.210526	1.157895	1.131579	1.105263	1.078947
10	1.444444	1.333333	1.277778	1.222222	1.166667
15	1.705882	1.529412	1.441176	1.352941	1.264706
20	2	1.75	1.625	1.5	1.375
25	2.333333	2	1.833333	1.666667	1.5
30	2.714286	2.285714	2.071429	1.857143	1.642857
35	3.153846	2.615385	2.346154	2.076923	1.807692
40	3.666667	3	2.666667	2.333333	2
45	4.272727	3.454545	3.045455	2.636364	2.227273
50	5	4	3.5	3	2.5
55	5.888889	4.666667	4.055556	3.444444	2.833333
60	7	5.5	4.75	4	3.25
65	8.428571	6.571429	5.642857	4.714286	3.785714
70	10.33333	8	6.833333	5.666667	4.5
75	13	10	8.5	7	5.5
80	17	13	11	9	7

(iii)

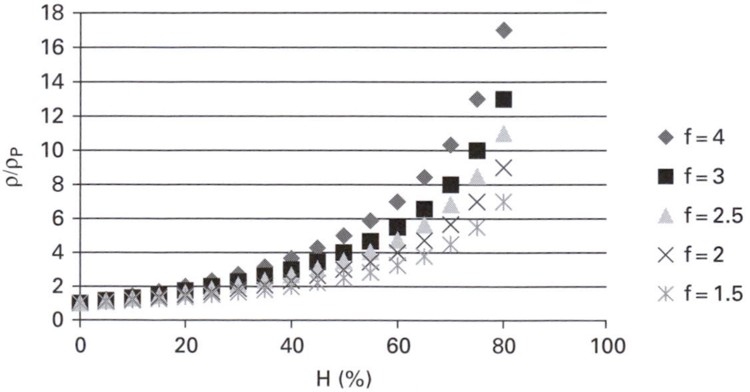

(7)

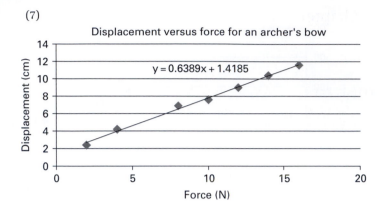

Displacement versus force for an archer's bow

$y = 0.6389x + 1.4185$

Displacement (cm) vs Force (N)

(8)

R(Ω)	V (V)	I = V/R (A)
900	7.75	0.008611
800	7.73	0.009663
700	7.71	0.011014
600	7.68	0.0128
500	7.64	0.01528
400	7.58	0.01895
300	7.48	0.024933

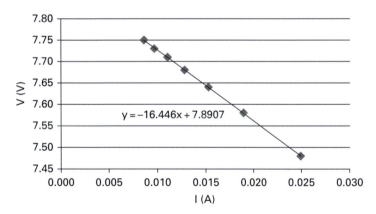

$y = -16.446x + 7.8907$

V (V) vs I (A)

(9)

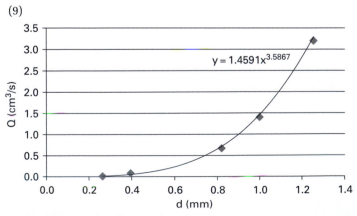

$y = 1.4591x^{3.5867}$

Q (cm³/s) vs d (mm)

Chapter 3

Exercises

A (1) (a) (ii) $A = \frac{1}{8}$, (iii) 0.4375, (iii) 4375.

(b) (ii) $A = \frac{3}{128}$, (iii) 0.2891, (iii) 2891.

(2) (ii) 0.4512, (iii) 0.5488.

B

x(g)	7.2	7.4	7.6	7.8	8.0	8.2	8.4	8.6	8.8	9.0
cdf	0.0000	0.0002	0.0062	0.0668	0.3085	0.6915	0.9332	0.9938	0.9998	1.0000

C 0.0476.

D

z	0.0	0.1	0.2	0.3	0.4	0.5	0.6	0.7	0.8	0.9	1.0
cdf	0.5000	0.5398	0.5793	0.6179	0.6554	0.6915	0.7257	0.7580	0.7881	0.8159	0.8413

E (1) (i) 0.2192, (ii) 7.9 (round to 8).

(2) (i) 0.02275, (ii) 0.9545, (iii) 0.0214, (iv) 0.02275.

(3) 0.01242.

(4) (i) 0.1587, (ii) 0.1359, (iii) 0.00135, (iv) 13.5 (round to 14).

F $\bar{x} = 18.41$ °C. The standard deviation, s, = 0.6283 °C (found using equation 1.16). We take this as a good estimate of σ.

Interval containing 50% of the data: 17.99 °C to 18.83 °C.
Interval containing 68% of the data: 17.79 °C to 19.03 °C.
Interval containing 90% of the data: 17.38 °C to 19.44 °C.
Interval containing 95% of the data: 17.18 °C to 19.64 °C.
Interval containing 99% of the data: 16.79 °C to 20.03 °C.

G 0.8416.

H 99% confidence interval: 893 kg/m³ to 917 kg/m³.

I Using equation 3.21, $s_{\bar{x}} = 0.014$ eV. Using equation 3.27, $s_{\bar{x}} = 0.014$ eV.

J 90% confidence interval: 4.64 mm to 4.72 mm.

K (i)

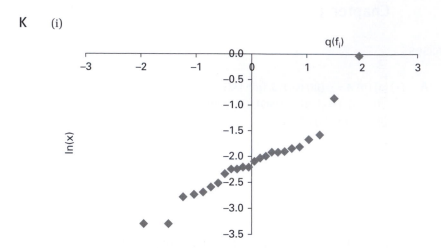

(ii) Comparing $\ln(x)$ versus $q(f_i)$ with the graph given in figure 3.26 indicates that the normality has been improved by the logarithmic transformation.

L 5

End of chapter problems

(1) (ii) $A = \frac{3}{2}$.

(iii) (a) 0.145, (b) 0.055.

(4) mean = 916.7 MPa, standard deviation = 20 MPa.

(5) $\bar{x} = 0.64086$ V, $s = 0.0027$ V. Number of diodes with voltage in excess of 0.6400 V would be 125.

(6) (i) (b)

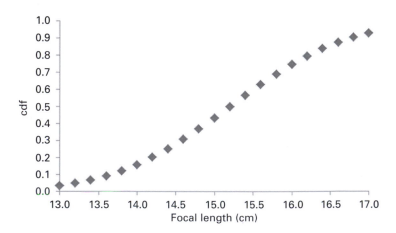

(ii) (a) 0.4004, (b) 0.1857.

(7)

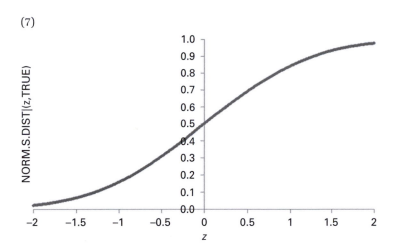

(8) 90% confidence interval: 2022 Hz to 2190 Hz.

(9) 95% confidence interval: 131.5 µPa to 144.5 µPa.

(10) (i) 0.147584, (ii) 0.02275, (iv) when $v = 61$.

(11) (i) 4.893 cm^3, 0.1670 cm^3. (ii) 4.743 cm^3 to 5.043 cm^3. (iii) Using equation 3.33, $\sigma_s \approx 0.71 \dfrac{0.1670}{\sqrt{12}} = 0.034$ cm^3.

(12) (i) Graph of x_i versus $q(f_i)$ is not linear suggesting data are not normally distributed.

(13) (iii) Normal quantile plot indicates that distribution is consistent with a log-normal distribution.

Chapter 4

Exercises

A   (i) $P(r = 0) = 0.8904$.
(ii) $P(r =1) = 0.1035$.
(iii) $P(r >1) = 0.0061$.

B   (i) $P(r = 20) = 0.09306$.
(ii) $P(r \le 20) = 0.7622$.
(iii) $P(r < 20) = 0.6692$.
(iv) $P(r > 20) = 0.2378$.
(v) $P(r \ge 20) = 0.3308$.

C   (i) $\mu = 8$.
(ii) $\sigma = 2.8$.
(iii) $P(r \ge 2) = 0.9972$.

D   (i)   Using binomial distribution, $P(290 \leq r \leq 320) = 0.6857$. Using Normal approximation, $P(289.5 \leq x \leq 320.5) = 0.6871$.

(ii)   Using binomial distribution, $P(r > 320) = 0.07923$. Using Normal approximation, $P(x > 320.5) = 0.07859$.

E   (i) $P(r = 0) = 0.6065$.
(ii) $P(r \leq 3) = 0.9982$.
(iii) $P(2 \leq r \leq 4) = 0.0900$.

F   (i) $P(r = 0) = 0.3535$.
(ii) $P(r = 1) = 0.3676$.
(iii) $P(r > 2) = 0.0878$.

G   (i) $P(r = 0) = 0.2209$.
(ii) $P(r = 1) = 0.3336$.
(iii) $P(r = 3) = 0.1267$.
(iv) $P(2 \leq r \leq 4) = 0.4265$.
(v) $P(r > 6) = 0.000962$.

H   (i)   (Using normal approximation to the Poisson distribution), $P(179.5 \leq r \leq 230.5) = 0.9109$.

End of chapter problems

(1) $C_{10,5} = 252$, $C_{15,2} = 105$, $C_{42,24} = 3.537 \times 10^{11}$, $C_{580,290} = 1.311 \times 10^{173}$.

(2) (i) $P(r = 1) = 0.0872$. (ii) $P(r > 1) = 0.004323$.

Number of screens expected to have one or more faulty transistors = 22.

(3) (i)   Number with three functioning ammeters = 18 (rounded).
(ii) Number with two functioning ammeters = 6 (rounded).
(iii) Number with less than two functioning ammeters = 1 (rounded).

(4) (i) $P(r \geq 20) = 0.8835$.
(ii) $n = 32$.

(5) (i)   Estimate of population mean = 123 counts.
(ii)   Estimate of population standard deviation = 11 counts.

(6) (i) 1.3167 flaws per metre.

(ii)

r	N(rounded)
0	16
1	21
2	14
3	6
4	2
5	1

(7) (i) 1.609.

(ii) $P(r > 4) = 0.02421$.

(iii) $P(r \leq 2) = 0.7809$.

(8) 0.9817.

Chapter 5

Exercises

A $R_{int} = 267\,M\Omega$ (beware of premature rounding in this problem).

B   17.27 s.

C   (i) $6153\,kg/m^3$, $26\,kg/m^3$.

(ii) $33.3\,N/m$, $1.4\,N/m$.

(iii) $0.986\,N{\cdot}s{\cdot}m^{-2}$, $0.055\,N{\cdot}s{\cdot}m^{-2}$.

D   (i) $0.2543\,cm^3/s$.

(ii) $0.003302\,cm^3/s$.

(iii) $0.0010\,cm^3/s$.

(iv) 9.

E   (i) $6497\,kg/m^3$.

(ii) $u = 14\,kg/m^3$, assuming a rectangular distribution is appropriate.

F   0.029.

G   (i) 100.063 g.

(ii) 0.0045 g.

H   (1) $f = 1.810 \times 10^3$, $u(f) = 0.039 \times 10^3$ Hz.
     (2) $A = 55.4$ mm^2, $u(A) = 7.9$ mm^2.
     (3) $k = -1.363$, $u(k) = 0.051$.

I   (1) $n = 1.467$, $u(n) = 0.064$.
     (2)   (i) $f = 104.6$ mm, $u(f) = 1.0$ mm, (ii) $m = 5.005$, $u(m) = 0.061$.
     (3)   (i) $A = 4.524 \times 10^{-8}$ m^2, (ii) $u(A) = 5.7 \times 10^{-9}$ m^2, (iii) 1.944×10^{-8} Ω·m, (iv) 2.6×10^{-9} Ω·m.
     (4) $\mu_s = 0.76$, $u(\mu_s) = 0.11$.

J   7.05325 g, 0.00038 g.

K   0.00033.

L   73 °C, 0.05.

M   (i)   Mean = 221.8 s, standard deviation, $s = 7.9$ s.
     (ii) Value furthest from mean is 235 s.
     (iii) Number expected to be at least as far from the mean as the 'suspect' value is 0.465. Note that this is very close to 0.5 and it would be sensible, if possible, to acquire more data rather than eliminate the 'outlier'.
     (iv) New mean = 218.5 s, new standard deviation, $s = 3.1$ s.

N   Weighted mean = 1.103 s.

O   Standard uncertainty of the weighted mean = 0.11 s.

End of chapter problems

(1) Relative standard uncertainty = 0.038.
(2)   (i)   Mean rebound height = 186.0 mm.
      (ii) Standard error in rebound height = 1.2 mm.
      (iii) 95% coverage interval = (186.0 ± 3.9) mm.
      (iv) Still 1.2 mm to two significant figures.
(3)   (i)   Best estimate of film thickness (which is the mean) = 328 nm.
      (ii) Standard uncertainty in the best estimate = 12 nm.
      (iii) 99% coverage interval = (328 ± 47) nm.
(4) $n = 1.461$, $u(n) = 0.033$.
(5) Best estimate of $r = 0.386$, standard uncertainty in best estimate = 0.031.
(6) Best estimate of $\theta_c = 5.52 \times 10^{-3}$ rad, standard uncertainty in best estimate $= 0.10 \times 10^{-3}$ rad.

(7) Best estimate of $R = 125.57\ \Omega$, standard uncertainty in best estimate $= 0.59\ \Omega$.

(8) Best estimate of $c = 0.6710$, standard uncertainty in best estimate $= 0.0073$.

(9) Best estimate of $H = -6.54\ \text{W}$, standard uncertainty $= 0.99\ \text{W}$.

(10)   (i)   Mean $= 47.83\ \text{cm}$.
      (ii)   Value furthest from mean is $42.7\ \text{cm}$.
      (iii)   Yes, reject outlier ($N = 0.064$).
      (iv)   New mean $= 48.4\ \text{cm}$.

(11)   (i) $166.0\ \text{mV}$.
      (ii) $1.4\ \text{mV}$.
      (iii) $163.1\ \text{mV}$ to $168.9\ \text{mV}$.

(12)   (i)   Mean mass $= 0.9656\ \text{g}$.
      (ii)   Standard uncertainty $= 0.0032\ \text{g}$.
      (iii)   95% coverage interval $= 0.9584\ \text{g}$ to $0.9729\ \text{g}$.

(13)   (i) $\bar{x} = 33.38\ \text{mL}$, $s = 0.28\ \text{mL}$.
      (ii)   Possible outlier is $33.9\ \text{mL}$. Applying Chauvenet's criterion indicates that outlier should be removed ($N = 0.38$).
      (iii)   New $\bar{x} = 33.28\ \text{mL}$, new $s = 0.13\ \text{mL}$.

(14) Weighted mean $= 1.0650\ \text{g/cm}^3$, standard uncertainty in weighted mean $= 0.0099\ \text{g/cm}^3$.

Chapter 6

Exercises

A $a = 332.1\ \text{m/s}$, $b = 0.6496\ \text{m/(s·°C)}$, $SSR = 15.41\ \text{m}^2/\text{s}^2$.

B $s_a = 1.5$, $s_b = 0.23$.

C   (1)   The 99% coverage interval for α is $(4 \pm 22) \times 10^{-3}$. The 99% coverage interval for β is $(2.51 \pm 0.12) \times 10^{-2}\ \text{mL/ng}$.
      (2)   (i) $a = 10.24\ \Omega$, $b = 4.324 \times 10^{-2}\ \Omega/\text{°C}$.
           (ii) $s_a = 0.066\ \Omega$, $s_b = 1.4 \times 10^{-3}\ \Omega/\text{°C}$.
           (iii)   The 95% coverage intervals for $A = (10.24 \pm 0.15)\ \Omega$ and for $B = (4.32 \pm 0.32) \times 10^{-2}\ \Omega/\text{°C}$.

D   (i) $a = 3.981$, $b = 16.48$, $s_a = 1.1$, $s_b = 1.9$.

E   (i)   Plot P versus h, (ii) $a = P_A$, $b = \rho g$.

F   (i) $a = 1.20046$ m, $b = 1.07818 \times 10^{-5}$ m/°C, $s_a = 5.0 \times 10^{-4}$ m, $s_b = 8.0 \times 10^{-7}$ m/°C .

   (ii) $\alpha = 8.981 \times 10^{-6}$ °C^{-1}.

   (iii) $u(\alpha) = 6.7 \times 10^{-7}$ °C^{-1}.

G   The 99% coverage interval for $\mu_{y|x_0}$ when $x_0 = 15$ is 4.5 ± 7.1.

H   The 95% prediction interval for y at $x_0 = 12$ is 12 ± 12.

I

   (i) $\hat{y} = 4677.89 + 14415.11 x_i$.

   (ii) $\hat{x}_0 = 3.64$ ppm, $s_{\hat{x}_0} = 0.17$ ppm.

J

   (i)   'Usual' least squares (error in y), $k_0 = 0.6842$ V, $k_1 = -2.391 \times 10^{-3}$ V/°C.

   (ii)   When errors are confined to x, $k_0 = 0.6847$ V, $k_1 = -2.401 \times 10^{-3}$ V/°C.

K   (i) $r = -0.8636$.

   (ii) $a = 31.794$ °C, $b = -0.5854$ °C/cm^3.

   (iv)   A plot of data indicates that the assumption of linearity is not valid.

L $r = 0.9667$.

M   (i) $r = 0.7262$.

   (ii)   Value of r is not significant.

N   (ii) $a = 0.4173$ s, $b = 0.4779$ s/kg.

   (iii)

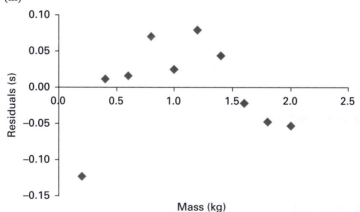

(iv) Yes, probably not a valid equation to fit to data (i.e. period of oscillation of a mass on a spring is not linearly related to mass).

O   Number of data expected to be at least as far from the best line as the outlier is 0.392. Based on Chauvenet's criterion, the outlier should be rejected and the intercept and slope recalculated.

P   (1) (i) Plot $\ln R$ versus T. Intercept $= \ln A$, slope $= -B$.

        (ii) Plot $\ln P$ versus $\ln V$. Intercept $= \ln C$, slope $= -\gamma$.

        (iii) Plot H versus T. Intercept $= -CT_0$, slope $= C$.

        (iv) Plot T_w versus R^2. Intercept $= T_c$, slope $= -k$.

        (v) Plot T^2 versus m. Intercept $= 0$, slope $= \dfrac{4\pi^2}{k}$.

        (vi) Plot $\dfrac{1}{I}$ versus R. Intercept $= \dfrac{r}{E}$, slope $= \dfrac{1}{E}$.

        (vii) Plot $\dfrac{1}{v}$ versus $\dfrac{1}{u}$. Intercept $= \dfrac{1}{f}$, slope $= -1$.

        (viii) Plot $\ln N$ versus $\ln C$. Intercept $= \ln k$, slope $= \dfrac{1}{n}$.

        (ix) Plot $\dfrac{1}{n^2 - 1}$ versus $\dfrac{1}{\lambda^2}$. Intercept $= \dfrac{1}{A}$, slope $= -\dfrac{B}{A}$.

        (x) Plot tD^2 versus D. Intercept $= A$, slope $= AB$.

        (xi) Plot $\ln D$ versus $\ln E$. Intercept $= \ln k$, slope $= n$.

        (xii) Plot v^2 versus l^2. Intercept $= kA^2$, slope $= -k$.

        (xiii) Plot g versus h. Intercept $= g_0$, slope $= -\dfrac{g_0}{2R}$.

        (xiv) Plot v^2 versus P. Intercept $= -\dfrac{2P_A}{\rho}$, slope $= \dfrac{2}{\rho}$.

    (2) $k = 2.311 \times 10^{-6}$ V^{-1}·pF^{-2}, $\phi = 1.064$ V.

Q   (i) $y = 7.483$, $u(y) = 0.13$.

    (ii) $y = 3136$, $u(y) = 220$.

    (iii) $y = 0.01786$, $u(y) = 6.4 \times 10^{-4}$.

    (iv) $y = 3.189 \times 10^{-4}$, $u(y) = 2.3 \times 10^{-5}$.

    (v) $y = 1.748$, $u(y) = 0.016$.

R $A = 1.998 \times 10^4$ counts, $\lambda = 6.087 \times 10^{-4}$ mm^{-2}, $u(A) = 1.3 \times 10^2$ counts, $u(\lambda) = 4.7 \times 10^{-6}$ mm^{-2}.

S $a = 2.311$, $b = -19.70$ V^{-1}, $s_a = 0.063$, $s_b = 0.82$ V^{-1}.

End of chapter problems

    (1) (ii) $a = 9.803$ m/s^2, $b = -2.915 \times 10^{-6}$ s^{-2}.

        (iii) $SSR = 0.0063$ (m/s^2)^2, $s = 0.028$ m/s^2.

(iv) $s_a = 0.019 \, \text{m/s}^2$, $s_b = 3.1 \times 10^{-7} \, \text{s}^{-2}$.

(v) $r = -0.9579$.

(vi) $a = 9.7927 \, \text{m/s}^2$, $b = -2.867 \times 10^{-6} \, \text{s}^{-2}$, $SSR = 0.0004533 \, (\text{m/s}^2)^2$, $s = 0.0075 \, \text{m/s}^2$, $s_a = 0.0051 \, \text{m/s}^2$, $s_b = 8.3 \times 10^{-8} \, \text{s}^{-2}$, $r = -0.9967$.

(2) $9.7927 \, \text{m/s}^2$, $6.832 \times 10^6 \, \text{m}$, $0.0051 \, \text{m/s}^2$, $1.9 \times 10^5 \, \text{m}$. Note that both intercept and slope are used to determine R_E, so this must be taken into account when finding the standard uncertainty in R_E (see section 6.4.1.1).

(3) Fitting $y = a + bx$ to data gives $k = 27.09 \, \text{MPa}$, $u(k) = 0.75 \, \text{MPa}$, but if we force the fitted line through the origin (which is not recommended) then $k = 29.75 \, \text{MPa}$, $u(k) = 0.49 \, \text{MPa}$.

(4) (ii) $K_1 = 2.556 \times 10^{-10} \, \text{m/(s·μA}^4)$, $K_2 = 5.583 \times 10^4 \, \text{μA}^2$.

(iii)   No discernable pattern in residuals – weighting seems appropriate.

(5) (ii) $k = 2.638$, $n = 2.346$, $u(k) = 0.036$, $u(n) = 0.028$.

(iii)   There is an indication that as $\ln C$ increases, so do the standardised residuals. A weighted fit is probably appropriate, but more points are needed to confirm this.

(iv)   When $C = 0.085 \, \text{mol/L}$, $Y = (0.923 \pm 0.017) \, \text{mol}$.

(6) (i) $a = 0.9591$, $b = 0.9039$.

(ii) $I_{max} = 1.863$, $I_{min} = 0.9591$.

(iii)   Note $I_{max} = a + b$. As a and b are correlated, replace a with $\bar{y} - b\bar{x}$ before proceeding to calculate $u(I_{max})$. $u(I_{max}) = 0.038$.

(7) (i)   Plot $1/X$ versus $1/P$. $a = -\frac{B}{A}$, $b = \frac{1}{A}$.

(ii) $a = 2228 \, \text{m}^2/\text{kg}$, $b = 1338 \, \text{N/kg}$, $u(A) = 31 \, \text{m}^2/\text{kg}$, $u(B) = 15 \, \text{N/kg}$.

(iii) $A = 7.473 \times 10^{-4} \, \text{kg/N}$, $B = -1.665 \, \text{m}^2/\text{N}$, $u(A) = 8.6 \times 10^{-6} \, \text{kg/N}$, $u(B) = 0.040 \, \text{m}^2/\text{N}$.

(8) (i) $b = 5.833 \, \text{m}^{-1}\cdot\text{Pa}^{-1}$, $s_b = 0.55 \, \text{m}^{-1}\cdot\text{Pa}^{-1}$.

(ii) $d = 1.748 \times 10^{-10} \, \text{m}$, $u(d) = 8.3 \times 10^{-12} \, \text{m}$.

(12) (ii) $r = 0.9220$.

(iii)   Here we have six points. Using table 6.20, the probability of having $r > 0.9$ when data are uncorrelated is 0.014, therefore we have evidence that the correlation is significant.

(13) (iv) $r = 2.85 \, \Omega$, $u(r) = 0.11 \, \Omega$, $E = 41.68 \, \text{mV}$, $u(E) = 0.54 \, \text{mV}$.

(v)   An examination of the residuals reveals no clear evidence that a weighted fit is required, but there is a possible model violation – more data needed to clarify this.

(14) (i) $a = 399.3 \, \text{ppm}$, $s_a = 2.5 \, \text{ppm}$, $b = 1.323 \, \text{ppm/min}$, $s_b = 0.011 \, \text{ppm/min}$.

(ii) $r = 0.9998$.

(iii)   When $t = 150 \, \text{min}$, concentration $= 597.9 \, \text{ppm}$. When $t = 400 \, \text{min}$, concentration $= 928.8 \, \text{ppm}$.

(iv)   At $t = 150 \, \text{min}$, 95% coverage interval is $(597.9 \pm 1.4) \, \text{ppm}$. At $t = 400 \, \text{min}$, 95% coverage interval is $(928.8 \pm 2.9) \, \text{ppm}$.

(15) (ii) $a = -97$ cps, $s_a = 128$ cps, $b = 2650$ cps/ppb, $s_b = 15$ cps/ppb.

    (iii) Best estimate of antimony concentration = 2.40 ppb, 95% coverage interval = (2.40 ± 0.17) ppb.

(16) (iii) $E_o = 455.002$ keV, $u(E_o) = 0.054$ keV, $v = 1.0077 \times 10^7$ m/s, $u(v) = 0.0061 \times 10^7$ m/s.

    (iv) A plot of residuals does show quite a strong trend, suggestive of a model violation.

(17) (ii) $k = 1.76$ °C, $D = 2.57$ °C.

    (iii) $u(k) = 0.35$ °C, $u(D) = 0.59$ °C.

(18) (ii) Increasing the number of points decreases the size of the standard uncertainties in both slope and intercept.

    (iv) The size of the standard uncertainty in the mean decreases as $\frac{1}{n^{1/2}}$, so that increasing the number of points by a factor of 5, will reduce the standard uncertainty by about 2.

    (v) Yes, expect a further decrease in the size of the standard uncertainties (by about a factor of 3 from when 51 points were used).

Chapter 7

Exercises

A

$$\begin{bmatrix} n & \sum T_i & \sum T_i \ln T_i \\ \sum T_i & \sum T_i^2 & \sum T_i^2 \ln T_i \\ \sum T_i \ln T_i & \sum T_i^2 \ln T_i & \sum (T_i \ln T_i)^2 \end{bmatrix} \begin{bmatrix} a \\ b \\ c \end{bmatrix} = \begin{bmatrix} \sum V_i \\ \sum V_i T_i \\ \sum V_i T_i \ln T_i \end{bmatrix}$$

B (i)

$$\begin{vmatrix} -0.0598 & -0.00427 & 0.134406 \\ 0.099696 & -0.14207 & 0.097605 \\ 0.024255 & 0.21152 & -0.19437 \end{vmatrix}$$

    (ii)

$$\begin{vmatrix} -0.02049 & 0.000461 & 0.025877 & -0.00427 \\ 0.02333 & -0.00381 & -0.03049 & 0.02004 \\ 0.02103 & -0.00537 & -0.00979 & -0.00087 \\ -0.01745 & 0.016652 & 0.014903 & -0.0125 \end{vmatrix}$$

    (iii)

$$\begin{vmatrix} -0.06572 & -0.00915 & 0.019853 & -0.0018 & 0.102176 \\ -0.17055 & -0.10107 & -0.57789 & 0.651764 & -0.01951 \\ 0.080788 & 0.050332 & 0.368402 & -0.2621 & -0.11095 \\ -0.11792 & -0.3062 & -0.79712 & 0.88727 & 0.06521 \\ 0.320455 & 0.414455 & 0.995139 & -1.21103 & -0.04705 \end{vmatrix}$$

C (i) $\begin{vmatrix} 9436.22 \\ 5547.23 \\ 7173.82 \end{vmatrix}$

   (ii) $\begin{vmatrix} 5529 \\ 6140 \\ 4428 \\ 6961 \end{vmatrix}$

D Best estimate of A, $a = -11.82\ \Omega$.
   Best estimate of B, $b = 0.4244\ \Omega/\text{K}$.
   Best estimate of C, $c = -5.928 \times 10^{-5}\ \Omega/\text{K}^2$.

E $a = 15.36$, $b = 2.408$, $c = 1.876$.

F Best estimate of A, $a = -6.776\ \mu\text{V}$.
   Best estimate of B, $b = 4.922 \times 10^{-3}\ \mu\text{V/K}$.
   Best estimate of C, $c = 9.121 \times 10^{-5}\ \mu\text{V/K}^2$.
   Best estimate of D, $d = -6.747 \times 10^{-8}\ \mu\text{V/K}^3$.

   $s_a = 0.083\ \mu\text{V}$.
   $s_b = 1.6 \times 10^{-3}\ \mu\text{V/K}$.
   $s_c = 9.0 \times 10^{-6}\ \mu\text{V/K}^2$.
   $s_d = 1.6 \times 10^{-8}\ \mu\text{V/K}^3$.

G $\alpha = (0.010 \pm 0.011)\ \text{N}$.
   $\beta = (3.9 \pm 3.4)\ \text{N/m}$.
   $\gamma = (427 \pm 24)\ \text{N/m}^2$.

H $a = 13.27$.
   $b = 3.628$.
   $c = 1.426$.

   $s_a = 1.1$.
   $s_b = 0.55$.
   $s_c = 0.095$.

I $R^2 = 0.9988$.

J (i) $a = 5.102 \times 10^{-3}$, $b = 1.028 \times 10^{-2}$, $c = 8.019 \times 10^{-3}$.
   (ii) $s_a = 7.7 \times 10^{-4}$, $s_b = 2.7 \times 10^{-3}$, $s_c = 2.0 \times 10^{-4}$.
   (iii) $R^2 = 0.9997$.
   (iv) $s = 1.2 \times 10^{-4}$.

K   (i)   0.875.

(ii)

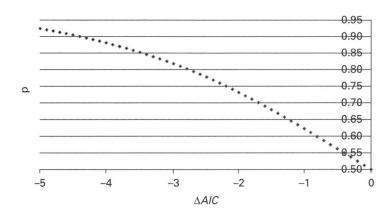

ΔAIC

End of chapter problems

(1) (ii)
$$\begin{bmatrix} n & \sum x_i & \sum \exp x_i \\ \sum x_i & \sum x_i^2 & \sum x_i \exp x_i \\ \sum \exp x_i & \sum x_i \exp x_i & \sum (\exp x_i)^2 \end{bmatrix} \begin{bmatrix} a \\ b \\ c \end{bmatrix} = \begin{bmatrix} \sum y_i \\ \sum y_i x_i \\ \sum y_i \exp x_i \end{bmatrix}$$

(iii) $a = 6.076$, $b = -0.9074$, $c = 1.492 \times 10^{-4}$.

(2) (i)
$$\begin{bmatrix} n & \sum \dfrac{1}{x_i} & \sum x_i \\ \sum \dfrac{1}{x_i} & \sum \dfrac{1}{x_i^2} & n \\ \sum x_i & n & \sum x_i^2 \end{bmatrix} \begin{bmatrix} a \\ b \\ c \end{bmatrix} = \begin{bmatrix} \sum y_i \\ \sum \dfrac{y_i}{x_i} \\ \sum x_i y_i \end{bmatrix}$$

(ii) $a = 1.740$ mm, $b = 26.87$ mm·mL/min, $c = 0.02366$ mm·min/mL.

(iii) $s_a = 0.13$ mm, $s_b = 0.89$ mm·mL/min, $s_c = 1.7 \times 10^{-3}$ mm·min/mL.

(3) Writing best estimates of s_0, u and $\tfrac{1}{2}g$ as a, b, and c respectively, we have

$a = 135.1$ m, $b = 14.32$ m/s, $c = -4.854$ m/s^2 (so $g = -9.71$ m/s^2).

$s_a = 0.85$ m, $s_b = 0.66$ m/s, $s_c = 0.11$ m/s^2.

$u(g) = 2s_c = 2 \times 0.11 = 0.22$ m/s^2.

(4) Let $y = \frac{PV}{RT}$, and $x = \frac{1}{V}$

Writing best estimates of A, B, C and D as a, b, c and d, respectively, we have

$a = 0.9230$, $b = -67.57$ cm^3, $c = 1977$ cm^6 and $d = 2.387 \times 10^4$ cm^9.

$s_a = 0.01$, $s_b = 5.2$ cm^3, $s_c = 430$ cm^6 and $s_d = 9.7 \times 10^3$ cm^9.

(5) (i)   For 2 parameters, $R_{ADJ}^2 = 0.9707$, $AIC = 121.5$, $AIC_c = 124.5$.

For 3 parameters, $R_{ADJ}^2 = 0.9675$, $AIC = 123.5$, $AIC_c = 129.2$.

(ii) 0.913

(6) (i) Writing best estimates of A, B, and C as a, b and c respectively, we have
   $a = 29.90\,\text{J·mol}^{-1}\text{·K}^{-1}$, $b = 4.304 \times 10^{-3}\,\text{J·mol}^{-1}\text{·K}^{-2}$, $c = -1.632 \times 10^{5}\,\text{J·mol}^{-1}\text{·K}$.

   (ii) $s_a = 0.22\,\text{J·mol}^{-1}\text{·K}^{-1}$, $s_b = 2.4 \times 10^{-4}\,\text{J·mol}^{-1}\text{·K}^{-2}$, $s_c = 1.8 \times 10^{4}\,\text{J·mol}^{-1}\text{·K}$.

   (iii) $A = (29.90 \pm 0.47)\,\text{J·mol}^{-1}\text{·K}^{-1}$, $B = (4.30 \pm 0.53) \times 10^{-3}\,\text{J·mol}^{-1}\text{·K}^{-2}$,
   $C = (-1.63 \pm 0.40) \times 10^{5}\,\text{J·mol}^{-1}\text{·K}$.

(7) Indicators of goodness of fit including AIC, AIC_c and residuals should show that equation 7.72 is a better fit to data than equation 7.73.

(8) $k = 0.028$, $u(k) = 0.094$, $r = 0.852$, $u(r) = 0.40$, $s = 3.48$, $u(s) = 0.19$, $t = -0.93$, $u(t) = 0.23$.

Chapter 8

Exercises

A $a = 1.11$, $b = 9.28$, $c = 1.77$. No, I arbitrarily tried starting values of $a = 10$, $b = 10$, $c = 10$. Solver found a local minimum in SSR for these values (as it would for many other starting values) – see section 8.3.1 for a discussion of local minima. To better assure finding the global minimum, I considered the function and recognised (for example) when $x = 0$, $y \approx a$, which gives a starting value for a of 1.77. By considering other properties of the function given in this exercise (including that the maximum of sin cx must be 1) I used the data to find approximate values for b and c of 9 and 2 respectively.

B $a = 24.50\,°\text{C}$, $s_a = 0.13\,°\text{C}$, $b = 3.061\,°\text{C}$, $s_b = 0.23\,°\text{C}$, $c = -0.068\,\text{min}^{-1}$, $s_c = 0.015\,\text{min}^{-1}$.

C   (ii)

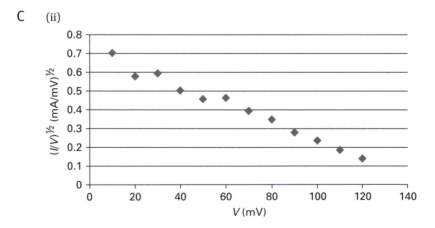

   (iii) $a = 2.30 \times 10^{-5}\,\text{mA/(mV)}^3$, $s_a = 2.0 \times 10^{-6}\,\text{mA/(mV)}^3$, $b = 149.8\,\text{mV}$, $s_b = 3.9\,\text{mV}$.

(iv) The normal distribution of errors (which is a basic assumption when applying least squares) is no longer valid after transforming the data (which is required by 'linearisation' in this problem). Non-linear least squares does not require data transformation, thereby avoiding this problem.

End of chapter problems

(1) (i) $0.0344\ ppm^{-1}$, $-7.01\ ppm$.

  (ii) $0.00027\ ppm^{-1}$, $0.16\ ppm$.

(2) (i) 0.2412, $0.03441\ ppm^{-1}$, $-7.01\ ppm$.

  (ii) 0.0038, $0.00028\ ppm^{-1}$, $0.16\ ppm$.

(3) (ii) $28.22\ ml$, $0.9906\ ml$, $0.0008469\ s^{-1}$.

  (iii) $0.38\ ml$, $0.22\ ml$, $3.0 \times 10^{-5}\ s^{-1}$.

  (iv) $(28.22 \pm 0.97)\ ml$, $(0.99 \pm 0.55)\ ml$, $(0.000847 \pm 0.000077)\ s^{-1}$.

(4) (ii) $0.0099\ mol/L$, $0.000662\ L/mol\cdot s$.

  (iii) $0.0017\ mol/L$, $2.0 \times 10^{-5}\ L/mol\cdot s$.

(5) $1.169871\ eV$, $0.0004832\ K^{-1}$, $662\ K$, $7.8 \times 10^{-5}\ eV$, $4.7 \times 10^{-6}\ K^{-1}$, $11\ K$.

(6) (i) -0.4021, $3.404\ (mg/L)^{-1}$, -0.5650, $3.581\ (mg/L)^{-0.979}$, 0.9790.

  (ii) 0.058, $0.013\ (mg/L)^{-1}$, 0.089, $0.079\ (mg/L)^{-0.979}$, 0.0090.

  (iii) 0.6652, 0.5074.

  (iv) -60.6, -62.8.

  Equation 8.67 better fits the data.

(7) (iv) $3199\ mA$, 11.7.

  (v) $17\ mA$, 1.8.

  (vi) $(3199 \pm 36)\ mA$, (11.7 ± 3.8).

(8) $8.689\ V$, $1.4593\ V$, -0.428, $0.019\ V$, $0.0074\ V$, 0.011.

(9) 55.4, 5.067, 1.8, 0.043.

(10) (iii) $63.1\ J$, $49.4\ J$, $16.9\ K$, $339.9\ K$.

  (iv) $1.3\ J$, $1.5\ J$, $2.1\ K$, $1.2\ K$.

Chapter 9

Exercises

A H_0: $\mu = 1.260\ V$; $H_a \neq 1.260\ V$. For data, $z = -3.0$. $z_{crit} = 1.96$. As $|z| > z_{crit}$, reject null hypothesis, i.e. the sample has not been drawn from a population with a mean of $1.260\ V$.

B p value $= 0.024$, therefore at $\alpha = 0.05$, there is a significant difference between the hypothesised population mean and the sample mean.

C (i)   When $\alpha = 0.2$, $z_{crit} = 0.84$.
   (ii) When $\alpha = 0.05$, $z_{crit} = 1.64$.
   (iii) When $\alpha = 0.01$, $z_{crit} = 2.33$.
   (iv) When $\alpha = 0.005$, $z_{crit} = 2.58$.

D $t = -2.119$, $t_{crit} = 2.365$. As $|t| < t_{crit}$ we cannot reject the null hypothesis, i.e. the mean of the values in table 9.8 is not significantly different from the published value of c.

E H_0: population intercept = 0. Carry out two tailed test at $\alpha = 0.05$. $t = 6.917$, $t_{crit} = 2.228$, therefore reject null hypothesis.
   H_0: population slope = 0. $|t| = 2.011$. $t_{crit} = 2.228$, therefore cannot reject null hypothesis.

F $t = 14.49$ and $t_{crit} = 2.228$, therefore reject the null hypothesis, i.e. the means of the coefficient of kinetic friction for the two contacts areas are significantly different at the $\alpha = 0.05$ level of significance.

G Using Excel's T.TEST() function, the p value is 0.046. As this is less than $\alpha = 0.05$ we reject the null hypothesis, i.e. there is a significant difference (at $\alpha = 0.05$) between the lead content at the two locations.

H Carry out t test for paired samples.
   $t = 2.762$. For a two tailed test at $\alpha = 0.05$ and with number of degrees of freedom = 7, $t_{crit} = 2.365$. As $t > t_{crit}$ we reject null hypothesis, i.e. the emfs of the batteries have changed over the period of storage.

I $p = 0.08037$. As $p > 0.05$, we would not reject the null hypothesis at $\alpha = 0.05$.

J $F = 2.207$, $F_{crit} = 7.15$. As $F < F_{crit}$, cannot reject null hypothesis at $\alpha = 0.05$.

K $F_{crit} = 5.285$.

L (i)

count	observed frequency	expected frequency
0	10	7.78
1	13	14.48
2	7	13.46
3	15	8.35
4	5	5.93

   (ii) $\chi^2 = 9.331$, $\chi^2_{crit} = 7.815$. Reject null hypothesis at $\alpha = 0.05$.

M (i)   10.60.
   (ii) 11.34.
   (iii) 11.07.
   (iv) 15.99.

N $F = 12.82$, $F_{crit} = 3.40$, as $F > F_{crit}$, the ANOVA indicates that, at $\alpha = 0.05$, the magnitude of the alpha wave *does* depends on light level.

End of chapter problems

(1) Two tailed t test required (samples not paired). $t = 1.831$, $t_{crit} = 2.228$. As $t < t_{crit}$ we cannot reject the hypothesis that both samples have the same population mean (at $\alpha = 0.05$).

(2) Paired sample t test required (two tailed). $t = 2.909$, $t_{crit} = 2.365$. As $t > t_{crit}$, we reject the hypothesis (at $\alpha = 0.05$) that there is no difference in the urea concentration as determined by the two laboratories.

(3) Paired sample t test required (single tailed). $t = 2.440$, $t_{crit} = 2.015$. As $t > t_{crit}$, we reject the hypothesis (at $\alpha = 0.05$) that the voltage generated when the heat sinks are vertically oriented is less than or equal to the voltage generated when the heat sinks are horizontally oriented.

(4) $t = 1.111$, $t_{crit} = 3.182$ (for $\alpha = 0.05$ and 3 degrees of freedom). As $t < t_{crit}$, we cannot reject the null hypothesis, i.e. the intercept is not significantly different from zero.

(5) Two tailed F test carried out at $\alpha = 0.05$. $F = 3.596$, $F_{crit} = 4.433$. As $F < F_{crit}$, we cannot reject the null hypothesis that both populations have the same variance.

(6) One tailed F test carried out at $\alpha = 0.05$. $F = 2.796$, $F_{crit} = 3.438$. As $F < F_{crit}$, we cannot reject a null hypothesis that both populations have the same variance.

(7) I chose a bin width of 0.1 s with a bin range beginning at 0.8 s and extending to 1.9 s. Where necessary, bins were combined to ensure that the frequencies were ≥ 5. A chi-squared test (carried out at $\alpha = 0.05$) indicates that the distribution of data in table 9.30 is consistent with the normal distribution.

(8) $F = 2.811$, $F_{crit} = 3.555$. As $F < F_{crit}$ we cannot reject a null hypothesis that the population means are equal.

(9) $F = 10.61$, $F_{crit} = 3.885$. As $F > F_{crit}$ we reject the null hypothesis and conclude that the population means are not equal.

References

Adler, H.A. and Roessler, E.B. *Introduction to Probability and Statistics 5th Edition* (1972) W. H. Freeman and Company, San Francisco.

Akaike, H. A new look at the statistical model identification (1974) *IEEE Transactions on Automatic Control* **19**, 716–723.

Barford, N.C. *Experimental Measurements: Precision, Error and Truth 2nd Edition* (1985) John Wiley and Sons, Chichester.

Bates, D.M. and Watts, D.G. *Nonlinear Regression Analysis and its Applications* (1988) Wiley, New York.

Bennett, C.A. *Principles of Physical Optics* (2008) John Wiley and Sons, New York.

Bentley, J.P. *Principles of Measurement Systems 4th Edition* (2004) Prentice Hall, New Jersey.

Berry, J., Norcliffe, A. and Humble, S. *Introductory Mathematics through Science Applications* (1989) Cambridge University Press, Cambridge.

Bevington, P.R. and Robinson, D.K. *Data Reduction and Error Analysis for the Physical Sciences 3rd Edition* (2002) McGraw-Hill, New York.

Blaisdell, E.A. *Statistics in Practice 2nd Edition* (1998) Saunders College Publishing, Fort Worth.

Blau, P.J. *Friction Science and Technology: From Concepts to Applications, 2nd Edition* (2009) CRC Press, Taylor and Francis Group, Florida.

Bloch, S.C. *Excel for Engineers and Scientists 2nd edition* (2003) John Wiley and Sons, New York.

Box, J.F. *R. A. Fisher: The Life of a Scientist* (1978) Wiley, New York.

Bube, R.H. *Photoconductivity of Solids* (1960) Wiley, New York.

Burnham, K.P. and Anderson, D. *Model Selection and Multi-Model Inference* (2002) Springer-Verlag, New York.

Cantrell, C.A. Review of methods for linear least-squares fitting of data and application to atmospheric chemistry problems (2008) *Atmospheric Chemistry and Physics* **8**, 5477–5487.

Cleveland, W.S. *The Elements of Graphing Data* (1994) Hobart Press, New Jersey.

Crow, E.L. and Shimizu, K. *Lognormal Distributions: Theory and Applications* (1988) Marcel Dekker, New York.

DeCoursey, W. *Statistics and Probability for Engineering Applications* (2003) Newnes.

Demas, J. N. *Excited State Lifetime Measurements* (1983) Academic Press, New York.

Denton, P. Analysis of first order kinetics using Microsoft Excel Solver (2000) *Journal of Chemical Education* **77**, 1524–1525.

Devore, J. L. *Probability and Statistics for Engineering and the Sciences 7th Edition* (2007) Brookes/Cole, California.

Dietrich, C. R. *Uncertainty, Calibration and Probability: Statistics of Scientific and Industrial Measurement 2nd Edition* (1991) Adam Hilger, Bristol.

Doebelin, E. O. *Measurement Systems: Applications and Design* (2003) McGraw-Hill, New York.

Feynman, R. P., Leighton, R. B. and Sands, M. *The Feynman lectures on Physics* (1963) Addison Wesley, Reading, Massachusetts.

Fylstra, D., Lasdon, L., Watson, J. and Waren, A. Design and use of Microsoft Excel Solver (1998) *Interfaces*, **28**, 29–55.

Ghilani, C. D. *Adjustment Computations: Spatial Data Analysis 5th Edition* (2010) John Wiley and Sons, New York.

Graham, R. C. *Data Analysis for the Chemical Sciences* (1993) VCH, New York.

Hamilton, L. C. *Modern Data Analysis: A First Course in Applied Statistics* (1990) Brookes/Cole, California.

Hayter, A. *Probability and Statistics for Engineers and Scientists* (2006) Duxbury Press, California.

Hoel, P. G. *Introduction to Mathematical Statistics 5th Edition* (1984) Wiley, New York.

Jacquez, J. A. *Compartmental Analysis in Biology and Medicine* (1996) Biomedware, Michegan.

Jelen, B. *Microsoft Excel 2010 in Depth* (2010) Que Publishing, Indianapolis.

Karlovsky, J. Simple method for calculating the tunneling current in an Esaki diode (1962) *Physical Review* **127**, 419.

Kennedy, J. B. and Neville, A. M. *Basic Statistical Methods for Engineers and Scientists 3rd Edition* (1986) Harper Row, New York.

Khazan, A. D. *Transducers and their Elements* (1994) Prentice Hall, New Jersey.

Kingman, J. F. C. *Poisson Processes* (1993) Oxford University Press, Oxford.

Kirkup, L. and Frenkel, R. *An Introduction to Uncertainty in Measurement* (2006) Cambridge University Press, Cambridge.

Kirkup, L. and Mulholland, M. Comparison of linear and non-linear equations for univariate calibration (2004) *Journal of Chromatography A*, **1029**, 1–11.

Kirkup, L. and Sutherland, J. Curve stripping and non-linear fitting of polyexponential functions to data using a microcomputer (1988) *Computers in Physics* **2**, 64–68.

Kraftmakher, Y. *Experiments and Demonstrations in Physics* (2007) World Scientific Publishing Company, Singapore.

Kutner, M. J., Nachtsheim, C. J. and Neter, J. *Applied Linear Regression Models 4th Edition* (2003) McGraw-Hill/Irwin, New York.

Limpert, E., Stahel, W. A. and Abbt, M. Log-normal distributions across the sciences: keys and clues (2001) *Bioscience* **51**, 341–352.

Lyon, A. J. Rapid statistical methods (1980) *Physics Education* **15**, 78–83.

Mathur, K. K., Needleman, A. and Tvergaard, V. 3D analysis of failure modes in the Charpy impact test (1994) *Modelling and Simulation in Materials Science and Engineering* **2**, 617–635.

McCullough, B. D. and Heiser, D. A. On the accuracy of statistical procedures in Microsoft Excel 2007 (2008) *Computational Statistics & Data Analysis* **52**, 4570–4578.

McPherson, G. *Statistics in Scientific Investigation: Its Basis, Application, and Interpretation* (1990) Springer-Verlag, New York.

Meyer, S. L. *Data Analysis for Scientists and Engineers* (1975) John Wiley and Sons, New York.

Middleton, M. R. *Data Analysis using Microsoft Excel 3rd Edition* (2003) South Western Educational Publishing.

Mohr, P. J., Taylor, B. N. and Newell, D. B. CODATA recommended values of the fundamental physical constants: 2006 (2008) *Reviews of Modern Physics* **80**, 633–730.

Moore, D. S. and McCabe, G. P. *Introduction to the Practice of Statistics* (1989) W H Freeman and Company, New York.

Motulsky, H. and Christopoulos, A. *Fitting Models to Biological Data Using Linear and Nonlinear Regression: A Practical Guide to Curve Fitting* (2005) GraphPad Software Inc., California.

Nielsen-Kudsk, F. A microcomputer program in Basic for iterative, non-linear data-fitting to pharmacokinetic functions (1983) *International Journal of Bio-Medical Computing* **14**, 95–107.

Nicholas, J. V. and White, D. R. *Traceable Temperatures: An Introduction to Temperature Measurement and Calibration* (2001) John Wiley and Sons, Chichester.

Patel, J. K. and Read, C. B. *Handbook of the Normal Distribution 2nd Edition* (1996) Marcel Dekker, New York.

Peck, R., Olsen, C. and Devore, J. *Introduction to Statistics and Data Analysis 3rd Edition* (2008) Duxbury Press, California.

Scheaffer, R. L., Mulekar, M. and McClave, J. T. *Probability and Statistics for Engineers 5th Edition* (2010) Duxbury Press, California.

Simpson, R. E. *Introductory Electronics for Scientists and Engineers 2nd Edition* (1987) Prentice-Hall, New Jersey.

Skoog, D. A. and Leary, J. J. *Principles of Instrumental Analysis 4th Edition* (1992) Harcourt Brace: Fort Worth.

Smith, S. and Lasdon, L. Solving large sparse nonlinear programs using GRG (1992) *ORSA Journal on Computing* **4**, 2–15.

Spiegel, M. R. and Stephens, L. J. *Statistics 4th Edition* (2007) McGraw-Hill, New York.

Student. The probable error of a mean (1908) *Biometrika* **6**, 1–25.

Telford, W. M., Geldart, L. P. and Sheriff, R. E. *Applied Geophysics 2nd Edition* (2010) Cambridge University Press, Cambridge.

Wagenmakers, E. and Farrel, S. AIC model selection using Akaike weights (2004) *Psychonomic Bulletin and Review* **11**, 192–196.

Walkenbach, J. *Excel 2010 Bible* (2010) Wiley Publishing, Indianapolis.

Walpole, R. E., Myers, R. H. and Myers, S. L. *Probability and Statistics for Engineers and Scientists 6th Edition* (1998) Prentice Hall, New Jersey.

Weisberg, S. *Applied Linear Regression 3rd Edition* (2005) John Wiley and Sons, New York.

Young, H. D., Freedman, R. A. and Ford, L. *University Physics with Modern Physics 12th Edition* (2007) Addison-Wesley, Reading, Massachusetts.

Zielinski, T. J. and Allendoerfer, R. D. Least squares fitting of nonlinear data in the undergraduate laboratory (1997) *Journal of Chemical Education* **74**, 1001–1007.

Index